Humanity in Crisis

A Comprehensive Reflection on Anthropogenic Challenges and Pathways to Solutions

RIVER PUBLISHERS SERIES IN SOCIAL, URBAN, ECONOMIC AND ENVIRONMENTAL SUSTAINABILITY

The River Series on Social, Urban, Economic and Environmental Sustainability is a series of comprehensive academic and professional books which focus on the societal side of sustainability. The series focuses on topics ranging from theory to policy and real-life case studies and applications.

Books published in the series include research monographs, edited volumes, handbooks and textbooks. The books provide professionals, researchers, educators, and advanced students in the field with an invaluable insight into the latest research and developments.

Topics included in this series are as follows:-

- Climate Change Mitigation
- Renewable Energy Policy
- Urban sustainability
- Strategic environmental planning
- Environmental Systems Monitoring and Analysis
- Greening the World Economy
- Sustainable Development

For a list of other books in this series, visit www.riverpublishers.com

Humanity in Crisis

A Comprehensive Reflection on Anthropogenic Challenges and Pathways to Solutions

Medani P. Bhandari
Akamai University, USA,
Sumy State University, Ukraine,
and, Gandaki University, Nepal

River Publishers

Routledge
Taylor & Francis Group
NEW YORK AND LONDON

Published 2026 by River Publishers
River Publishers
Broagervej 10, 9260 Gistrup, Denmark
www.riverpublishers.com

Distributed exclusively by Routledge
605 Third Avenue, New York, NY 10017, USA
4 Park Square, Milton Park, Abingdon, Oxon OX14 4RN

Humanity in Crisis: A Comprehensive Reflection on Anthropogenic Challenges and Pathways to Solutions / Medani P. Bhandari.

Routledge is an imprint of the Taylor & Francis Group, an informa business

ISBN 978-87-438-0877-0 (hardback)
ISBN 978-87-438-0882-4 (paperback)
ISBN 978-87-438-0879-4 (online)
ISBN 978-8-743-80878-7 (ebook master)

DOI: 10.1201/9788743808787

To my friend, wife, coauthor Mrs. Prajita Bhandari, without your help and support this book was impossible.

Together We Have Miles To Go

You have assumed a pivotal role akin to that of a true teacher, actively fostering an environmentally conscious atmosphere within your home, workplace, and community. Your impactful actions and exemplary behavior serve as an inspiration, motivating others to embrace sustainable practices and embrace a greener way of life. Your endeavors encompass a wide spectrum, from waste reduction to energy conservation and the advocacy for eco-friendly initiatives. Through these contributions, you actively participate in the creation of a better world for all. Your unwavering dedication to the stewardship of the environment shines as a guiding beacon, urging others to join the collective effort toward a healthier and more sustainable future. Amidst life's challenges, you provide me with a reassuring sense of comfort and a valuable lesson in resilience. For this, I extend my heartfelt gratitude to you.

Contents

Preface

Today's world stands at a crossroads where our choices will determine the trajectory of humanity and the planet. *Humanity in Crisis: A Comprehensive Reflection on Anthropogenic Challenges and Pathways to Solutions* is born from decades of observation, experience, and analysis of the urgent, multi-faceted crises that define our era. This book is not just a compilation of facts or an academic exercise; it is a heartfelt exploration of the challenges we face and the pathways to solutions within our grasp.

This work inspires me from a deep-seated concern for the interconnected crises threatening our existence: environmental degradation, social inequality, political instability, and cultural fragmentation. These challenges are not abstract concepts but realities that impact communities, ecosystems, and future generations. Through this book, I aim to shed light on the known, seen, and experienced realities of these crises, weaving together insights from history, science, and human experience to present a comprehensive understanding of the state of our world.

At its core, this book is also a message of hope. Despite the enormity of the challenges, the human spirit has consistently demonstrated remarkable resilience, creativity, and the ability to transform adversity into opportunity. While daunting, the crises we face today also present an unprecedented chance to rethink, reimagine, and rebuild our world. By embracing sustainability, equity, and solidarity, we can address the root causes of these crises and lay the foundation for a brighter, more sustainable future.

Each chapter of this book delves into a specific dimension of the global crisis—environmental, social, political, economic, cultural, and ethical—offering an in-depth analysis of the problems and potential pathways to solutions. The concluding sections emphasize the power of collective action, innovative thinking, and ethical leadership in overcoming these challenges. The vision presented here is not utopian but grounded in the belief that humanity's best days lie ahead if we act with urgency, compassion, and foresight as part of a powerful collective movement.

This book is a call to action for individuals, communities, and policymakers alike. It seeks to inspire dialogue, collaboration, and transformative change, urging readers to reflect on their roles in shaping the future. Whether you are an activist, scholar, policymaker, or concerned global citizen, this work will resonate with you and empower you to contribute to building a peaceful and healthy planet filled with the potential for transformative change.

Let us not view today's crises as insurmountable obstacles but as catalysts for a profound shift in how we live, think, and interact with one another and the Earth. Together, we can redefine humanity's trajectory, ensuring that the challenges of our time become the steppingstones to a better tomorrow.

Thank you!

Prof. Medani P. Bhandari, PhD

Acknowledgements

Writing *Humanity in Crisis: A Comprehensive Reflection on Anthropogenic Challenges and Pathways to Solutions* has been an extraordinary journey, and it would not have been possible without the support, inspiration, and contributions of many individuals and organizations. This work is a culmination of decades of learning, observing, and collaborating, and I am deeply grateful to all those who have contributed to its realization.

First and foremost, I extend my heartfelt gratitude to my family. To my beloved wife, Prajita Bhandari, my daughter, Manaslu Bhandari, and my son-in-law, Abhimanyu Iyer, your unwavering love, patience, and encouragement have been my constant source of strength. Your belief in my vision and the sacrifices you made, putting up with the countless hours I dedicated to this book, have been invaluable. I sincerely appreciate your sacrifices and understand the time and effort I poured into this project.

I also honor my mother, Heema Devi, whose wisdom and resilience have been a guiding light, and my late father, Loknath Bhandari, whose values and teachings laid the foundation for my journey. To my brothers and sisters, your support and camaraderie have enriched my life and strengthened my resolve to pursue this work.

My deepest gratitude extends to my son, Prameya Bhandari, and my daughter-in-law, Kelsey Bhandari, for their love and encouragement. To my wonderful grandchildren—Prasmin, Aahana, Neena, and Neelok—your innocence, joy, and boundless potential remind me of the importance of creating a better world for future generations.

I am profoundly thankful to my mentors, colleagues, and friends who have guided and supported me throughout my career. Your insights, critiques, and encouragement have shaped my understanding of the complex issues addressed in this book and have been a significant part of my journey. I am especially grateful to the global community of environmentalists, activists, scholars, and policymakers I have worked with. Your tireless efforts to address the world's most pressing challenges have not gone unnoticed and have been a constant source of inspiration.

I also wish to acknowledge the countless individuals and communities worldwide who have shared their stories, struggles, and triumphs with me. Your lived experiences have enriched this book and provided human context for our crises. This work is dedicated to you and the hope that your voices will resonate far and wide, inspiring action and change.

I owe a debt of gratitude to the organizations and institutions that have supported my work over the years. Your resources, platforms, and collaborations have enabled me to explore, document, and advocate for solutions to the crises of our time. I am particularly grateful to the Rivers Publishers and editorial team for their professionalism, dedication, and belief in the importance of this project.

Finally, I want to express my heartfelt thanks to you, the readers of this book. Your curiosity, commitment, and willingness to engage with these critical issues are appreciated and vital to building a sustainable and equitable future. Your engagement is what makes this work meaningful. This book will inspire you to act, foster dialogue, and contribute to the solutions we urgently need.

This book is not just my work; it results from a shared journey with countless individuals who have contributed their knowledge, perspectives, and support. To each of you, I offer my deepest thanks. Together, we can turn today's crises into opportunities for transformation and hope for future generations.

Thank you!

Prof. Medani P. Bhandari, PhD

List of Tables

List of Abbreviations

AI	Artificial intelligence
AMR	Antimicrobial resistance
ARRA	American Recovery and Reinvestment Act
AU	African Union
CBD	Convention on Biological Diversity
CBDR	Common but differentiated responsibilities
CCS	Carbon capture and storage
CITES	Convention on International Trade in Endangered Species
COPD	Chronic obstructive pulmonary disease
CSOs	Civil society organizations
EPR	Extended producer responsibility
ESD	Education for sustainable development
FAO	Food and Agriculture Organization
FGM	Female genital mutilation
FOI	Freedom of information
GCF	Green Climate Fund
GDI	Genuine progress indicator
GEF	Global Environment Facility
GHG	Greenhouse gases
GHSA	Global Health Security Agenda
GIS	Geographic information systems
GPI	Genuine progress indicator
HRC	Human Rights Campaign
HDI	Human Development Index
IAEA	International Atomic Energy Agency
ICC	International Criminal Court
ICRC	International Committee of the Red Cross
IDPs	Internally displaced persons
IEA	International Energy Agency
IHL	International humanitarian law
IHR	International health regulations
IMF	International Monetary Fund

IOM	International Organization for Migration
IRENA	International Renewable Energy Agency
IPV	Intimate partner violence
LDCs	Least developed countries
MDGs	Millennium Development Goals
MPAs	Marine protected areas
NCDs	Non-communicable diseases
NDCs	Nationally determined contributions
NGOs	Non-governmental organizations
NAFTA	North American Free Trade Agreement
OECD	Organisation for Economic Co-operation and Development
OGP	Open Government Partnership
PTSD	Post-traumatic stress disorder
SDGs	Sustainable Development Goals
SEL	Social-emotional learning
SIDS	Small island developing states
TARP	Troubled Asset Relief Program
TRC	Truth and Reconciliation Commission
UHC	Universal health coverage
UNEP	UN Environment Programme
UNESCO	United Nations Educational, Scientific and Cultural Organization
UNFCCC	United Nations Framework Convention on Climate Change
UNHCR	United Nations High Commissioner for Refugees
UNMIL	UN Peacekeeping Mission in Liberia
WFP	World Food Programme
WMO	World Meteorological Organization
WTE	Waste-to-energy
WTO	World Trade Organization
WWF	World Wildlife Fund

1

Humanity in Crisis – A Comprehensive Reflection on Anthropogenic Challenges and Pathways to Solutions

This book, Humanity in Crisis: A Comprehensive Reflection on Anthropogenic Challenges and Pathways to Solutions, critically examines humanity and the planet's profound challenges due to anthropogenic greed and mismanagement. It poignantly reflects the crises that have emerged due to unsustainable practices, unchecked consumption, and a disregard for the interconnectedness of all life forms. Through its seven-part structure, the book provides a holistic analysis of the crises and offers actionable pathways to address them, emphasizing compassion, cooperation, and ethical stewardship.

The book begins by exploring humanity's overarching crises, presented in Understanding the Crisis. Chapters such as "Humanity at the Brink" and "The Anthropocene and Its Consequences" delve into the unprecedented changes brought about by human activity. The Anthropocene, marked by human dominance over natural systems, has triggered systemic crises that affect every aspect of life on Earth (Steffen et al., 2015). These initial chapters provide a foundation for understanding the global interconnectedness of ecological, social, and economic systems, highlighting how disruptions in one area can have cascading effects.

Part II: Environmental Crises examines the ecological consequences of human actions. Climate change, biodiversity loss, and water scarcity are presented as emblematic of the Anthropocene's destructive potential. The chapter on climate change underscores the urgency of reducing greenhouse gas emissions to mitigate global warming (IPCC, 2021). Similarly, the chapter on biodiversity and ecosystem collapse highlights the critical role of ecosystems in maintaining planetary health, emphasizing the need for conservation and restoration efforts (Dasgupta, 2021). The global water crisis explored in

DOI: 10.1201/9788743808787-1

depth reveals the dire consequences of over-extraction, pollution, and inequitable access to freshwater resources (Gleick, 2014). These chapters clarify that environmental degradation is not an isolated issue but a deeply interconnected crisis with social and economic implications.

Part III: Social and Humanitarian Crises addresses the human toll of these challenges. Issues such as poverty, public health, and displacement are analyzed through the lens of systemic inequality and injustice. For instance, the chapter on economic inequality explores how wealth disparities exacerbate vulnerabilities to environmental and social shocks (Piketty, 2014). The chapters on gender inequality and education in crisis highlight the need for inclusive policies that empower marginalized communities and ensure equitable access to resources and opportunities.

The book also examines the structural dimensions of the crises in Part IV: Political and Economic Crises and Part V: Cultural and Ethical Crises. Chapters on governance, misinformation, and ethical dilemmas in technology reveal how political instability and the erosion of trust in institutions hinder collective action. The chapter on the crisis of meaning and purpose reflects on the existential challenges of modernity, urging readers to consider the ethical and philosophical dimensions of humanity's trajectory.

The concluding sections, Part VI: Pathways to Solutions and Part VII: Hope and Vision for Humanity, offer a roadmap for addressing these crises. These chapters emphasize sustainability, global cooperation, and ethical leadership as cornerstones of a transformative agenda. The book calls for innovative solutions, such as regenerative agriculture, circular economies, and transformative education, to build resilience and foster long-term sustainability (Raworth, 2017). The final chapters, "Lessons from History and Nature" and "A Vision for Humanity's Future," inspire hope by drawing on historical examples of resilience and the wisdom of natural systems.

This book is a critical analysis and a call to action. It invites readers to reflect on their roles in shaping a sustainable and equitable future. By weaving together insights from diverse disciplines, it offers a comprehensive framework for understanding and addressing the crises of our time, making it an essential resource for policymakers, scholars, and concerned citizens alike.

1.1 Introduction

Humanity stands at a pivotal juncture in history, facing a convergence of crises threatening our interconnected world's stability and well-being. These crises – environmental degradation, economic inequality, political instability, technological disruption, cultural fragmentation, and wars – are not isolated phenomena but deeply intertwined challenges that exacerbate one another. Addressing them requires a holistic, inclusive, collaborative approach transcending traditional policy, geography, and ideology boundaries.

1.2 Environmental Degradation

Environmental degradation, driven by climate change, deforestation, biodiversity loss, and pollution, represents an existential threat to human survival. According to the Intergovernmental Panel on Climate Change (IPCC, 2023), global temperatures are on track to exceed the critical 1.5°C threshold, leading to catastrophic impacts on ecosystems, food security, and human livelihoods. The depletion of natural resources has intensified conflicts over water, land, and energy, often fueling broader geopolitical instability (Gleick, 2014).

1.3 Economic Inequality

Economic inequality exacerbates marginalized populations' vulnerability to environmental and social crises. The World Inequality Report (2022) highlights that the richest 10% of the global population controls over 76% of the world's wealth, while billions live on less than $5.50 per day. Such disparities are morally indefensible and destabilizing, as they fuel social unrest, reduce trust in institutions, and hinder collective action on global challenges.

1.4 Political Instability

Political instability, marked by the erosion of democratic norms, the rise of authoritarian regimes, and widespread corruption, undermines global governance at a time when coordinated action is most needed. The Edelman Trust Barometer (2024) reports a historic decline in public trust in governments, media, and institutions, reflecting a crisis of legitimacy. This erosion of trust hampers international cooperation on pressing issues such as climate change, migration, and conflict resolution.

1.5 Technological Disruption

Technological disruption, while a driver of progress, also poses significant risks. Advances in artificial intelligence (AI), biotechnology, and digital surveillance have outpaced the development of ethical frameworks, leading to privacy violations, algorithmic bias, and widening inequalities in access to technology (Zuboff, 2019). The digital divide further marginalizes underrepresented communities, while the misuse of technology amplifies misinformation and polarizes societies.

1.6 Cultural Fragmentation

Cultural fragmentation, fueled by globalization, identity politics, and the erosion of shared narratives, threatens social cohesion. As societies become more interconnected, cultural differences have often been exploited to sow division rather than foster understanding. The rise of nationalism and xenophobia has deepened polarization, undermining collaborative efforts to address global challenges (Appadurai, 2006).

1.7 Wars and Armed Conflicts: A Deep Dive into Their Global Impact

Wars and armed conflicts exacerbate all these crises, displacing millions, destroying ecosystems, and diverting resources from critical global challenges. The United Nations High Commissioner for Refugees (UNHCR, 2024) estimates that over 110 million people are currently displaced due to conflict and persecution, creating a humanitarian crisis of unprecedented scale. Wars also accelerate environmental destruction, deepen economic inequalities, and destabilize political systems, creating a vicious cycle of crises.

1.7.1 The prevalence and scale of wars

Wars and armed conflicts continue to plague humanity, leaving devastating impacts on societies, economies, and ecosystems. The Stockholm International Peace Research Institute (SIPRI, 2024) reports that global military spending exceeded $2.2 trillion in 2023, reflecting a world heavily invested in militarization rather than peacebuilding. Current conflicts in regions such as Ukraine, Sudan, Yemen, and the Sahel highlight the persistent nature of warfare and its evolving dynamics, including proxy wars, cyber warfare, and insurgencies.

1.7.1.1 Humanitarian consequences

Wars inflict profound suffering on civilian populations, with millions displaced, injured, or killed. The United Nations High Commissioner for Refugees (UNHCR, 2024) estimates that over 110 million people are forcibly displaced worldwide, the highest number in recorded history. Displacement leads to overcrowded refugee camps, limited access to basic necessities, and long-term challenges in integration and resettlement.

Children are particularly vulnerable, with many losing access to education, healthcare, and family support. According to UNICEF (2023), over 222 million children living in conflict zones lack access to quality education, jeopardizing their future and the development of their communities.

1.7.1.2 Economic devastation

Wars severely disrupt local and global economies. Infrastructure, including roads, schools, and hospitals, is often destroyed, setting back development by decades. For example, the Syrian conflict has caused an estimated $120 billion in economic losses (World Bank, 2021). Additionally, wars divert resources from critical areas such as healthcare, education, and climate action, creating long-term economic instability.

Global supply chains are also impacted. The war in Ukraine, for instance, disrupted global grain supplies, exacerbating food insecurity in vulnerable regions such as Africa and the Middle East (FAO, 2023).

1.7.1.3 Environmental destruction

The environmental toll of wars is immense and often overlooked. Armed conflicts destroy ecosystems, pollute water sources, and degrade arable land. Bombing campaigns, deforestation for military purposes, and the use of chemical weapons leave long-lasting scars on the environment. For example, the Gulf War's oil spills caused extensive damage to marine ecosystems, and the use of Agent Orange during the Vietnam War resulted in deforestation and soil contamination that persists to this day.

Additionally, military activities contribute significantly to greenhouse gas emissions. A 2022 study by CEOBS and Scientists for Global Responsibility estimated that the U.S. Department of Defense alone emits more CO_2 annually than many small nations, highlighting the environmental cost of militarization.

1.7.1.4 Political instability and governance collapse

Wars often result in the collapse of governance structures, creating power vacuums that lead to lawlessness and further conflict. Failed states, such as Libya and Somalia, struggle to rebuild, leaving citizens vulnerable to exploitation by armed groups and criminal networks. This instability often spills over into neighboring regions, creating regional insecurity.

1.7.2 The role of proxy wars and geopolitical rivalries

Modern conflicts are increasingly shaped by geopolitical rivalries and proxy wars, where major powers back opposing sides for strategic advantage. For example, the Syrian Civil War has been influenced by the involvement of the United States, Russia, Turkey, and Iran, each pursuing its agenda. Such conflicts prolong suffering and complicate peace negotiations as external actors prioritize their interests over local stability.

1.7.2.1 The intersection of war with other crises

Wars exacerbate other global crises, creating a vicious cycle of instability:

Economic inequality: Wars deepen poverty and inequality as resources are diverted to military spending, and wealth becomes concentrated in the hands of war profiteers.

Environmental degradation: Conflicts over natural resources, such as water and oil, are both a cause and consequence of war.

Cultural fragmentation: Wars often exploit ethnic and religious divisions, eroding social cohesion and fostering long-term animosity.

Technological disruption: Cyber warfare and the militarization of AI introduce new dimensions of conflict with potentially catastrophic consequences for global security.

The book's central argument is that humanity is facing a convergence of six critical crises – environmental degradation, economic inequality, political instability, technological disruption, cultural fragmentation, and wars and conflicts – that threaten the stability and well-being of our world. These crises are deeply interconnected and require a holistic, inclusive, and collaborative approach.

1.8 Bridging Knowledge Gaps and Contributing New Insights

Humanity in Crisis addresses contemporary society's pressing challenges by exploring the convergence of five critical crises: environmental degradation, economic inequality, political instability, technological disruption, and cultural fragmentation, focusing on the role of wars and armed conflicts. These crises are interconnected and cannot be tackled in isolation. The book seeks to bridge the knowledge gap by offering a holistic, interdisciplinary approach integrating insights from environmental science, economics, political theory, sociology, and technology studies. By synthesizing these diverse perspectives, the book provides a comprehensive understanding of the crises and offers practical pathways for resolution.

The primary audience for this book includes policymakers, academics, and engaged citizens seeking a deeper understanding of the interconnected nature of global crises. It is designed to be accessible to a broad audience while maintaining academic rigor, making it valuable to general readers and those in specialized fields. Policymakers will find the book helpful in shaping more effective, integrated strategies for addressing these crises. At the same time, academics will benefit from its interdisciplinary approach, incorporating recent research across various fields. Additionally, the book is aimed at the growing number of activists, students, and individuals interested in social justice, environmental sustainability, and global peace who seek a more explicit framework for understanding and addressing the crises that threaten the stability of our world.

One of the key ways this book bridges the knowledge gap is by offering a holistic framework that recognizes the interdependencies between various global crises. Traditionally, these issues have been addressed in isolation – environmental degradation has been dealt with in environmental science, economic inequality in economics, political instability in political science, and so on. However, this book argues that these crises are not isolated but deeply interconnected and mutually reinforcing. For example, environmental degradation often exacerbates economic inequality, which fuels political instability and social fragmentation. This interconnectedness demands a comprehensive, cross-disciplinary approach to understanding and addressing the crises. The book provides a complete picture of humanity's challenges and the necessary solutions by bringing together diverse academic disciplines and real-world case studies.

The book adds new knowledge by proposing a collaborative, inclusive approach to crisis resolution. While many academic works focus on individual crises or solutions, this book emphasizes the need for global cooperation and multi-stakeholder engagement. It challenges the traditional, siloed approach to addressing global issues and calls for a more integrated, holistic response. For instance, in the context of climate change, the book discusses how environmental degradation affects ecosystems and exacerbates social inequalities, leading to conflicts over resources, migration, and displacement. Similarly, it explores how, if not managed responsibly, technological disruption can worsen cultural fragmentation and deepen economic divides. The book proposes that addressing these crises requires technological and policy solutions and a fundamental shift in how we view global cooperation, governance, and the role of institutions.

Another important contribution of this book is its focus on the role of wars and armed conflicts in amplifying these crises. While much literature on global crises emphasizes environmental or economic issues, this book highlights the often-overlooked intersection of warfare and global instability. Wars contribute to environmental degradation, economic collapse, and political fragmentation, creating a vicious cycle that is difficult to break. The book provides new insights into how modern warfare, including proxy wars and cyber warfare, has evolved and has devastatingly impacted nations' and regions' immediate and long-term stability. By including this critical perspective, the book offers a more nuanced understanding of the crises and proposes solutions that account for the complex dynamics of global conflict.

The book also emphasizes the importance of inclusivity and equity in addressing global crises. It argues that solutions to these crises must prioritize the voices and needs of marginalized communities, particularly those in the Global South, who are disproportionately affected by environmental degradation, economic inequality, and political instability. This focus on inclusion is a key aspect of the book's contribution to the field, as it challenges traditional top-down approaches to crisis resolution and advocates for more participatory, grassroots strategies. The book's focus on equity is informed by the principles of social justice and the capabilities approach, which emphasizes expanding individuals' freedoms and opportunities to live a life they value (Sen, 1999; Nussbaum, 2011).

Humanity in Crisis bridges the knowledge gap by offering a comprehensive, interdisciplinary analysis of today's interconnected crises. It challenges conventional approaches to crisis resolution and proposes a more integrated, inclusive, and collaborative framework for addressing these challenges. The book provides a fresh perspective on the path forward by adding new insights into the role of war in exacerbating global crises and emphasizing the need for equity and inclusivity. Its theoretical and practical contributions offer actionable solutions for policymakers, activists, and individuals seeking to make a difference in the world.

1.9 Humanity at a Crossroads: A Critical Moment in History

Humanity stands at a pivotal moment in history, facing a convergence of multiple crises that threaten the very foundations of our societies, economies, and ecosystems. The world is grappling with unprecedented challenges, including environmental degradation, economic inequality, political instability, technological disruption, and cultural fragmentation. These crises are not isolated;

they are deeply interconnected, creating a complex web of issues requiring a holistic response. At this crossroads, humanity must decide whether to continue with unsustainable practices and escalating conflict or embrace transformative change that fosters a more equitable, sustainable, and peaceful world.

One of humanity's most pressing crises is environmental degradation, which has reached alarming levels in recent decades. Climate change, deforestation, biodiversity loss, and pollution are all symptoms of a larger environmental crisis that is rapidly accelerating. According to the Intergovernmental Panel on Climate Change (IPCC, 2023), global temperatures are on track to rise by 1.5°C above pre-industrial levels by as early as 2030, leading to devastating impacts on ecosystems, food security, and human health. This environmental crisis is not just an ecological issue; it is deeply intertwined with social, economic, and political systems. For example, the most vulnerable populations – particularly those in the Global South – are disproportionately affected by environmental degradation, exacerbating existing inequalities and contributing to social unrest (Schlosberg, 2012). Addressing climate change and other environmental issues requires a paradigm shift in how we relate to the natural world, moving away from exploitation and towards sustainability, equity, and stewardship.

Economic inequality is another critical challenge that has reached unprecedented levels in the 21st century. The gap between the rich and the poor continues to widen, with wealth concentrated in the hands of a small elite while billions of people live in poverty or struggle to meet basic needs. According to Oxfam (2023), the richest 1% of the global population now owns more than half of the world's wealth, while millions are denied access to education, healthcare, and other essential services. This inequality is not just a matter of income disparity; it is a systemic issue that permeates every aspect of society, from access to resources and opportunities to political power and influence. Economic inequality fuels social fragmentation, as marginalized communities are excluded from decision-making processes and denied the opportunities to flourish. The widening wealth gap also exacerbates political instability as people lose faith in institutions, perpetuating inequality and leading to rising populism, authoritarianism, and social unrest (Piketty, 2014). To address economic inequality, we must focus on redistributive policies and reimagine global economic systems to prioritize human well-being and environmental sustainability.

Political instability is another manifestation of the interconnected crises facing humanity. In recent years, we have witnessed the rise of populist movements, the erosion of democratic institutions, and the increasing polarization of societies. These political shifts are often fueled by economic inequality, social fragmentation, and the erosion of trust in government institutions. Geopolitical rivalries, regional conflicts, and the rise of authoritarian regimes further complicate the political landscape. The war in Ukraine, for example, has not only caused immense human suffering but also triggered a global energy crisis, disrupted food supplies, and deepened geopolitical divisions. According to the United Nations (2023), conflicts and political instability are now the leading drivers of displacement, with over 110 million people forcibly displaced worldwide. Political instability also exacerbates environmental and economic crises, as governments are often unable or unwilling to address long-term challenges like climate change or poverty. To stabilize political systems, we must strengthen democratic institutions, promote inclusive governance, and address the root causes of conflict and instability.

Technological disruption is another force that is shaping the future of humanity. While technology can potentially solve many of the world's problems, it also poses significant risks. The rapid pace of technological advancement, particularly in artificial intelligence (AI), automation, and biotechnology, reshapes labor markets, economies, and social structures. According to Brynjolfsson and McAfee (2014), automation is expected to displace millions of jobs in the coming decades, particularly in manufacturing, transportation, and retail. This disruption can potentially exacerbate economic inequality, as those displaced by automation may struggle to find new employment opportunities. Additionally, the rise of AI and other emerging technologies raises concerns about privacy, surveillance, and the concentration of power in the hands of a few large corporations. While technology can offer solutions to pressing global challenges, it also requires careful regulation and ethical considerations to ensure that it benefits all of humanity and does not deepen existing inequalities or create new forms of exploitation.

Finally, cultural fragmentation is a growing concern in an increasingly interconnected yet deeply divided world. The rise of nationalism, religious extremism, and identity politics has led to a breakdown in social cohesion, with communities becoming more polarized and less willing to engage in constructive dialogue. According to Huntington (1996), cultural and civilizational divides are becoming more pronounced, with conflicts arising between nations and within them. This fragmentation is exacerbated by the rise of social media, which often amplifies divisive rhetoric and reinforces echo chambers. As societies become more fragmented, building the trust and cooperation necessary to address global challenges like climate change, inequality, and conflict becomes increasingly difficult. To overcome cultural fragmentation, we must promote cross-cultural understanding, inclusivity, and dialogue, recognizing the value of diverse perspectives and fostering a sense of shared humanity.

At this crossroads, humanity faces a choice: continue down a path of fragmentation, inequality, and environmental destruction or embrace a transformative shift towards sustainability, equity, and global cooperation. The crises we face are not insurmountable but require bold, collective action. By addressing these interconnected crises in a holistic, inclusive, and collaborative manner, we can create a more just and sustainable future for all.

1.10 Humanity in Crisis: A Call to Action for Global Transformation

Humanity is facing an unprecedented crisis, with multiple interwoven challenges threatening the fabric of our societies, economies, and ecosystems. These crises – environmental degradation, economic inequality, political instability, technological disruption, and cultural fragmentation – are not isolated problems but are deeply interconnected, creating a complex web of issues that require a comprehensive and integrated response. The urgency of addressing these crises cannot be overstated, as their consequences are already being felt globally, and their impacts will only deepen without collective action. This is a critical juncture in human history, and the time for transformative change is now.

The environmental crisis is perhaps the most pressing of all. Climate change, biodiversity loss, deforestation, and pollution threaten the planet's ability to sustain life. According to the Intergovernmental Panel on Climate Change (IPCC, 2023), global temperatures are on track to rise by 1.5°C above pre-industrial levels by as early as 2030, with catastrophic consequences for ecosystems, agriculture, water resources, and human health. Environmental degradation disproportionately affects the most vulnerable populations, particularly those in the Global South, exacerbating poverty, displacement, and social inequality (Schlosberg, 2012). Moreover, the environmental crisis is intrinsically linked to economic systems prioritizing short-term profit over long-term sustainability. The transition to a sustainable global economy requires rethinking how we produce, consume, and distribute resources, shifting away from exploitative practices towards a model of regenerative development that respects both people and the planet.

Economic inequality is another fundamental crisis that has deepened in recent decades. The gap between the rich and the poor continues to widen, with wealth and power concentrated in the hands of a small elite while billions of people are left behind, struggling to meet basic needs. Oxfam (2023) states that the richest 1% controls more than half of the world's wealth, while nearly 10% of the global population lives in extreme poverty. This economic disparity is a matter of income inequality and reflects systemic issues in global economic structures, including unequal access to education, healthcare, and opportunities. The COVID-19 pandemic further exacerbated these inequalities, highlighting the vulnerability of marginalized communities and the failures of current economic systems to address basic human needs. To achieve global justice, we must radically rethink our economic systems, promoting policies prioritizing equity, human well-being, and environmental sustainability over profit maximization (Piketty, 2014).

Political instability is another critical issue that threatens global stability. Across the world, we are witnessing the erosion of democratic institutions, the rise of authoritarianism, and the increasing polarization of societies. These political shifts are often fueled by economic inequality, social fragmentation, and the erosion of trust in government institutions. The war in Ukraine, for example, has not only caused immense human suffering but also triggered a global energy crisis, disrupted food supplies, and deepened geopolitical divisions. As the United Nations (2023) reports, conflicts and political instability are now the leading drivers of displacement, with over 110 million people forcibly displaced worldwide. Political instability also exacerbates other crises, such as climate change and economic inequality, as governments struggle to address long-term challenges while focusing on short-term survival. Strengthening democratic institutions, promoting inclusive governance, and addressing the root causes of conflict and instability are essential steps towards global transformation.

Technological disruption, while offering many opportunities, also presents significant risks. The rapid advancement of technologies such as artificial intelligence (AI), automation, and biotechnology is reshaping labor markets, economies, and social structures. As Brynjolfsson and McAfee (2014) argue, automation is expected to displace millions of jobs in the coming decades, particularly in industries like manufacturing and transportation, leading to greater economic inequality and social unrest. Moreover, the rise of AI and other emerging technologies raises concerns about privacy, surveillance, and the concentration of power in the hands of a few large corporations. While technology has the potential to solve many of the world's problems, its rapid development must be accompanied by careful regulation and ethical considerations to ensure that it serves the common good rather than exacerbating existing inequalities or creating new forms of exploitation.

Cultural fragmentation is another key challenge that undermines social cohesion and global cooperation. In an increasingly interconnected world, societies are becoming more polarized, with identity politics, nationalism, and religious extremism on the rise. According to Huntington (1996), cultural and civilizational divides are becoming more pronounced, with conflicts arising between nations and within them. This fragmentation is further exacerbated by the rise of social media, which amplifies divisive rhetoric and reinforces echo chambers. As societies become more fragmented, building the trust and cooperation necessary to address global challenges like climate change, inequality, and conflict becomes increasingly difficult. Overcoming cultural fragmentation requires promoting cross-cultural understanding, inclusivity, and dialogue, recognizing the value of diverse perspectives, and fostering a sense of shared humanity.

At this critical juncture, humanity faces a choice: continue down a path of division, inequality, and environmental destruction or embrace a transformative shift towards sustainability, equity, and global cooperation. The interconnectedness of these crises means that solutions cannot be piecemeal or isolated; they must be holistic, inclusive, and collaborative. This call to action for global transformation requires reimagining our political, economic, and social systems and a deep commitment to shared values of justice, equity, and sustainability. By addressing the root causes of these crises and fostering a culture of global solidarity, we can create a more just, peaceful, and sustainable world for future generations.

1.11 Wars and Their Role in Escalating the Humanity Crises: A Historical Perspective

Wars have long been a central force in shaping the trajectory of human societies, often exacerbating crises that extend far beyond the battlefield. From the ancient world to the present day, the consequences of war have been profound, contributing to political instability, economic collapse, social upheaval, and environmental degradation. Understanding the historical role of wars in bringing about crises is essential to recognizing the interconnectedness of global challenges today and the urgent need for collective action to prevent future conflicts.

1.11.1 Ancient and medieval wars: The roots of conflict

In the ancient and medieval periods, wars were often fought for territorial expansion, resources, and power. The consequences of these conflicts were felt by the combatants and the broader populations who suffered from the disruption of agricultural production, loss of life, and the destruction of infrastructure. For example, the conquests of Alexander the Great in the 4th century BCE led to the spread of Hellenistic culture across vast regions but also caused widespread devastation and displacement of populations. Similarly, the Roman Empire's wars of expansion in the Mediterranean region, while facilitating economic growth and cultural exchange, also resulted in the enslavement of millions and the destabilization of local societies (Goldsworthy, 2000).

The medieval period, marked by frequent wars between kingdoms and empires, also saw the rise of crusades driven by religious ideologies and territorial ambitions. The Crusades, which spanned from the 11th to the 13th centuries, resulted in the loss of life, the displacement of communities, and the weakening of local economies. The destruction of cities and the disruption of trade routes contributed to the economic decline of many regions in the Middle East and Europe (Tyerman, 2006). These wars also intensified cultural and religious divisions, laying the groundwork for future conflicts based on identity and belief systems.

1.11.2 Early modern and industrial age wars: The rise of total warfare

The early modern period ushered in new forms of warfare, particularly with the advent of gunpowder, which allowed for large-scale conflicts. The Thirty Years' War (1618–1648), one of the most devastating conflicts of the early modern period, not only involved a series of battles between European powers but also led to a humanitarian crisis across the continent. The war resulted in the deaths of millions, primarily due to violence, famine, and disease. The destruction of infrastructure and the economic collapse that followed further deepened the crisis, leaving much of Central Europe in ruins (Parker, 1997).

The Industrial Revolution in the 19th century introduced new technologies of war, such as railways, steamships, and more advanced weaponry, which transformed the scale and intensity of conflicts. The Napoleonic Wars (1803–1815) were among the first to demonstrate the full potential of industrialized warfare, and the consequences of Napoleon's expansionist ambitions were felt across Europe. These wars not only caused significant loss of life but also disrupted economies, leading to food shortages and the displacement of millions. The collapse of empires and the reordering of political boundaries after the Napoleonic Wars set the stage for future conflicts in Europe and beyond (Browning, 1993).

1.11.3 The world wars: Global crises of the 20th century

The two world wars of the 20th century are the most catastrophic conflicts in human history regarding the sheer scale of destruction and the long-term crises they triggered. World War I (1914–1918) resulted in the collapse of four major empires – the Austro-Hungarian, Ottoman, Russian, and German Empires – leading to political instability and the redrawing of national borders. The Treaty of Versailles, which ended the war, imposed harsh reparations on Germany, creating economic turmoil that contributed to the rise of extremist political movements, including Nazism (Tooze, 2006). The war also caused millions of deaths, widespread famine, and the displacement of populations, particularly in Eastern Europe and the Middle East.

World War II (1939–1945) had an even more profound impact, with an estimated 70–85 million people killed, including six million Jews in the Holocaust. The war led to the destruction of entire cities, particularly in Europe and Asia, and caused immense economic hardship. The post-war world was marked by the division of Europe into East and West, the onset of the Cold War, and the establishment of new international institutions, such as the United Nations, designed to prevent future conflicts. However, the legacy of World War II continues to shape global politics today, as many of the political and economic crises that emerged in its wake, such as the Israeli and Palestinian conflict and the Korean War, are still unresolved (Gaddis, 2005).

1.11.4 The Cold War and proxy wars: Political instability and proxy conflicts

The Cold War (1947–1991) between the United States and the Soviet Union was characterized by ideological, political, and military competition. However, it was also marked by a series of proxy wars fought in the developing world. These conflicts, often fueled by Cold War superpower rivalries, led to widespread suffering, political instability, and the displacement of millions. The Korean War (1950–1953), the Vietnam War (1955–1975), and the Soviet invasion of Afghanistan (1979–1989) are just a few examples of conflicts that devastated entire regions, leading to the loss of life, economic destruction, and the destabilization of political systems (Herring, 2014).

The Cold War also saw the rise of nuclear weapons, which brought the threat of global annihilation to the forefront of international relations. While the Cold War did not escalate into a direct military confrontation between the superpowers, the specter of nuclear war created an atmosphere of fear and uncertainty that permeated global politics. The arms race between the United States and the Soviet Union diverted vast resources into military spending, while issues like poverty, education, and healthcare were neglected (Sagan, 1993).

1.11.5 Contemporary conflicts: The war on terror and ongoing instability

In the post-Cold War era, warfare has continued to evolve, with new forms of conflict emerging in the 21st century. The 9/11 attacks and the subsequent War on Terror have led to prolonged military engagements in Afghanistan, Iraq, and other parts of the Middle East. These conflicts, often characterized by asymmetric warfare, have caused significant loss of life, economic devastation, and political instability. The invasion of Iraq in 2003, for example, destabilized the entire region, leading to the rise of extremist groups like ISIS and the ongoing Syrian Civil War, which has displaced millions and created a humanitarian crisis (Gerges, 2016).

In addition to traditional state-based conflicts, contemporary warfare increasingly involves non-state actors, cyber warfare, and information warfare. These new forms of conflict present unique challenges for global security, as they blur the lines between combatants and civilians and have far-reaching consequences for political stability and social cohesion (Arquilla & Ronfeldt, 1997). Moreover, the proliferation of weapons, including nuclear, biological, and chemical weapons, continues to pose a significant threat to global peace and security.

Throughout history, wars have been a major catalyst for crises, leading to widespread destruction, displacement, and suffering. The consequences of war are not limited to the immediate aftermath of conflict but extend to long-term political, economic, and social instability. As humanity grapples with contemporary challenges, including climate change, inequality, and technological disruption, it is essential to recognize war's deep and lasting impact on global stability. Addressing the root causes of conflict – such as resource competition, political instability, and ideological extremism – will be crucial to preventing future wars and mitigating their catastrophic effects.

1.12 Humanity in Crisis: The Need for Global Harmony

As we witness the unfolding crises of the 21st century, it becomes increasingly clear that the root causes of many of these challenges – wars, conflicts, and environmental degradation – are deeply interconnected. The competition for resources, development, and comforts often exacerbates tensions between nations and peoples, creating a global crisis that transcends borders. Humanity's struggle for dominance, security, and economic growth has led to significant human suffering, environmental destruction, and political instability. However, in the face of these crises, there lies an urgent need for reflection, understanding, and, most importantly, the pursuit of a new path forward – one that fosters harmony between humanity and the planet.

1.13 The Interconnectedness of Global Crises

Wars and conflicts have long been the driving forces behind the suffering and instability in human societies. From ancient civilizations to contemporary times, the struggle for power, resources, and territorial control has often led to violence and destruction. Today, this dynamic is more pronounced than ever, as conflicts are not only waged between nation-states but also involve non-state actors, insurgencies, and proxy wars. These conflicts often have global ramifications, as they spill over borders, displace millions of people, and disrupt economies. The Syrian Civil War, for instance, has caused one of the largest refugee crises in modern history, with millions of displaced Syrians seeking refuge in neighboring countries and Europe (Maciulaitis & Tolk, 2018).

In addition to armed conflict, the competition for development and comforts – often manifesting as economic inequality and environmental degradation – has exacerbated global tensions. The race for technological advancements, resource extraction, and economic dominance often leads to environmental harm, social inequality, and the depletion of natural resources. The extraction of fossil fuels, for example, has contributed significantly to climate change, while the overconsumption of resources by affluent nations has created a growing divide between the Global North and South. These dynamics contribute to a cycle of poverty, inequality, and environmental degradation, all of which further fuel conflicts and instability (Piketty, 2014; Sachs, 2015).

1.14 The Need for a New Path: Harmony between Humanity and the Planet

In response to these crises, one of the major objectives of this book is to propose a new path forward – one that seeks to harmonize humanity's relationship with the planet. The concept of global harmony is not a utopian ideal, but a practical and necessary approach to resolving the crises we face. Achieving global harmony requires recognizing the deep interdependence of all life on Earth, understanding that human well-being is inextricably linked to the health of the planet, and embracing a holistic approach to addressing the challenges of our time.

The idea of harmonizing humanity involves fostering a collective sense of responsibility and shared purpose. It means prioritizing cooperation over competition, empathy over hostility, and mutual respect over domination. The concept of "harmonizing humanity" is rooted in the belief that we must work together – across nations, cultures, and ideologies – to address the crises that threaten our survival. This approach requires recognizing the dignity and rights of all individuals, promoting social equity, and ensuring that the benefits of development and progress are shared more equitably across the globe. It also involves rethinking the traditional notion of development, moving away from a model that prioritizes material growth at the expense of social and environmental well-being, towards one that emphasizes sustainability, equity, and human flourishing (Sen, 1999; Nussbaum, 2011).

1.14.1 The role of global cooperation and collaboration

A key aspect of achieving global harmony is the recognition that no single nation or group can address these crises alone. Global challenges such as climate change, pandemics, and economic inequality require collective action. The Paris Agreement on climate change, for instance, represents a global effort to address the environmental crisis, with countries committing to reduce greenhouse gas emissions and limit global warming. However, the success of such agreements depends on the willingness of all nations to cooperate and work towards common goals. The global community must move beyond nationalistic and protectionist tendencies and embrace a shared vision of a sustainable and peaceful world.

International organizations, such as the United Nations, play a crucial role in fostering global cooperation and diplomacy. The UN's Sustainable Development Goals (SDGs) provide a framework for addressing the interconnected challenges of poverty, inequality, and environmental degradation. These goals emphasize the need for collaborative action, with a focus on inclusivity, justice, and sustainability. By aligning national policies with the SDGs and promoting global cooperation, we can begin to address the root causes of conflict and instability, paving the way for a more harmonious and equitable world (UN, 2015).

1.14.2 The role of education and awareness

In addition to political and economic solutions, the pursuit of global harmony requires a shift in societal values and attitudes. Education and awareness play a critical role in this process. By fostering a culture of empathy, respect, and shared responsibility, we can cultivate a global citizenry that is committed to working for the common good. Transformative education, which emphasizes critical thinking, empathy, and global citizenship, is essential for creating a generation of leaders and activists who are equipped to tackle the complex challenges of the modern world.

Moreover, the integration of diverse knowledge systems – such as indigenous knowledge, local ecological wisdom, and scientific expertise – can provide a more holistic understanding of the world and its problems. This epistemological pluralism can help bridge the gap between different cultural perspectives and offer innovative solutions to global crises. By embracing multiple ways of knowing and learning from each other, we can create a more inclusive and sustainable future (Escobar, 2018; Nussbaum, 2011).

1.15 A Vision for the Future

Humanity is at a crossroads. The crises we face – wars, conflicts, environmental degradation, and inequality – are deeply interconnected and threaten the stability and well-being of our world. However, there is hope. By harmonizing humanity's relationship with the planet and fostering global cooperation, we can address these challenges and create a more just, peaceful, and sustainable world. This book aims to show that achieving global harmony is not only possible but necessary for the survival and flourishing of all life on Earth. It is a call to action for individuals, communities, and nations to come together and work for the common good, guided by the principles of empathy, equity, and sustainability.

References

Appadurai, A. (2006). *Fear of small numbers: An essay on the geography of anger*. Duke University Press.

Arquilla, J., & Ronfeldt, D. (1997). *The emergence of noopolitik: Toward an American information strategy*. RAND Corporation.

Browning, R. (1993). *The Napoleonic Wars: The rise and fall of an empire*. The University of Chicago Press.

Brynjolfsson, E., & McAfee, A. (2014). *The second machine age: Work, progress, and prosperity in a time of brilliant technologies*. W. W. Norton & Company.

CEOBS & Scientists for Global Responsibility. (2022). *Military emissions gap: The military's contribution to climate change*. https://ceobs.org

Dasgupta, P. (2021). *The economics of biodiversity: The dasgupta review*. HM Treasury.
Escobar, A. (2018). *Designs for the pluriverse: Radical interdependence, autonomy, and the making of worlds*. Duke University Press.
FAO. (2023). *The impact of the war in Ukraine on global food security*. Food and Agriculture Organization of the United Nations. https://www.fao.org
FAO. (2023). *The state of food security and nutrition in the world 2023*. Food and Agriculture Organization of the United Nations.
Gaddis, J. L. (2005). *The Cold War: A new history*. Penguin Press.
Gerges, F. A. (2016). *ISIS: A history*. Princeton University Press.
Gleick, P. H. (2014). *The world's water volume 8: The biennial report on freshwater resources*. Island Press.
Goldsworthy, A. (2000). *The complete Roman army*. Thames & Hudson.
Herring, G. C. (2014). *The American war in Vietnam: A short history*. Oxford University Press.
Huntington, S. P. (1996). *The clash of civilizations and the remaking of world order*. Simon & Schuster.
IPCC. (2021). *Climate change 2021: The physical science basis*. Cambridge University Press.
IPCC. (2023). *Climate change 2023: Synthesis report*. Intergovernmental Panel on Climate Change.
IPCC. (2023). *Climate change 2023: The physical science basis*. Intergovernmental Panel on Climate Change.
Maciulaitis, M., & Tolk, A. (2018). *The humanitarian crisis in Syria: An analysis*. Springer.
Nussbaum, M. (2011). *Creating capabilities: The human development approach*. Harvard University Press.
Nussbaum, M. C. (2011). *Creating capabilities: The human development approach*. Harvard University Press.
Oxfam. (2023). *The inequality virus: The impact of COVID-19 on inequality*. Oxfam International.
Parker, G. (1997). *The thirty years' war*. Routledge.
Piketty, T. (2014). *Capital in the twenty-first century*. Harvard University Press.
Raworth, K. (2017). *Doughnut economics: Seven ways to think like a 21st-century economist*. Chelsea Green Publishing.
Sachs, J. D. (2015). *The age of sustainable development*. Columbia University Press.
Sagan, S. D. (1993). *The limits of safety: Organizations, accidents, and nuclear weapons*. Princeton University Press.

Schlosberg, D. (2012). *Environmental justice and the new global politics of difference*. Oxford University Press.
Sen, A. (1999). *Development as freedom*. Oxford University Press.
Sen, A. (1999). *Development as freedom*. Oxford University Press.
SIPRI. (2024). *SIPRI yearbook 2024: Armaments, disarmament, and international security*. Stockholm International Peace Research Institute.
SIPRI. (2024). *World military expenditure reaches a new high*. Stockholm International Peace Research Institute. https://www.sipri.org
Steffen, W., Crutzen, P. J., & McNeill, J. R. (2015). The Anthropocene: Are humans now overwhelming the great forces of nature? *Ambio, 36*(8), 614–621. https://doi.org/10.1579/0044-7447(2007)36[614:taaah]2.0.co;2
The Edelman Trust Barometer. (2024). *2024 Edelman Trust Barometer: Global report*. Edelman.
The World Inequality Report. (2022). *World Inequality Report 2022*. World Inequality Lab.
Tooze, A. (2006). *The wages of destruction: The making and breaking of the Nazi economy*. Viking.
Tyerman, C. (2006). *God's war: A new history of the Crusades*. Belknap Press.
UN. (2015). *Transforming our world: The 2030 agenda for sustainable development*. United Nations.
UNHCR. (2024). *Global trends: Forced displacement in 2023*. United Nations High Commissioner for Refugees. https://www.unhcr.org
UNICEF. (2023). *Education in emergencies: A global crisis*. United Nations Children's Fund. https://www.unicef.org
UNICEF. (2023). *The state of the world's children 2023: For every child, vaccination*. United Nations Children's Fund.
United Nations. (2023). *Global trends: Forced displacement in 2023*. United Nations High Commissioner for Refugees.
World Bank. (2021). *The economic impact of the Syrian conflict*. https://www.worldbank.org
World Bank. (2021). *World development report 2021: Data for better lives*. World Bank Group.
Zuboff, S. (2019). *The age of surveillance capitalism: The fight for a human future at the new frontier of power*. PublicAffairs.

PART I

Understanding the Crisis

2

Understanding the Crisis: Theories and Epistemology

The contemporary crises facing humanity – environmental degradation, economic inequality, political instability, technological disruption, and conflicts – are deeply complex and interconnected. To address these crises effectively, it is essential to explore the underlying theories and epistemological frameworks that shape our understanding of the world and guide our responses to these challenges. Theories of crisis and epistemology help us make sense of the multifaceted nature of these global issues and offer potential pathways for resolution. By examining different theoretical perspectives and epistemological approaches, we can better understand the root causes of these crises and the most effective ways to address them.

2.1 Theories of Crisis: A Multifaceted Perspective

A crisis can be defined as a critical turning point or a period of intense difficulty, disruption, or danger. In the context of contemporary global challenges, crises are not isolated events but rather interconnected phenomena that affect multiple dimensions of human life – social, economic, political, and environmental. Several theoretical frameworks help us understand the nature of these crises and their impact on society.

One prominent theory is Systems Theory, which views crises as disruptions in complex systems. According to systems theory, societies, economies, and ecosystems are composed of interconnected components that interact with one another. When one part of the system is disrupted – whether through environmental degradation, economic collapse, or political instability – it can trigger a cascade of effects that destabilize the entire system (Meadows, 2008). This framework highlights the importance of understanding the

DOI: 10.1201/9788743808787-3

interconnectedness of global crises and the need for holistic, systemic approaches to problem-solving.

Another important theoretical approach is Critical Theory, which emphasizes the role of power, inequality, and social structures in shaping crises. Scholars like Habermas (1984) and Marcuse (1964) have argued that societal crises often arise from structural injustices, including economic exploitation, political repression, and cultural domination. Critical theory suggests that the root causes of contemporary crises are deeply embedded in the social, economic, and political systems that perpetuate inequality and injustice. To address these crises, it is necessary to challenge these power structures and promote more equitable and inclusive social arrangements.

Additionally, ecological economics provides a framework for understanding the environmental dimensions of global crises. Ecological economists argue that traditional economic models, which prioritize growth and profit maximization, fail to account for the ecological limits of the planet (Daly, 1996). These models often contribute to environmental degradation, resource depletion, and climate change. Ecological economics calls for a shift towards a steady-state economy, where economic activity is aligned with the carrying capacity of the Earth. This perspective emphasizes the need for sustainable development and the importance of integrating ecological principles into economic decision-making.

2.2 Epistemology and the Crisis of Knowledge

Epistemology, the study of knowledge and how we come to know the world, plays a crucial role in understanding the crises we face. The way we perceive and interpret the world shapes our responses to global challenges. Traditional epistemological approaches, which prioritize objective, scientific knowledge, have been instrumental in advancing technological innovation and solving many practical problems. However, in the context of contemporary crises, it is increasingly clear that these approaches are insufficient for addressing the complexity and interconnectedness of global challenges.

One of the limitations of traditional epistemology is its emphasis on reductionism – the idea that complex phenomena can be understood by breaking them down into simpler components. While reductionism has been successful in many scientific fields, it often fails to capture the complexity of global crises, which involve multiple, interrelated factors. For example, the climate crisis cannot be understood solely in terms of carbon emissions or deforestation; it also involves social, economic, and political dimensions, such as the role of global inequality, the political economy of fossil fuel industries, and the cultural values that drive consumption patterns (Latour, 2018).

In response to the limitations of reductionism, epistemological pluralism has emerged as an important framework for understanding global crises. Epistemological pluralism acknowledges that there are multiple ways of knowing and understanding the world, including scientific knowledge, indigenous knowledge, local ecological wisdom, and experiential knowledge. This approach recognizes that different cultures and communities may have different ways of interpreting and responding to crises, and that these diverse perspectives can offer valuable insights and solutions (Escobar, 2018). For example, indigenous knowledge systems often emphasize the interconnectedness of humans and nature, offering alternative models for sustainable living that challenge dominant Western paradigms of development and progress (Nash, 2001).

Moreover, postcolonial epistemology provides a critical lens for understanding the role of colonialism and imperialism in shaping global crises. Postcolonial scholars argue that the historical exploitation of colonized peoples and resources has contributed to the contemporary patterns of inequality, environmental degradation, and conflict that we see today (Said, 1978). Postcolonial epistemology calls for decolonization of knowledge and the recognition of marginalized voices and perspectives in addressing global crises. This approach seeks to challenge the dominance of Western epistemological frameworks and promote a more inclusive, diverse understanding of the world.

2.3 Integrating Theories and Epistemologies for Global Transformation

To effectively address the crises facing humanity, it is essential to integrate multiple theories and epistemologies. A holistic approach that draws on systems theory, critical theory, ecological economics, and epistemological pluralism can help us better understand the root causes of global crises and identify pathways for resolution. This integration requires a shift in how we approach problem-solving, moving away from siloed, disciplinary perspectives towards more interdisciplinary and collaborative approaches.

For example, addressing the climate crisis requires not only scientific knowledge about carbon emissions and renewable energy technologies but also an understanding of the social, economic, and political factors that drive environmental degradation. It also requires an appreciation of indigenous knowledge and local ecological practices that have sustained communities for centuries. Similarly, addressing global inequality requires not only economic theories of wealth distribution but also a critical understanding of power dynamics, historical injustices, and the cultural values that perpetuate social hierarchies.

By integrating diverse theories and epistemologies, we can develop more comprehensive and inclusive solutions to the global crises we face. This approach encourages collaboration across disciplines, cultures, and knowledge systems, fostering a more inclusive and equitable global response to the challenges of the 21st century.

Understanding the crises facing humanity requires a multifaceted approach that draws on diverse theories and epistemologies. Systems theory, critical theory, ecological economics, and epistemological pluralism offer valuable insights into the interconnectedness of global crises and the need for holistic, collaborative solutions. By integrating these perspectives, we can develop more effective and inclusive strategies for addressing the challenges of our time. The integration of diverse knowledge systems, including indigenous and local knowledge, is essential for fostering global cooperation and creating a more sustainable, equitable, and peaceful world.

2.4 Humanity at the Brink: Understanding the Global Crisis

The phrase "humanity at the brink" conveys a sense of urgency and alarm about the state of the world. It reflects the growing recognition that humanity is facing a series of interrelated crises that threaten not only the stability of societies but the very survival of the planet itself. From environmental degradation to economic inequality, political instability, technological disruptions, and ongoing conflicts, the world is witnessing a convergence of challenges that demand immediate and collective action. These crises are not isolated events but rather interconnected phenomena that feed into one another, creating a cycle of destruction and suffering.

2.4.1 Environmental crisis: The planet in peril

One of the most pressing crises humanity faces is the environmental degradation of the planet. Climate change, driven largely by human activities such as the burning of fossil fuels and deforestation, is causing rising temperatures, extreme weather events, and sea-level rise. Biodiversity loss, pollution, and resource depletion are further exacerbating the ecological imbalance. According to the Intergovernmental Panel on Climate Change (IPCC), we are already witnessing the effects of climate change, with projections indicating that the situation will worsen unless immediate action is taken to reduce greenhouse gas emissions and shift towards sustainable practices (IPCC, 2021). The environmental crisis is not only a matter of ecological concern but also has profound social, economic, and political implications, particularly for vulnerable populations who are most affected by environmental degradation.

2.4.2 Economic inequality: The growing divide

Economic inequality is another critical crisis that has reached alarming levels in recent decades. While global wealth has increased, the distribution of that wealth has become increasingly uneven. The World Inequality Report (2022) highlights that the richest 1% of the global population now controls more than a quarter of the world's wealth, while billions of people continue to live in poverty. The gap between the rich and poor has widened, with many developing countries facing systemic challenges in addressing inequality. Economic inequality fuels social unrest and political instability, and contributes to the erosion of social cohesion. It also hampers economic growth, as a large portion of the population is excluded from opportunities for education, healthcare, and employment.

2.4.3 Political instability: The fragility of governance

Political instability is another crisis that is threatening global peace and security. From authoritarian regimes to failed states, the world is witnessing a breakdown of governance in many regions. Political instability often leads to civil unrest, violence, and the displacement of millions of people. The erosion of democratic institutions and the rise of populist movements have further exacerbated tensions within and between countries. Conflicts in the Middle East, Africa, and Latin America, along with the growing political polarization in established democracies, underscore the fragility of political systems in the face of economic and social challenges. Political instability is not only a threat to national security but also global peace, as it often leads to the escalation of conflicts and the spread of extremism.

2.4.4 Technological disruption: A double-edged sword

Technological advancements have brought significant benefits to humanity, but they have also introduced new challenges and risks. The rise of artificial intelligence, automation, and biotechnology has the potential to revolutionize industries and improve quality of life. However, these technologies also pose significant ethical, social, and economic risks. Automation, for example, threatens to displace millions of jobs, particularly in manufacturing and service industries, leading to widespread unemployment and social unrest. The development of AI raises concerns about privacy, surveillance, and the concentration of power in the hands of a few corporations and governments. Additionally, the rapid pace of technological change is outpacing the ability of regulatory frameworks to keep up, leading to a regulatory vacuum that leaves society vulnerable to unforeseen consequences.

2.4.5 Conflict and war: The persistence of violence

Conflict and war have been persistent features of human history, and they continue to be a major source of suffering and instability in the contemporary world. From ongoing wars in Syria, Yemen, and Ukraine to the rise of armed conflicts in Africa and Asia, violence continues to ravage societies and disrupt global peace. The proliferation of weapons, the rise of extremist ideologies, and the breakdown of international diplomacy have contributed to the persistence of war in the 21st century. The humanitarian consequences of war are staggering, with millions of people displaced, injured, or killed, and entire societies torn apart. War not only threatens the lives of those directly involved but also has long-term social, economic, and environmental consequences that reverberate across generations.

2.5 The Need for Global Transformation

These crises – environmental, economic, political, technological, and military – are deeply interconnected and cannot be addressed in isolation. The challenges we face today reflect the complex and interdependent nature of global systems. To address these crises, humanity must embrace a holistic, inclusive, and collaborative approach that recognizes the interconnectedness of all aspects of human life and the planet. This requires a fundamental shift in how we think about development, governance, technology, and peace.

The idea of global transformation is central to addressing the crises we face. This transformation involves rethinking our systems, structures, and values to create a more sustainable, equitable, and peaceful world. It requires collective action at the global, national, and local levels, with a focus on cooperation rather than competition, solidarity rather than division, and sustainability rather than short-term profit. It also involves a deep reimagining of our relationship with the planet, recognizing that human well-being is inseparable from the health of the Earth.

Humanity is indeed at the brink, facing a series of crises that threaten our future. However, this moment of crisis also presents an opportunity for global transformation. By understanding the interconnectedness of the challenges, we face and embracing a holistic approach to solutions, we can chart a new path forward – one that prioritizes sustainability, equity, and peace. The time for action is now, and it is up to all of us to work together to create a more harmonious and resilient world for future generations.

2.6 Overview of Contemporary Crises

Humanity is currently facing a series of interconnected crises that threaten the stability of societies and the well-being of the planet. These crises, which span across environmental, economic, political, technological, and social domains, are not isolated from one another but are deeply intertwined, creating a complex web of challenges that require urgent attention and action. Understanding these crises in their entirety is crucial for developing effective solutions. The following sections provide an overview of the most pressing contemporary crises and their interconnected nature.

2.6.1 Environmental crisis: Climate change and ecological degradation

The environmental crisis is perhaps the most urgent and widespread crisis facing humanity. Climate change, driven by human activities such as the burning of fossil fuels, deforestation, and industrial agriculture, is causing global temperatures to rise, leading to extreme weather events, rising sea levels, and the loss of biodiversity. The Intergovernmental Panel on Climate Change (IPCC) (2021) reports that the world is on track to exceed the critical threshold of 1.5°C of warming, which will have catastrophic impacts on ecosystems, food security, water availability, and human health. Additionally, deforestation, pollution, and the over-exploitation of natural resources are pushing the planet's ecosystems to the brink of collapse, threatening not only the survival of species but also the ability of the Earth to support human life.

2.6.2 Economic inequality: Widening gaps and social injustice

Economic inequality has reached alarming levels, both within and between countries. The wealthiest individuals and corporations control an ever-increasing share of global resources, while billions of people remain trapped in poverty. According to the World Inequality Report (2022), the top 1% of the global population controls more than a quarter of the world's wealth, while the poorest half of humanity owns less than 10%. This growing divide contributes to social unrest, instability, and a lack of access to necessities such as education, healthcare, and housing. Economic inequality also hinders overall economic growth, as large portions of the population are excluded from opportunities for development and participation in the global economy.

2.6.3 Political instability: Erosion of democracy and governance

Political instability is another major crisis facing contemporary society. The erosion of democratic institutions, the rise of authoritarianism, and the weakening of the rule of law have led to increasing political polarization, civil unrest, and even the breakdown of governance in some regions. The United Nations (2021) has highlighted the rise of political violence, the suppression of civil liberties, and the decline of democratic freedoms in many countries. This instability is compounded by corruption, weak institutions, and the failure of governments to address pressing social and economic issues. The result is a loss of public trust in government, increased political polarization, and the rise of extremist ideologies, all of which contribute to social fragmentation and conflict.

2.6.4 Technological disruption: Opportunities and risks

While technological advancements have brought many benefits, they have also introduced new challenges and risks. The rise of artificial intelligence, automation, and biotechnology has the potential to revolutionize industries and improve quality of life. However, these technologies also threaten to displace millions of jobs, exacerbate economic inequality, and raise ethical concerns. For example, automation in manufacturing and service sectors could lead to widespread unemployment, particularly among low-skilled workers, while AI and machine learning technologies could lead to the concentration of power in the hands of a few tech giants and governments. Moreover, the rapid pace of technological innovation often outpaces the development of regulatory frameworks, leading to a lack of oversight and the potential for unintended consequences.

2.6.5 Social fragmentation: Cultural and identity conflicts

Social fragmentation, driven by cultural, ethnic, and religious divisions, is another critical crisis facing the world today. The rise of populism, nationalism, and identity politics has deepened social divisions and fueled conflict in many countries. Social media platforms have amplified these divisions, created echo chambers and polarized public discourse. The refugee crisis, exacerbated by conflict, climate change, and economic hardship, has also led to tensions over immigration, national identity, and cultural assimilation. These social fractures not only undermine social cohesion but also contribute to the rise of hate speech, discrimination, and violence. The lack of understanding and empathy between different groups has made it increasingly difficult to build inclusive societies that respect diversity and promote social justice.

2.6.6 Conflict and war: Persistent violence and human suffering

Conflict and war remain among the most devastating crises humanity faces. The 21st century has seen a continuation of violent conflicts in regions such as the Middle East, Africa, and Eastern Europe, with devastating consequences for civilian populations. The war in Syria, the ongoing conflict in Yemen, and the Russian invasion of Ukraine are just a few examples of how warfare continues to cause immense human suffering, displacement, and economic destruction. United Nations High Commissioner for Refugees (UNHCR) (2023) reports that over 100 million people are currently displaced worldwide due to conflict, violence, and persecution. Wars not only lead to loss of life but also disrupt economies, destroy infrastructure, and create long-term psychological and social impacts on affected populations.

2.7 Interconnected Crises and the Need for Global Solutions

These crises – environmental, economic, political, technological, social, and military – are not isolated from one another but are deeply interconnected. Climate change exacerbates economic inequality and social unrest, while political instability and conflict disrupt efforts to address environmental and economic challenges. Technological disruption can both contribute to and alleviate these crises, depending on how it is managed. The growing fragmentation of societies further complicates efforts to find common ground and work together to address global challenges.

The complexity and interconnectedness of these crises make it clear that no single solution or approach will suffice. Instead, addressing these crises requires a holistic, collaborative, and inclusive approach that recognizes the interdependence of all aspects of human life and the planet. The path forward lies in building global cooperation, fostering social and economic justice, and embracing sustainable development practices that prioritize the well-being of both people and the planet.

2.8 Interconnected Challenges in the Modern World

In the contemporary era, the challenges humanity faces are no longer isolated or confined to individual regions or sectors. Instead, they are deeply interconnected, with environmental, economic, political, technological, and social crises influencing one another in complex ways. This interconnectedness amplifies the severity of these challenges, making it increasingly difficult to address them in isolation. Understanding the interlinkages between these crises is crucial for developing comprehensive solutions that can foster long-term stability, sustainability, and peace.

2.8.1 Environmental and economic interdependence

One of the most prominent examples of interconnected challenges is the relationship between environmental degradation and economic inequality. Climate change, deforestation, and biodiversity loss are not only environmental issues but also economic ones. The poorest communities, who contribute least to environmental degradation, are often the most vulnerable to its impacts, including extreme weather events, crop failures, and water scarcity. According to the Intergovernmental Panel on Climate Change (IPCC) (2021), climate change disproportionately affects developing countries, exacerbating existing economic inequalities.

Conversely, economic systems that prioritize short-term profit and growth over sustainability often contribute to environmental degradation. The extraction of natural resources for industrial purposes, such as mining and fossil fuel extraction, accelerates environmental damage while also perpetuating economic inequality by concentrating wealth and power in the hands of a few corporations and individuals. The World Bank (2022) highlights how economic policies that disregard environmental sustainability lead to long-term costs, such as the loss of biodiversity, which undermines the foundation of many industries, including agriculture and tourism.

2.8.2 Political instability and social fragmentation

Political instability and social fragmentation are also intricately linked. Political instability, often fueled by economic inequality, corruption, and poor governance, leads to social unrest, which in turn destabilizes countries and regions. The rise of populism and nationalism in many parts of the world has exacerbated political divisions, creating environments where social fragmentation thrives. For example, in countries like Brazil, Hungary, and the United States, political polarization has led to social fragmentation, as different groups compete for power, resources, and recognition.

This political fragmentation is often amplified by technological platforms, where social media can exacerbate divisions by creating echo chambers and enabling the spread of misinformation. As Zeynep Tufekci (2017) argues in *Twitter and Tear Gas*, social media platforms can both mobilize social movements and deepen political divides, contributing to political instability and social unrest. The erosion of trust in democratic institutions, coupled with rising inequality, creates fertile ground for political violence and extremism, undermining social cohesion and peace.

2.8.3 Technological disruption and inequality

Technological advancements, while offering tremendous potential for innovation and development, also contribute to widening inequality. Automation, artificial intelligence (AI), and digital technologies are transforming industries, but they are also displacing jobs, particularly in sectors that rely on manual labor. According to a report by the World Economic Forum (2020), AI and automation could displace up to 85 million jobs globally by 2025, while simultaneously creating 97 million new roles. However, the new roles require different skills, and many workers, particularly those in low-income or low-skilled jobs, may not have the resources or opportunities to reskill.

This technological divide is further compounded by unequal access to technology, with poorer communities and countries often lacking the infrastructure to benefit from technological advancements. The OECD (2021) highlights how digital inequality exacerbates existing disparities, limiting access to education, healthcare, and economic opportunities. Moreover, the concentration of power in the hands of a few tech giants raises concerns about data privacy, surveillance, and the monopolization of digital markets, creating new forms of economic and political control.

2.8.4 Global conflict and social instability

Conflict and war are other critical areas where interconnected challenges are evident. Armed conflicts, whether in Syria, Yemen, or Ukraine, not only cause immediate human suffering but also destabilize entire regions, leading to refugee crises, economic collapse, and environmental destruction. The ongoing war in Ukraine, for example, has exacerbated global food insecurity by disrupting grain exports, particularly affecting countries in Africa and the Middle East that rely on Ukrainian wheat. The United Nations World Food Programme (2022) warns that conflicts around the world are pushing millions into hunger, further entrenching poverty and inequality.

Moreover, the environmental impacts of war, such as the destruction of infrastructure, pollution, and the displacement of populations, contribute to the long-term instability of affected regions. As Robert D. Kaplan (2012) discusses in *The Revenge of Geography*, the geopolitical consequences of conflict are often intertwined with environmental factors, such as access to water, energy, and fertile land. The competition for these resources can escalate tensions and fuel further conflict, creating a vicious cycle of war and environmental degradation.

2.8.5 Health crises and global inequality

The COVID-19 pandemic is a stark example of how global health crises are interconnected with economic, political, and social issues. The pandemic exposed deep inequalities in healthcare systems, with wealthier nations having greater access to vaccines and medical resources, while poorer nations struggled to secure even basic supplies. The World Health Organization (WHO) (2021) reported that the pandemic exacerbated existing health disparities, particularly in low-income countries, where healthcare systems were already fragile.

Additionally, the economic impact of the pandemic has been severe, with millions losing their jobs, particularly in sectors such as tourism, hospitality, and retail. The pandemic also disrupted education systems, exacerbating inequalities in access to quality education and digital learning. As Amartya Sen (1999) argues in *Development as Freedom*, the inability to access healthcare, education, and economic opportunities limits human capabilities, perpetuating cycles of poverty and inequality.

2.9 The Need for Holistic Solutions

The interconnectedness of contemporary crises underscores the need for holistic solutions that address the root causes of these challenges. Environmental degradation, economic inequality, political instability, technological disruption, and social fragmentation are not isolated problems but are part of a complex, interwoven system. Addressing these challenges requires a global, collaborative approach that recognizes their interdependencies and fosters cooperation across borders, sectors, and disciplines. By understanding the interconnected nature of these crises, we can begin to develop solutions that promote sustainability, equity, and peace for future generations.

2.10 The Urgency of Collective Action

In the face of the numerous crises confronting humanity – climate change, economic inequality, political instability, technological disruption, and social fragmentation – the urgency of collective action has never been more apparent. These challenges are not only interconnected but also transcend national borders, affecting people across the globe in profound and often devastating ways. To address these crises effectively, it is essential that humanity acts together, drawing on the strength of collective efforts rather than relying on fragmented, individual responses.

2.10.1 Global crises demand global solutions

The complexity and scale of the crises we face require solutions that are equally vast and interconnected. Climate change, for example, is a global phenomenon that cannot be tackled by any one nation acting alone. The Paris Agreement, adopted in 2015, represents a significant step toward collective action, with nearly every country in the world committing to limit global warming to well below 2°C. However, as Hegerl et al. (2017) emphasize, the lack of full implementation and the need for greater ambition highlight the challenges of global cooperation in addressing environmental crises. Climate change mitigation requires sustained global cooperation, investment in clean technologies, and a rethinking of economic models that prioritize growth over sustainability.

Similarly, the economic inequality that has been exacerbated by the COVID-19 pandemic cannot be solved by isolated national policies or market-based solutions alone. As Piketty (2014) argues in *Capital in the Twenty-First Century*, inequality is a global issue that requires international coordination in areas such as tax policy, wealth redistribution, and the regulation of multinational corporations. Collective action to address economic inequality includes not only government interventions but also the engagement of civil society, businesses, and international organizations to create a more equitable global economy.

2.10.2 Political instability and the need for global governance

Political instability, particularly in fragile states and conflict zones, exacerbates global challenges and undermines efforts to build a stable and sustainable world. In many cases, political instability is a direct result of systemic inequality, corruption, and the concentration of power in the hands of a few. Acemoglu and Robinson (2012) argue in *Why Nations Fail* that inclusive political and economic institutions are essential for sustainable development and peace. However, achieving such institutions requires collective action from both national governments and the international community to support democratic governance, human rights, and the rule of law.

The role of global governance in addressing political instability is crucial. International organizations such as the United Nations and regional bodies like the European Union play critical roles in mediating conflicts, providing humanitarian aid, and promoting peace. However, as Kofi Annan (2003) noted, the international community must be more proactive in preventing conflict and addressing the root causes of instability, such as poverty, inequality, and environmental degradation. The need for effective global governance mechanisms is more pressing than ever, as political instability in one region can quickly ripple out to affect global security, trade, and migration patterns.

2.10.3 Technological disruption and the need for ethical standards

The rapid pace of technological change has created both opportunities and challenges for societies worldwide. Automation, artificial intelligence (AI), and biotechnology have the potential to transform industries, improve quality of life, and solve complex global problems. However, these technologies also present significant risks, including job displacement, privacy violations, and the potential for widening inequality. As Brynjolfsson and McAfee (2014) argue in *The Second Machine Age*, technological disruption requires a collective response that includes both governments and the private sector working together to ensure that the benefits of innovation are shared equitably and that the risks are mitigated.

Moreover, the ethical implications of new technologies – ranging from data privacy concerns to the potential for surveillance capitalism – demand global cooperation. International frameworks for regulating emerging technologies, such as AI and biotechnology, are necessary to ensure that these innovations are used for the common good and do not exacerbate existing inequalities or create new forms of exploitation. As Floridi (2019) discusses in *The Fourth Industrial Revolution*, the ethical challenges posed by technological disruption require a global, collaborative effort to establish norms and standards that prioritize human dignity, equity, and sustainability.

2.10.4 Social fragmentation and the role of solidarity

Social fragmentation, fueled by rising nationalism, populism, and polarization, is a significant barrier to collective action. In an increasingly interconnected world, the inability to work together across borders and cultures undermines efforts to address global challenges. As Putnam (2000) argues in *Bowling Alone*, social capital – trust, cooperation, and solidarity – is essential for building resilient communities and societies. However, the erosion of social trust, exacerbated by misinformation and political polarization, makes it more difficult to build the solidarity necessary for collective action.

To overcome social fragmentation, it is essential to foster a sense of shared responsibility and common purpose. This involves promoting dialogue, understanding, and cooperation across cultural and ideological divides. Global movements such as Fridays for Future, led by Greta Thunberg, demonstrate the power of collective action in addressing climate change, with people from diverse backgrounds coming together to demand action. The success of such movements highlights the potential for solidarity to transcend national borders and create a united front in the fight against global crises.

2.10.5 The role of education and public awareness

Education and public awareness are fundamental to mobilizing collective action. As Sen (1999) argues in *Development as Freedom*, education is not only a tool for personal empowerment but also a means of fostering collective responsibility and social change. Public awareness campaigns, informed by scientific research and grounded in the principles of justice and equity, can help build the social and political will needed to tackle global crises.

Moreover, the role of education in fostering global citizenship is critical. As Nussbaum (2011) discusses in *Creating Capabilities*, education should aim to cultivate individuals who can think critically about global issues and who are motivated to act in ways that promote the common good. By integrating global challenges such as climate change, inequality, and conflict into educational curricula, we can empower future generations to engage in collective action and work toward a more sustainable and just world.

The urgency of collective action in addressing the interconnected crises facing humanity cannot be overstated. Global challenges such as climate change, economic inequality, political instability, technological disruption, and social fragmentation require coordinated efforts from governments, international organizations, businesses, and civil society. Only through collective action can we hope to build a sustainable, equitable, and peaceful future. As we confront these challenges, it is essential to recognize our shared responsibility and the power of solidarity in creating a better world for all.

References

Acemoglu, D., & Robinson, J. A. (2012). *Why nations fail: The origins of power, prosperity, and poverty*. Crown Business.

Brynjolfsson, E., & McAfee, A. (2014). *The second machine age: Work, progress, and prosperity in a time of brilliant technologies*. W. W. Norton & Company.

Daly, H. E. (1996). *Beyond growth: The economics of sustainable development*. Beacon Press.

Escobar, A. (2018). *Designs for the pluriverse: Radical interdependence, autonomy, and the making of worlds*. Duke University Press.

Floridi, L. (2019). *The fourth industrial revolution*. Springer.

Habermas, J. (1984). *The theory of communicative action* (Vol. 1). Beacon Press.

Hegerl, G. C., et al. (2017). *Understanding climate change: A guide for decision-makers*. Cambridge University Press.

Intergovernmental Panel on Climate Change (IPCC). (2021). *Climate change 2021: The physical science basis*. Cambridge University Press.

Kaplan, R. D. (2012). *The revenge of geography: What the map tells us about coming conflicts and the battle against fate*. Random House.

Latour, B. (2018). *Down to earth: Politics in the new climatic regime*. Polity Press.

Meadows, D. H. (2008). *Thinking in systems: A primer*. Chelsea Green Publishing.

Nash, R. (2001). *American Indian environmental ethics: An overview*. Environmental Ethics, 23(1), 25–44.

Nussbaum, M. (2011). *Creating capabilities: The human development approach*. Harvard University Press.

OECD. (2021). *The digital divide and its impact on economic inequality*. OECD Publishing.

Piketty, T. (2014). *Capital in the twenty-first century*. Harvard University Press.

Putnam, R. D. (2000). *Bowling alone: The collapse and revival of American community*. Simon & Schuster.

Said, E. W. (1978). *Orientalism*. Pantheon Books.

Sen, A. (1999). *Development as freedom*. Oxford University Press.

Tufekci, Z. (2017). *Twitter and tear gas: The power and fragility of networked protest*. Yale University Press.

United Nations High Commissioner for Refugees (UNHCR). (2023). *Global trends: Forced displacement in 2023*. UNHCR.

United Nations World Food Programme. (2022). *Global food security report 2022*. United Nations.

United Nations. (2021). *The state of democracy and governance in the world*. United Nations.

World Bank. (2022). *The economics of climate change*. World Bank.

World Economic Forum. (2020). *The future of jobs report 2020*. World Economic Forum.

World Health Organization (WHO). (2021). *The impact of COVID-19 on global health systems*. WHO.

World Inequality Report. (2022). *World Inequality Report 2022*. World Inequality Lab.

3

The Anthropocene and its Consequences

3.1 Introduction – Anthropocene

The Anthropocene is a proposed geological epoch that highlights the significant and lasting impact of human activities on the Earth's ecosystems, geology, and climate. While not yet officially recognized as a formal geological epoch, it is widely used in scientific, environmental, and cultural discussions to describe the profound changes humans have caused to the planet.

The term "Anthropocene" combines "anthropo" (human) and "cene" (new) and was popularized by atmospheric chemist Paul Crutzen and ecologist Eugene Stoermer in the early 2000s. It represents a proposed geological epoch that emphasizes the profound and lasting impact of human activities on Earth's systems. The Anthropocene signifies a departure from the Holocene epoch, which began approximately 11,700 years ago after the last Ice Age, to an era where human influence is the dominant force shaping the planet.

Key markers of the Anthropocene include several interconnected phenomena. Climate change is a central feature, characterized by accelerated global warming due to greenhouse gas emissions from industrial and agricultural activities. Another hallmark is biodiversity loss, with mass extinctions driven by habitat destruction, overexploitation of natural resources, and pollution.

The Anthropocene is also defined by altered biogeochemical cycles, as human activities have drastically changed the natural flows of nitrogen, carbon, and phosphorus, leading to environmental imbalances. Land use changes, such as urbanization, deforestation, and large-scale agriculture, have reshaped Earth's landscapes, further impacting ecosystems. Additionally, pollution has left a lasting imprint, with the accumulation of plastics, radioactive isotopes, and other pollutants becoming ubiquitous in the environment.

The Anthropocene is marked by the presence of technofossils, which are physical remnants of human activity embedded in geological strata. These include materials such as plastics, concrete, and electronic waste, serving as enduring evidence of humanity's transformative influence on the planet.

DOI: 10.1201/9788743808787-4

The Anthropocene refers to a proposed geological epoch in which human activities have become the dominant force shaping Earth's systems (Crutzen & Stoermer, 2000). This concept underscores the unprecedented impact of human actions on the planet, including changes to the climate, biodiversity, land use, and the chemical composition of the atmosphere, oceans, and soils. Unlike previous epochs, which were driven by natural processes, the Anthropocene highlights the extent to which human activities – industrialization, deforestation, urbanization, and fossil fuel combustion – are altering Earth's systems on a global scale (Steffen et al., 2015). The consequences of the Anthropocene are profound, with implications for ecological balance, human societies, and the planet's future habitability.

3.2 Consequences of Anthropocene

One of the most significant consequences of the Anthropocene is climate change, driven by the excessive release of greenhouse gases such as carbon dioxide (CO_2), methane (CH_4), and nitrous oxide (N_2O) into the atmosphere. The *Intergovernmental Panel on Climate Change* (IPCC, 2023) reports that anthropogenic emissions have led to a rise in global temperatures, resulting in extreme weather events, sea-level rise, and disruptions to ecosystems. The melting of polar ice caps, ocean acidification, and biodiversity loss further underscore the interconnected nature of these changes. The Anthropocene signifies a shift where human activities are no longer localized but have planetary-scale consequences, threatening the stability of Earth's climate system and the survival of countless species.

Another consequence of the Anthropocene is the loss of biodiversity and ecological degradation. Human activities such as habitat destruction, overexploitation of resources, and pollution have accelerated species extinction rates to levels not seen since the last mass extinction event 65 million years ago (IPBES, 2019). The *Food and Agriculture Organization* (FAO, 2023) highlights that unsustainable agricultural practices and deforestation are key drivers of biodiversity loss, reducing the resilience of ecosystems that provide essential services such as pollination, water purification, and climate regulation. The degradation of ecosystems has direct consequences for human well-being, particularly for vulnerable communities that depend on natural resources for their livelihoods.

The Anthropocene also exacerbates global inequalities, as its consequences are not evenly distributed. While industrialized nations are responsible for the majority of historical greenhouse gas emissions, developing countries disproportionately bear the brunt of climate change impacts, including extreme weather, food insecurity, and displacement (UNHCR, 2024). The *World Bank* (2021) emphasizes that climate change has the potential to push millions into poverty, further widening the gap between the rich and the poor. Vulnerable populations, particularly in the Global South, face challenges in adapting to environmental changes due to limited resources and infrastructure. This unequal burden raises critical questions about environmental justice and the ethical responsibilities of nations to address the consequences of the Anthropocene equitably (Nussbaum, 2011; Sen, 1999).

Furthermore, the Anthropocene has profound implications for human health and well-being. Environmental changes such as air and water pollution, rising temperatures, and extreme weather events have increased the prevalence of respiratory diseases, heat-related illnesses, and vector-borne diseases (UNICEF, 2023). For instance, urbanization and industrialization have led to the deterioration of air quality in many parts of the world, contributing to millions of premature deaths annually (Zuboff, 2019). Additionally, the displacement of communities due to climate-related disasters poses significant challenges to mental health, social stability, and access to basic services such as healthcare, education, and shelter (UNHCR, 2024).

The Anthropocene marks a transformative period in Earth's history, characterized by the dominance of human activity in shaping planetary systems. Its consequences – climate change, biodiversity loss, ecological degradation, global inequalities, and human health challenges – demand urgent and coordinated global responses. Addressing the Anthropocene requires a paradigm shift toward sustainability, equity, and resilience, as well as a recognition of humanity's collective responsibility to safeguard the planet for future generations. As Steffen et al. (2015) argue, the Anthropocene challenges us to rethink our relationship with nature and adopt practices that promote ecological balance and social justice.

3.3 Human Dominance and Planetary Impact

Human dominance over Earth's systems has reached an unprecedented scale, marking a transformative period in planetary history. This dominance, often framed within the concept of the Anthropocene, underscores humanity's ability to reshape Earth's natural systems through industrial, technological, and socio-economic activities. Since the onset of the Industrial Revolution, human actions – such as fossil fuel combustion, deforestation, urbanization, and large-scale agriculture – have accelerated environmental changes that were once driven primarily by natural processes (Crutzen & Stoermer, 2000). These activities have led to profound and often irreversible impacts on the planet, including climate change, biodiversity loss, land degradation, and ocean acidification.

One of the most significant planetary impacts of human dominance is climate change, driven by the massive release of greenhouse gases into the atmosphere. The *Intergovernmental Panel on Climate Change* (IPCC, 2023) highlights those anthropogenic emissions of carbon dioxide (CO_2), methane (CH_4), and nitrous oxide (N_2O) have caused a measurable rise in global temperatures, leading to extreme weather events, melting ice caps, rising sea levels, and disruptions to ecosystems. Industrialization and reliance on fossil fuels have accelerated the warming of the planet, with long-term consequences for both human societies and natural systems. This disruption to Earth's climate equilibrium highlights the scale of human influence on planetary processes and the urgent need for mitigation strategies to limit further harm.

In addition to climate change, human dominance has caused a drastic loss of biodiversity and the degradation of ecosystems. The *Intergovernmental Science-Policy Platform on Biodiversity and Ecosystem Services* (IPBES, 2019) reports that species extinction rates are now hundreds to thousands of times higher than natural background rates due to human activities. Habitat destruction, pollution, overexploitation, and the introduction of invasive species have resulted in the loss of critical ecosystems such as forests, wetlands, and coral reefs. These ecosystems are essential for providing ecosystem services, including clean air, water purification, food production, and climate regulation. The loss of biodiversity undermines ecological resilience and threatens human well-being, particularly for communities that depend directly on natural resources for survival (FAO, 2023).

Human dominance has also profoundly altered landscapes and oceans, transforming the Earth's surface and its biogeochemical cycles. Large-scale deforestation for agriculture and urbanization has led to soil degradation, desertification, and the loss of carbon sinks, exacerbating climate change (World Bank, 2021). Similarly, industrial and agricultural runoff, combined with overfishing and plastic pollution, has degraded marine ecosystems, leading to ocean acidification and the collapse of fish populations (UNEP, 2021). These changes demonstrate the interconnectedness of human actions and natural systems, where disruptions in one area often have cascading effects on others.

The consequences of human dominance are not evenly distributed across the globe, exacerbating global inequalities and environmental injustices. While industrialized nations have historically contributed the most to greenhouse gas emissions and resource extraction, developing countries and marginalized populations disproportionately bear the brunt of climate change, pollution, and resource depletion (UNHCR, 2024). The *World Inequality*

Report (2022) highlights the disparity between the environmental impacts of affluent societies and the vulnerabilities faced by poorer communities, who often lack the resources and infrastructure to adapt to environmental changes. This unequal burden raises ethical questions about responsibility, equity, and the need for global cooperation to address planetary challenges.

In conclusion, human dominance over Earth's systems has brought both remarkable progress and significant planetary impacts. The Anthropocene reflects a period in which human activities have become the primary drivers of environmental change, leading to climate disruption, biodiversity loss, and ecological degradation. Addressing these challenges requires a paradigm shift toward sustainable practices, technological innovation, and equitable solutions, recognizing humanity's collective responsibility to protect the planet. Steffen et al. (2015) argue that humanity must transition from being a dominant force of planetary disruption to a steward of Earth's systems, ensuring a balanced and resilient future for all life forms.

3.4 The Age of Overexploitation

The Age of Overexploitation is the current period in human history characterized by the unsustainable use of natural resources, driven by economic growth, industrialization, and increasing human consumption. Overexploitation involves extracting resources – such as forests, fisheries, minerals, and freshwater – at rates exceeding Earth's regeneration capacity. This phenomenon, fueled by rapid population growth and the demands of modern economies, has led to severe environmental degradation, biodiversity loss, and the disruption of planetary systems that sustain life (IPBES, 2019).

One of the most evident forms of overexploitation is deforestation. Forests, which are vital for carbon sequestration, biodiversity conservation, and the livelihoods of millions of people, are being cleared at alarming rates for agriculture, logging, and infrastructure development. According to the *Food and Agriculture Organization* (FAO, 2023), approximately 10 million hectares of forest are lost annually, with tropical rainforests experiencing the most significant decline. The expansion of industrial-scale agriculture, particularly for commodities such as palm oil, soy, and cattle ranching, is a primary driver of deforestation. This overexploitation contributes to climate change by releasing stored carbon and destroying critical habitats, pushing many species toward extinction (World Bank, 2021).

Another significant manifestation of overexploitation is the depletion of marine resources through overfishing. Global fish stocks have been exploited at unsustainable levels due to industrial fishing practices and the growing demand for seafood. The *Food and Agriculture Organization* (FAO, 2023) reports that over one-third of the world's fish stocks are overfished, jeopardizing marine ecosystems and the livelihoods of coastal communities dependent on fishing. Unsustainable practices, such as bottom trawling and illegal, unreported, and unregulated (IUU) fishing, exacerbate the decline of fish populations and damage marine habitats, including coral reefs and seafloors. The loss of marine biodiversity disrupts ecological balance, affecting food chains and the ocean's ability to regulate climate (IPCC, 2023).

Freshwater resources are also under significant strain due to overexploitation. Human activities, including agriculture, industry, and urbanization, consume vast amounts of freshwater, leading to water scarcity in many regions. The *United Nations Children's Fund* (UNICEF, 2023) highlights that billions of people lack access to clean water, and the overuse and mismanagement of freshwater sources such as rivers, lakes, and aquifers exacerbate this crisis. Overextraction of groundwater for irrigation, particularly in arid regions, has led to the depletion of aquifers faster than they can be naturally replenished. The consequences of freshwater overexploitation include desertification, loss of wetlands, and increased vulnerability to droughts and climate variability (UNEP, 2021).

The overexploitation of mineral and energy resources has intensified environmental degradation and contributed to the global climate crisis. The extraction of fossil fuels, metals, and minerals is central to industrial economies, but it often comes at the cost of habitat destruction, pollution, and greenhouse gas emissions. Mining activities, particularly in developing countries, have led to deforestation, soil erosion, and contaminating water sources with toxic chemicals (Zuboff, 2019). Additionally, burning fossil fuels for energy remains the primary driver of climate change, releasing carbon dioxide and other greenhouse gases into the atmosphere (IPCC, 2023). This overreliance on finite resources highlights the unsustainable nature of current economic models and the urgent need for a transition toward renewable energy and circular economies.

The consequences of overexploitation are disproportionately borne by vulnerable populations, exacerbating global inequalities and environmental injustices. Indigenous communities, for instance, are often displaced or marginalized due to resource extraction projects despite their traditional knowledge of sustainable resource management (Appadurai, 2006). The *World Inequality Report* (2022) emphasizes that while industrialized nations and wealthy corporations are the primary beneficiaries of resource exploitation, developing countries and marginalized communities suffer the most from environmental degradation, pollution, and resource depletion.

In conclusion, the Age of Overexploitation represents a critical turning point for humanity, as the unsustainable use of natural resources threatens ecological stability, human well-being, and planetary health. Addressing this crisis requires a fundamental shift toward sustainable resource management, equitable economic systems, and global cooperation. Strategies such as promoting renewable energy, implementing conservation initiatives, and adopting circular economy principles are essential for mitigating the impacts of overexploitation and ensuring a resilient future for both humans and the natural world (Steffen et al., 2015).

3.5 Navigating the Balance Between Progress and Sustainability

In the 21st century, balancing economic progress and environmental sustainability has emerged as one of humanity's most pressing challenges. While economic growth has led to significant advancements in technology, infrastructure, and global living standards, it has often come at the cost of environmental degradation, climate change, and resource depletion. Achieving harmony between progress and sustainability requires a shift in perspective – moving beyond traditional economic development models to embrace practices that prioritize long-term ecological health, equity, and resilience (UNEP, 2021).

Economic progress, driven by industrialization, globalization, and technological innovation, has undoubtedly improved human well-being. Over the past century, billions of people have gained access to education, healthcare, and improved living conditions. According to the *World Bank* (2021), global poverty rates have declined significantly, and technological advancements have facilitated unprecedented levels of communication, connectivity, and productivity. However, this progress has been accompanied by unsustainable consumption and production patterns, leading to rising greenhouse gas emissions, deforestation, pollution, and biodiversity loss. The IPCC (2023) warns that these trends have accelerated climate change, threatening planetary systems that sustain life and disproportionately impacting vulnerable populations.

The challenge lies in redefining progress to align with sustainability principles. Economic systems rooted in linear growth models, which emphasize resource extraction, production, and waste, are no longer viable. Instead, adopting circular economy frameworks can help minimize environmental impacts while fostering innovation and growth. A circular economy emphasizes reducing, reusing, and recycling resources to create a closed-loop system, thereby reducing waste and conserving natural capital (Ellen MacArthur Foundation, 2020). For example, transitioning from fossil fuels to renewable energy sources – such as solar, wind, and hydropower – can drive economic growth while mitigating climate change and reducing dependence on finite resources (IPCC, 2023).

Sustainability also requires addressing social and economic inequalities often exacerbated by unsustainable development practices. The *World Inequality Report* (2022) highlights that the benefits of economic progress have been unevenly distributed, with wealthier nations and individuals consuming a disproportionate share of global resources. At the same time, marginalized communities bear the brunt of environmental degradation, resource scarcity, and climate-related disasters. Achieving a balance between progress and sustainability necessitates inclusive development strategies prioritizing equity, resource access, and resilience-building for vulnerable populations (Sen, 1999; Nussbaum, 2011).

Technological innovation plays a critical role in navigating this balance. Emerging technologies, such as artificial intelligence (AI), clean energy solutions, and sustainable agriculture practices, offer opportunities to decouple economic growth from environmental harm. For instance, precision agriculture uses data and technology to optimize water and fertilizer use, reducing resource waste and increasing crop yields. Similarly, advancements in energy storage and electric vehicles enable a transition toward low-carbon transportation systems (Zuboff, 2019). However, the deployment of such technologies must be guided by ethical considerations and policies that ensure equitable access and minimize unintended consequences (UNEP, 2021).

Education, awareness, and governance are equally critical in achieving this balance. Governments, businesses, and civil society must work together to implement policies that incentivize sustainable practices, such as carbon pricing, green taxation, and subsidies for renewable energy. International agreements, like the Paris Agreement, provide frameworks for global cooperation on climate action, emphasizing shared but differentiated responsibilities (IPCC, 2023). Also, fostering a sustainability culture through education and awareness can empower individuals to make informed choices that contribute to a more sustainable future (Mezirow, 2000).

In conclusion, navigating the balance between progress and sustainability requires a transformative shift in how societies define and pursue development. It involves integrating environmental stewardship, social equity, and economic growth into a cohesive framework that prioritizes long-term planetary health. By embracing circular economies, advancing technological innovation, reducing inequalities, and fostering global cooperation, humanity can achieve a future where progress and sustainability coexist in harmony (Steffen et al., 2015).

References

Appadurai, A. (2006). *Fear of small numbers: An essay on the geography of anger*. Duke University Press.

Crutzen, P. J., & Stoermer, E. F. (2000). The "Anthropocene." *Global change newsletter, 41*(1), 17-18.

Ellen MacArthur Foundation. (2020). *The circular economy: A transformative pathway to sustainability*. Ellen MacArthur Foundation.

FAO. (2023). *The state of food security and nutrition in the world 2023*. Food and Agriculture Organization of the United Nations.

IPBES. (2019). *Global assessment report on biodiversity and ecosystem services*. Intergovernmental Science-Policy Platform on Biodiversity and Ecosystem Services.

IPCC. (2023). *Climate change 2023: Synthesis report*. Intergovernmental Panel on Climate Change.

Mezirow, J. (2000). *Learning as transformation: Critical perspectives on a theory in progress*. Jossey-Bass.

Nussbaum, M. C. (2011). *Creating capabilities: The human development approach*. Harvard University Press.

Sen, A. (1999). *Development as freedom*. Oxford University Press.

Steffen, W., Broadgate, W., Deutsch, L., Gaffney, O., & Ludwig, C. (2015). The trajectory of the Anthropocene: The great acceleration. *The Anthropocene Review, 2*(1), 81–98.

UNEP. (2021). *From pollution to solution: A global assessment of marine litter and plastic pollution*. United Nations Environment Programme.

UNEP. (2021). *Global environmental outlook 6: Healthy planet, healthy people*. United Nations Environment Programme.

UNHCR. (2024). *Global trends: Forced displacement in 2024*. United Nations High Commissioner for Refugees.

UNICEF. (2023). *Water scarcity: The challenge for children*. United Nations Children's Fund.

World Bank. (2021). *World development report 2021: Data for better lives*. World Bank Group.

World Inequality Report. (2022). *World inequality report 2022*. World Inequality Lab.

Zuboff, S. (2019). *The age of surveillance capitalism: The fight for a human future at the new frontier of power*. PublicAffairs.

4

Systemic Crises in a Globalized World

In an increasingly interconnected and interdependent globalized world, the frequency, complexity, and far-reaching impact of systemic crises have reached a critical level. These crises, from economic recessions and pandemics to climate change and geopolitical conflicts, are no longer isolated to individual nations or regions. Instead, they ripple across borders, impacting global economies, ecosystems, and societies. The interconnected nature of modern systems amplifies vulnerabilities, making it urgent to address systemic risks through collective action, resilience-building, and sustainable governance (UNEP, 2021).

One of the most notable examples of systemic crises is the 2008 global financial crisis, which highlighted the fragility of interconnected financial systems. Originating in the U.S. housing market, the crisis quickly spread across the globe due to the intricate web of financial institutions and international markets. The collapse of major banks and financial institutions led to widespread economic downturns, unemployment, and loss of livelihoods. As Stiglitz (2010) notes, the crisis underscored the dangers of deregulated markets, unsustainable financial practices, and a lack of accountability. It also demonstrated how systemic failures in one part of the world could destabilize the entire global economy, emphasizing the need for stronger regulatory frameworks and international cooperation.

The COVID-19 pandemic is another stark example of a systemic crisis with global repercussions. The rapid spread of the virus, facilitated by globalization and increased human mobility, overwhelmed healthcare systems, disrupted supply chains and led to unprecedented economic and social challenges. According to the World Bank (2021), the pandemic pushed millions into poverty, exacerbated inequalities, and revealed systemic weaknesses in healthcare infrastructure, particularly in low- and middle-income countries. The pandemic also highlighted the interconnectedness of human and planetary health, as zoonotic diseases are increasingly linked to environmental degradation, deforestation, and biodiversity loss (UNEP, 2021).

DOI: 10.1201/9788743808787-5

Climate change represents the most profound and long-term systemic crisis facing humanity. Driven by industrialization, unsustainable consumption, and fossil fuel dependence, climate change poses existential risks to ecosystems, economies, and communities worldwide. The IPCC (2023) warns that rising global temperatures, extreme weather events, and sea-level rise threaten food security, water resources, and human settlements. The impacts of climate change are disproportionately felt by vulnerable populations in developing nations, further exacerbating global inequalities (Nussbaum, 2011). Climate change is a systemic crisis because it affects multiple interconnected economic, environmental, and social systems, requiring coordinated global action to mitigate its effects and adapt to its consequences.

Globalization has also intensified geopolitical and social crises. Issues such as conflict, forced migration, and resource competition have become increasingly globalized, with local conflicts often producing far-reaching consequences. For instance, the war in Syria triggered one of the largest refugee crises in modern history, with millions of people displaced and seeking asylum in neighboring countries and Europe (UNHCR, 2024). Such crises strain host countries' social, economic, and political systems, highlighting the interconnectedness of conflict, migration, and global stability. Furthermore, competition over scarce resources, such as freshwater and arable land, is expected to intensify in the face of climate change, potentially leading to further conflicts and displacement (SIPRI, 2024).

Addressing systemic crises requires a multidimensional and collaborative approach prioritizing resilience, equity, and sustainability. Governments, international organizations, and civil society must work together to strengthen global governance frameworks, improve early warning systems, and build adaptive capacity. For example, initiatives like the Paris Agreement on climate change demonstrate the importance of collective action in addressing systemic environmental risks (IPCC, 2023). Additionally, promoting sustainable development through the United Nations' Sustainable Development Goals (SDGs) offers a roadmap for addressing interconnected crises by integrating economic, social, and environmental priorities (UNEP, 2021).

Systemic crises in a globalized world underscore the need for a paradigm shift toward resilience, equity, and sustainability. As interconnected systems amplify vulnerabilities, solutions must transcend national boundaries and prioritize collective action. By addressing root causes, strengthening governance, and fostering international cooperation, humanity cannot only navigate systemic risks but also build a more resilient and equitable future. This potential for positive change should inspire hope and optimism in our collective ability to overcome these challenges (Steffen et al., 2015; Stiglitz, 2010).

4.1 Understanding Complex Systems and Feedback Loops

Understanding the behavior of complex systems is of utmost importance. These intricate networks of interconnected components interact dynamically, often giving rise to emergent behaviors that cannot be predicted by analyzing individual parts alone. These systems exist across various domains, including ecosystems, climate systems, economies, and societies. Recognizing their non-linear nature, interdependencies, and sensitivity to small changes is crucial, as it can lead to significant and sometimes unintended consequences. Feedback loops are central to the functioning of complex systems, which play a critical role in maintaining stability or driving change (Meadows, 2008).

Feedback loops are mechanisms within complex systems that either reinforce or counteract change, often categorized as positive and negative feedback loops. Positive feedback loops amplify change and can lead to exponential growth or collapse. For example, in climate change, the melting of Arctic ice reduces the Earth's albedo (reflectivity), causing more solar radiation to be absorbed by the ocean, accelerating ice melt. This self-reinforcing loop exacerbates global warming and highlights the destabilizing effects of positive feedback (IPCC, 2023). Similarly, in economic systems, speculative bubbles represent positive feedback, where rising asset prices fuel further investment, eventually leading to unsustainable growth and collapse (Minsky, 1986).

In contrast, negative feedback loops work to stabilize systems by counteracting deviations and maintaining equilibrium. An example in natural systems is homeostasis in living organisms, where mechanisms like sweating or shivering regulate body temperature to ensure survival (Odum & Barrett, 2005). In socioeconomic systems, regulatory policies, such as interest rate adjustments by central banks, act as negative feedback loops to stabilize inflation and economic growth. These stabilizing forces are critical for maintaining resilience in systems under stress.

Complex systems often exhibit emergent properties, where the collective behavior of components produces outcomes that are not reducible to the actions of individual elements. For instance, ecosystems exhibit emergent properties such as biodiversity, nutrient cycling, and resilience arising from the interactions between species and their environment. Similarly, social systems display emergent phenomena like cultural norms, economic markets, and technological innovations shaped by collective actions and feedback within societies (Holling, 2001). Recognizing these emergent properties is essential for managing complex systems effectively, as interventions in one part of the system can produce unintended consequences elsewhere.

The interplay of feedback loops within complex systems often leads to tipping points, where a small change triggers a significant and usually irreversible shift in the system's state. For example, in climate systems, the collapse of the Amazon rainforest or the disintegration of the West Antarctic Ice Sheet represents potential tipping points with far-reaching global consequences (Lenton et al., 2019). Such thresholds underscore the importance of understanding feedback dynamics to anticipate and mitigate risks before systems reach points of no return.

Understanding complex systems and feedback loops in socio-economic and environmental contexts is critical for addressing global challenges such as climate change, biodiversity loss, and economic inequality. Systems thinking – an approach that emphasizes holistic analysis and the interconnections within systems – has become an essential tool for developing sustainable solutions. For instance, the Sustainable Development Goals (SDGs) integrate multiple interdependent targets, recognizing that progress in one area, such as poverty alleviation, depends on addressing other areas like education, gender equality, and environmental sustainability (UNEP, 2021).

To navigate complex systems effectively, policymakers, scientists, and stakeholders must adopt adaptive and flexible strategies that account for uncertainty and dynamic interactions. Tools like systems dynamics modeling help visualize feedback loops, simulate scenarios, and identify leverage points where interventions can produce the most significant impact (Meadows, 2008). For example, in environmental management, restoring wetlands can act as a leverage point by enhancing carbon sequestration, improving water quality, and supporting biodiversity, thereby creating stabilizing feedback loops within ecosystems.

In conclusion, understanding complex systems and feedback loops is essential for addressing the multifaceted challenges of the modern world. Whether in natural systems, economies, or societies, feedback mechanisms drive change, maintain stability, and shape emergent behaviors. By adopting systems thinking and recognizing the dynamic interactions within these systems, humanity can develop strategies to enhance resilience, avoid tipping points, and promote sustainability in a rapidly changing world (Lenton et al., 2019; Meadows, 2008).

4.2 Global Interdependencies and Vulnerabilities

In an increasingly interconnected world, global interdependencies have become a defining characteristic of modern societies, economies, and ecosystems. These interdependencies refer to the complex web of relationships between nations, industries, and environments, where actions or disruptions in one part of the world can have cascading effects on other regions and systems. While global interdependencies have fostered economic growth, technological advancements, and cultural exchange, they have also heightened vulnerabilities, making the world more susceptible to systemic risks. Understanding these interdependencies and vulnerabilities is critical for addressing contemporary global challenges such as climate change, pandemics, economic instability, and social inequality (UNEP, 2021).

One of the most evident examples of global interdependencies is in the global supply chain. The movement of goods, services, and raw materials across borders has created a highly integrated and efficient global economy. However, as the COVID-19 pandemic demonstrated, such interdependencies also create vulnerabilities. The pandemic disrupted global supply chains, leading to shortages of essential goods, including medical supplies, food, and electronics. The closure of factories, restrictions on international trade, and labor shortages in key industries exposed the fragility of a system that relies on just-in-time production and the free movement of goods (World Bank, 2021). Moreover, the pandemic revealed how global interdependencies could exacerbate inequalities, as low-income countries faced disproportionate challenges in accessing vaccines and healthcare resources (UNHCR, 2024).

Similarly, climate change represents a critical area where global interdependencies and vulnerabilities intersect. The impacts of climate change, such as rising sea levels, extreme weather events, and shifting agricultural patterns, are not confined to one region but have global repercussions. For example, droughts in one part of the world can lead to food shortages that affect international markets. At the same time, extreme weather events in one country can disrupt global supply chains and lead to economic instability. The IPCC (2023) warns that the interconnectedness of global systems means that climate-related risks are compounded by socio-economic vulnerabilities, with poorer nations and marginalized communities bearing the brunt of the impacts. These vulnerabilities are exacerbated by historical inequalities in the distribution of resources and emission responsibilities, highlighting the need for equitable solutions to global environmental challenges.

Economic interdependencies also expose the world to vulnerabilities, particularly in financial markets. The 2008 global financial crisis exemplifies how interconnected financial systems can lead to widespread economic turmoil. The collapse of Lehman Brothers in the United States triggered a global recession, affecting countries across all continents. As Stiglitz (2010) argues, the crisis directly resulted from the over-leveraging of financial institutions, lack of regulation, and the global nature of financial markets. The interconnectedness of economies meant that a failure in one country could quickly spread across borders, destabilizing financial systems and leading to long-term economic hardship for millions of people. The crisis also exposed the vulnerabilities of global trade networks, where disruptions in one sector or country can lead to ripple effects in others, highlighting the need for more resilient economic structures.

In addition to economic and environmental vulnerabilities, social vulnerabilities are another critical aspect of global interdependencies. Migration, for example, has become a global issue, with millions of people displaced due to conflict, climate change, and economic instability. The United Nations High Commissioner for Refugees (UNHCR, 2024) reports that the number of displaced persons worldwide has reached unprecedented levels, with refugees and migrants often facing hostile political environments, limited access to resources, and inadequate support systems. The interconnectedness of global economies and societies means that migration flows can create tensions within host countries, particularly in regions already struggling with social and economic challenges. Furthermore, the rise of nationalism and protectionism in many countries has led to increased barriers to migration and a reduction in international cooperation on refugee issues, exacerbating the vulnerabilities of displaced populations.

Global governance has become increasingly important in addressing the interdependencies and vulnerabilities of the modern world. Effective governance requires international cooperation, transparent decision-making, and the ability to respond to global challenges in a coordinated manner. The Sustainable Development Goals (SDGs), adopted by the United Nations, provide a framework for addressing the interconnected issues of poverty, inequality, climate change, and peace (UNEP, 2021). However, achieving these goals requires overcoming the barriers posed by national interests, geopolitical conflicts, and the unequal distribution of power and resources. As Holling (2001) suggests, building resilience in the face of global vulnerabilities requires adaptive governance systems that can respond to changing circumstances, anticipate risks, and foster collaboration across borders.

The global interdependencies and vulnerabilities are central to understanding the complexities of contemporary global challenges. The interconnectedness of economies, ecosystems, and societies has led to significant benefits, but it has also created a web of vulnerabilities that can lead to cascading risks. Addressing these vulnerabilities requires a holistic approach that recognizes the interdependencies between different systems and promotes global cooperation, equitable solutions, and resilience-building. By understanding and managing these interdependencies, the world can better navigate the challenges of the 21st century and build a more sustainable and equitable future (Stiglitz, 2010; UNEP, 2021).

4.3 Lessons from Past Crises: Major Crises and Their Impact

Throughout history, major crises have shaped human development, offering valuable lessons about vulnerability, resilience, and the need for proactive, coordinated responses. From financial collapses to environmental disasters, these crises have exposed systemic weaknesses and highlighted the interconnectedness of global systems. By examining past crises, we can better understand the dynamics that lead to societal breakdowns and learn how to mitigate their impacts in the future. This analysis is crucial in addressing contemporary global challenges such as climate change, economic inequality, and geopolitical instability (Stiglitz, 2010; IPCC, 2023).

One of the most significant crises of the 20th century was the Great Depression of the 1930s. This economic downturn, triggered by the stock market crash of 1929, led to widespread poverty, unemployment, and social unrest. The crisis exposed the fragility of the global financial system, where speculative investments and inadequate regulation created a bubble that eventually burst, leading to a global economic collapse (Kindleberger, 2005). One of the key lessons from the Great Depression was the importance of financial regulation and the role of government intervention in stabilizing economies. The establishment of institutions like the International Monetary Fund (IMF) and the World Bank after World War II directly responded to the economic instability of the interwar period to prevent future financial crises (Stiglitz, 2003). Additionally, the crisis underscored the need for social safety nets as unemployment and poverty levels reached unprecedented heights, highlighting the importance of social welfare programs in protecting vulnerable populations.

Another major crisis that has left lasting lessons is the 2008 global financial crisis. Triggered by the collapse of the housing market in the United States, the crisis quickly spread to other financial markets, leading to a global recession. The crisis exposed the dangers of financialization – where financial markets, institutions, and financial instruments dominate the economy – and the consequences of inadequate regulation in the banking sector. Minsky's (1986) theory of financial instability, which posits that financial markets tend to move from periods of stability to instability through speculative lending and borrowing, proved highly relevant in explaining the causes of the 2008 crisis. One of the primary lessons from this crisis was the need for stronger financial oversight and the implementation of regulatory frameworks that limit risky behavior in financial markets. In response to the crisis, governments worldwide implemented bailouts for failing banks and introduced quantitative easing measures to stabilize economies. However, the crisis also revealed the growing inequality between the financial elite and the general population, highlighting the need for more equitable economic policies (Piketty, 2014).

In addition to financial crises, environmental disasters have also provided important lessons about the vulnerability of human societies to ecological changes. One of the most devastating examples is the Chernobyl nuclear disaster of 1986, which resulted in the release of large amounts of radioactive material into the atmosphere, causing widespread health and environmental damage. The crisis revealed the dangers of unchecked technological advancement and the need for rigorous safety protocols in industries with potentially catastrophic risks. The Chernobyl disaster also highlighted the importance of transparency and accountability in managing technological risks, as government secrecy and misinformation marred the initial response to the disaster (Gorbachev, 2009). In the aftermath, there was a global reevaluation of nuclear energy policies, with many countries reassessing nuclear power plants' safety and environmental impacts. This crisis underscored the need for global cooperation in managing technological risks, particularly in an era of increasing interconnectedness and environmental degradation.

The COVID-19 pandemic is another major crisis that has provided valuable lessons about global vulnerabilities. The pandemic exposed the fragility of public health systems worldwide as countries struggled to manage the surge in infections and hospitalizations. It also highlighted the interconnectedness of global supply chains, as disruptions in one part of the world had ripple effects across industries and economies. The pandemic underscored the importance of preparedness and early intervention in mitigating the spread of infectious diseases. The global response to the pandemic also revealed the critical role of science and technology in addressing global health crises, with the rapid development of vaccines providing a crucial tool in combating the disease (WHO, 2021). However, the pandemic also exacerbated existing inequalities, as poorer countries faced challenges accessing vaccines and medical resources while wealthier nations could secure supplies through bilateral agreements (UNICEF, 2023). This disparity highlighted the need for global solidarity and equitable access to healthcare resources in future crises.

In conclusion, major crises throughout history have provided critical insights into human societies' vulnerabilities and the need for proactive, coordinated responses. Whether in finance, technology, or public health, these crises have exposed systemic weaknesses and highlighted the importance of resilience, regulation, and global cooperation in mitigating their impacts. By learning from past crises, policymakers and global leaders can better prepare for future challenges and work toward creating more sustainable, equitable, and resilient societies (IPCC, 2023; Stiglitz, 2010).

References

Gorbachev, M. (2009). *The new political thinking: A global perspective*. European University Press.

Holling, C. S. (2001). Understanding the complexity of economic, ecological, and social systems. *Ecosystems, 4*(5), 390–405.

IPCC. (2023). *Climate change 2023: Synthesis report*. Intergovernmental Panel on Climate Change.

Kindleberger, C. P. (2005). *The world in depression, 1929–1939* (2nd ed.). University of California Press.

Lenton, T. M., Rockström, J., Gaffney, O., Rahmstorf, S., Richardson, K., Steffen, W., & Schellnhuber, H. J. (2019). Climate tipping points—Too risky to bet against. *Nature, 575*(7784), 592–595.

Meadows, D. H. (2008). *Thinking in systems: A primer*. Chelsea Green Publishing.

Minsky, H. P. (1986). *Stabilizing an unstable economy*. Yale University Press.

Nussbaum, M. C. (2011). *Creating capabilities: The human development approach*. Harvard University Press.

Odum, E. P., & Barrett, G. W. (2005). *Fundamentals of ecology* (5th ed.). Thomson Brooks/Cole.

Piketty, T. (2014). *Capital in the twenty-first century*. Harvard University Press.

SIPRI. (2024). *SIPRI yearbook 2024: Armaments, disarmament, and international security*. Stockholm International Peace Research Institute.

Steffen, W., Broadgate, W., Deutsch, L., Gaffney, O., & Ludwig, C. (2015). The trajectory of the Anthropocene: The great acceleration. *The Anthropocene Review, 2*(1), 81–98.

Stiglitz, J. E. (2003). *Globalization and its discontents*. W.W. Norton & Company.

Stiglitz, J. E. (2010). *Freefall: America, free markets, and the sinking of the world economy*. W.W. Norton & Company.

Stiglitz, J. E. (2010). *Freefall: America, free markets, and the sinking of the world economy*. W.W. Norton & Company.

UNEP. (2021). *Making peace with nature: A scientific blueprint to tackle climate, biodiversity and pollution emergencies*. United Nations Environment Programme.

UNHCR. (2024). *Global trends: Forced displacement in 2024*. United Nations High Commissioner for Refugees.

UNICEF. (2023). *The state of the world's children 2023: Children, food, and nutrition*. UNICEF.

WHO. (2021). *World health statistics 2021: Monitoring health for the SDGs, sustainable development goals*. World Health Organization.

World Bank. (2021). *World development indicators 2021*. World Bank Group.

PART II

Environmental Crises

5

Environmental Crises: Causes, Consequences, and Responses

Environmental crises are a defining feature of the Anthropocene, a geological epoch characterized by significant human impact on the Earth's ecosystems and climate. These crises, including climate change, biodiversity loss, deforestation, and pollution, are increasingly recognized as existential threats to human civilization and the planet's ecological balance. Understanding the causes, consequences, and potential responses to these environmental crises is critical for fostering sustainable solutions and mitigating future harm.

5.1 Causes of Environmental Crises

Human activities primarily drive environmental crises, particularly industrialization, urbanization, and unsustainable consumption patterns. Climate change, for instance, is primarily caused by burning fossil fuels, deforestation, and agricultural practices that release greenhouse gases into the atmosphere. According to the Intergovernmental Panel on Climate Change (IPCC, 2023), human activities have caused unprecedented warming of the Earth, leading to rising sea levels, extreme weather events, and disruptions to ecosystems. Biodiversity loss, another major environmental crisis, is driven by habitat destruction, pollution, overexploitation of natural resources, and climate change. The World Wildlife Fund (WWF, 2022) reports that human-induced pressures have dramatically declined wildlife populations, with some species facing extinction.

The deforestation crisis is also a major contributor to environmental degradation. Large-scale logging, agricultural expansion, and infrastructure development deplete forests, critical for carbon sequestration, water regulation, and biodiversity. The Food and Agriculture Organization (FAO, 2023) estimates that approximately 10 million hectares of forests are lost yearly,

DOI: 10.1201/9788743808787-7

exacerbating climate change and threatening the livelihoods of communities dependent on forest ecosystems.

Pollution, in its various forms, is another significant driver of environmental crises. Plastic pollution in oceans, air pollution from industrial activities, and water pollution from agricultural runoff and industrial waste profoundly affect ecosystems and human health. According to the United Nations Environment Programme (UNEP, 2022), plastic waste in oceans has reached alarming levels, with an estimated 8 million tons entering the seas annually, threatening marine life and food security.

5.2 Consequences of Environmental Crises

The consequences of environmental crises are far-reaching and multifaceted, affecting both the natural world and human societies. Climate change, for example, has led to more frequent and severe weather events, such as hurricanes, droughts, and wildfires, which disrupt communities, agriculture, and infrastructure. These events disproportionately affect vulnerable populations, including low-income communities, Indigenous peoples, and developing countries, exacerbating existing social inequalities (IPCC, 2023).

Biodiversity loss has equally severe consequences. Ecosystems that depend on diverse species for pollination, pest control, and nutrient cycling are being destabilized, leading to declines in agricultural productivity and the collapse of essential ecosystem services. The World Economic Forum (2023) highlights that biodiversity loss directly threatens food security, as many crops depend on pollinators such as bees and butterflies. Additionally, the extinction of species diminishes the genetic diversity necessary for resilience in ecosystems and human health.

Deforestation contributes to climate change and biodiversity loss, as forests act as carbon sinks, absorbing significant amounts of CO_2 from the atmosphere. The destruction of forests accelerates global warming and disrupts the livelihoods of millions of people who depend on forests for food, medicine, and shelter. According to the FAO (2023), forest loss is a major driver of rural poverty and social conflict, particularly in regions where forests are integral to local economies and cultures.

Pollution has wide-ranging effects on both ecosystems and human health. Air pollution is responsible for millions of premature deaths each year, with respiratory and cardiovascular diseases being the leading causes of mortality. The World Health Organization (WHO, 2021) estimates that 7 million people die annually from exposure to air pollution. On the other hand, water pollution contaminates drinking water sources, leading to waterborne diseases and harming aquatic ecosystems, reducing fish populations and damaging coral reefs.

5.3 Responses to Environmental Crises

Addressing environmental crises requires coordinated global efforts, systemic changes, and transformative policies. International agreements such as the Paris Agreement on climate change aim to limit global warming and mitigate its effects through emissions reductions and adaptation strategies. The IPCC (2023) stresses the importance of urgent action to reduce greenhouse gas emissions and transition to renewable energy sources, such as solar, wind, and hydropower, to avoid the worst impacts of climate change.

Efforts to combat biodiversity loss have led to the establishment of international frameworks like the Convention on Biological Diversity (CBD), which seeks to conserve biodiversity and promote sustainable use of natural resources. The WWF (2022) advocates for creating protected areas, restoring degraded ecosystems, and implementing sustainable agricultural practices to safeguard biodiversity.

Initiatives like the REDD+ (Reducing Emissions from Deforestation and Forest Degradation) program have been launched to address deforestation. These programs incentivize forest conservation and sustainable land use. The FAO (2023) also promotes sustainable forestry practices, such as agroforestry and community-based forest management, which integrate environmental conservation with local economic development.

Pollution control is another area where global cooperation is crucial. The UNEP (2022) calls for stronger regulations on plastic production and waste management and efforts to reduce emissions from industrial and transportation sectors. Circular economy principles, which emphasize reducing, reusing, and recycling materials, are gaining traction as a way to minimize pollution and conserve resources.

In addition to these global efforts, local and grassroots movements are critical in addressing environmental crises. Indigenous communities, for example, have long advocated for protecting their lands and natural resources, emphasizing the importance of traditional ecological knowledge in conservation efforts (UNICEF, 2023). Similarly, youth-led movements, such as those spearheaded by Greta Thunberg, have mobilized millions of people worldwide to demand action on climate change and environmental justice.

Environmental crises are a pressing challenge that requires immediate and sustained action at all levels of society. The causes of these crises are rooted in unsustainable human activities, and their consequences are profound, affecting ecosystems, human health, and global stability. However, through international cooperation, policy reforms, and community engagement, there is potential to mitigate these crises and build a more sustainable future. As we navigate the complexities of the Anthropocene, it is essential to recognize the interconnectedness of environmental, social, and economic systems and work collectively to address the root causes of environmental degradation.

5.4 Climate Change and Global Warming: Causes, Impacts, and Solutions

Climate change and global warming are among the most pressing environmental challenges of the 21st century. These phenomena are driven by human activities, particularly the burning of fossil fuels, deforestation, and industrial processes, which accumulate greenhouse gases (GHGs) in the atmosphere. The rising global temperature has far-reaching consequences for ecosystems, human societies, and the planet. Addressing climate change requires a comprehensive understanding of its causes, impacts, and potential solutions.

5.4.1 Causes of climate change and global warming

At the heart of global warming lies the enhanced greenhouse effect, which occurs when human activities release large amounts of GHGs, such as carbon dioxide (CO_2), methane (CH_4), and nitrous oxide (N_2O), into the atmosphere. These gases trap heat from the sun, preventing it from escaping back into space and thus increasing the Earth's average temperature. According to the Intergovernmental Panel on Climate Change (IPCC, 2023), burning fossil fuels for energy production, transportation, and industrial processes is the largest contributor to GHG emissions, accounting for over 70% of global emissions. Deforestation, agricultural practices, and waste management also contribute significantly to the increase in GHGs, as forests act as carbon sinks, absorbing CO_2 from the atmosphere.

Global temperatures have risen particularly pronounced since the mid-20th century, with the World Meteorological Organization (WMO, 2022) reporting that the last five years have been the warmest on record. The Paris Agreement (2015), a landmark international treaty, aims to limit global warming to well below 2°C above pre-industrial levels, with an aspiration to limit it to 1.5°C. However, the current emissions trajectory suggests that these targets are increasingly challenging to meet without significant and immediate action.

5.4.2 Impacts of climate change and global warming

The impacts of climate change and global warming are widespread and multifaceted, affecting ecosystems, human health, and economies. Rising temperatures have led to more frequent and intense heatwaves, which pose direct risks to human health, particularly for vulnerable populations such as the elderly and those with pre-existing health conditions (WHO, 2021). Additionally, extreme weather events, such as hurricanes, droughts, and floods, are becoming more common and severe, devastatingly affecting infrastructure, agriculture, and communities. The IPCC (2023) warns that climate change will exacerbate these events, leading to more frequent and intense storms, flooding, and wildfires, especially in regions already vulnerable to such phenomena.

Rising sea levels are another critical consequence of global warming. As global temperatures increase, polar ice caps and glaciers are melting accelerated, contributing to higher sea levels. According to the National Aeronautics and Space Administration (NASA, 2022), global sea levels have risen by approximately 20 centimeters since 1880, and the rate of rise is accelerating. This poses a significant threat to coastal communities, where millions of people live and work, and threatens to displace large populations through climate-induced migration.

Biodiversity loss is another severe consequence of climate change. Many species cannot adapt quickly enough to the changing conditions, leading to ecosystem shifts and even extinction. The World Wildlife Fund (WWF, 2022) reports that climate change is one of the primary drivers of biodiversity loss, alongside habitat destruction and pollution. As temperatures rise, ecosystems such as coral reefs, wetlands, and forests are particularly vulnerable, and the loss of biodiversity undermines ecosystem services essential for human survival, including pollination, water purification, and soil fertility.

5.4.3 Solutions to climate change and global warming

Addressing climate change requires a multi-faceted approach that includes reducing GHG emissions, transitioning to renewable energy sources, and adapting to the inevitable changes that are already occurring. The most urgent solution is reducing emissions through policy, technology, and behavior change. The IPCC (2023) emphasizes the need for rapid decarbonization of the global economy, particularly in energy, transportation, and industry.

Transitioning from fossil fuels to renewable energy sources, such as solar, wind, and hydropower, is crucial for achieving emissions reduction targets. The International Renewable Energy Agency (IRENA, 2022) highlights that renewable energy technologies are key to reducing emissions and offer economic opportunities, such as job creation in green industries.

Energy efficiency is another critical solution. Improving the efficiency of buildings, transportation systems, and industrial processes can significantly reduce energy consumption and emissions. The Global Alliance for Energy Productivity (2023) reports that energy efficiency improvements could account for more than 40% of the emissions reductions needed to meet global climate targets.

Carbon capture and storage (CCS) technologies also play an important role in mitigating climate change. These technologies involve capturing CO_2 emissions from power plants and industrial sources, storing them underground, or using them in industrial processes. While CCS can potentially reduce emissions from hard-to-abate sectors, its large-scale implementation is still early and faces significant technical and economic challenges (IEA, 2022).

In addition to mitigation strategies, adaptation to climate change impacts is essential. This includes building resilient infrastructure, improving water management, and protecting vulnerable communities. The United Nations Framework Convention on Climate Change (UNFCCC, 2023) emphasizes the need for adaptation plans, particularly in developing countries most vulnerable to climate impacts. Nature-based solutions, such as reforestation and wetland restoration, can also help mitigate the effects of climate change by sequestering carbon and reducing the risks of flooding and storm surges.

Climate change and global warming are among humanity's most significant challenges. The causes are rooted in human activities, particularly the burning of fossil fuels and deforestation, and the consequences are far-reaching, affecting ecosystems, human health, and economies. However, through coordinated global efforts, technological innovation, and policy interventions, it is possible to mitigate the worst effects of climate change and build a more sustainable future. The situation's urgency requires immediate action to reduce emissions, transition to renewable energy, and adapt to the already underway changes. Only through collective action and global cooperation can we hope to address the climate crisis and safeguard the future of our planet.

5.4.4 Causes and consequences of climate change

Climate change is one of the most critical environmental challenges of the 21st century, driven primarily by human activities that release greenhouse gases (GHGs) into the atmosphere. Its consequences are widespread, affecting ecosystems, human health, and the global economy. Understanding the causes and consequences of climate change is essential for developing effective strategies to mitigate its impacts and adapt to its inevitable effects.

5.4.4.1 Causes of climate change

The primary cause of climate change is the enhanced greenhouse effect, which occurs when certain atmospheric gases trap heat. While the greenhouse effect is a natural phenomenon that helps maintain the planet's temperature, human activities have significantly amplified this effect, leading to global warming. The key drivers of climate change are:

Burning of fossil fuels: The combustion of fossil fuels, such as coal, oil, and natural gas, for energy production, transportation, and industrial processes is the largest source of greenhouse gas emissions. The Intergovernmental Panel on Climate Change (IPCC, 2023) estimates that fossil fuel use accounts for about 75% of global CO_2 emissions. As fossil fuels are burned, carbon dioxide (CO_2) is released into the atmosphere, increasing the concentration of GHGs and contributing to global warming.

Deforestation and land use changes: Forests act as carbon sinks, absorbing CO_2 from the atmosphere. However, deforestation for agriculture, urban development, and logging reduces the Earth's ability to sequester carbon. According to the Food and Agriculture Organization (FAO, 2023), land-use changes, including deforestation, contribute to approximately 23% of global greenhouse gas emissions. The loss of forests not only releases CO_2 but also disrupts biodiversity and the regulation of water cycles.

Agriculture and livestock: Agricultural practices, particularly livestock farming, significantly contribute to climate change. Livestock, such as cattle, produce methane (CH_4), a potent greenhouse gas. Using synthetic fertilizers in agriculture also releases another GHG of nitrous oxide (N_2O). The United Nations Environment Programme (UNEP, 2022) highlights that agriculture is responsible for around 25% of global emissions, with livestock and rice cultivation being major contributors.

Industrial processes: Certain industrial activities, such as cement production, steel manufacturing, and the chemical industry, release GHGs, including CO_2, methane, and fluorinated gases. These industrial processes contribute to the accumulation of GHGs in the atmosphere, exacerbating global warming. According to the International Energy Agency (IEA, 2022), industrial processes account for around 20% of global emissions.

Waste management: Improper waste management, such as decomposing organic waste in landfills, releases methane into the atmosphere. Inadequate waste disposal practices, particularly in rapidly growing urban areas, contribute to the buildup of GHGs. The World Bank (2021) reports that waste management is responsible for a significant share of global methane emissions, especially in low- and middle-income countries.

5.4.4.2 Consequences of climate change

The consequences of climate change are profound and affect every aspect of life on Earth. These impacts are already being felt and are expected to intensify in the coming decades, with the most vulnerable populations and ecosystems being hit hardest. The primary consequences of climate change include:

Rising global temperatures: One of the most visible effects of climate change is the increase in global temperatures. According to the World Meteorological Organization (WMO, 2022), global temperatures have risen by approximately 1.2°C since the pre-industrial era, with the past decade being the warmest on record. This increase in temperature has led to more frequent and intense heatwaves, particularly in regions already vulnerable to extreme heat. Heatwaves pose significant health risks, including heatstroke, dehydration, and cardiovascular problems, especially for vulnerable populations such as the elderly and those with pre-existing health conditions (WHO, 2021).

Sea-level rise: As global temperatures rise, polar ice caps and glaciers are melting, contributing to rising sea levels. According to NASA (2022), global sea levels have risen by about 20 cm since 1880, and the rate of increase is accelerating. Rising sea levels threaten coastal communities, leading to flooding, erosion, and population displacement. Low-lying island nations and coastal cities are particularly at risk, with millions of people facing the prospect of climate-induced migration.

Extreme weather events: Climate change contributes to more frequent and severe extreme weather events, such as hurricanes, droughts, floods, and wildfires. The IPCC (2023) reports that extreme weather events are becoming more intense due to the warming of the atmosphere and oceans. Hurricanes and typhoons are growing in intensity, while droughts are becoming longer and more severe, particularly in regions like sub-Saharan Africa and parts of South Asia. These extreme weather events disrupt agriculture, damage infrastructure, and lead to significant economic losses.

Disruption of ecosystems and biodiversity: Climate change is having a profound impact on ecosystems and biodiversity. Rising temperatures, altered precipitation patterns, and changing seasons disrupt habitats and migration patterns. The World Wildlife Fund (WWF, 2022) reports that climate change is one of the primary drivers of biodiversity loss, alongside habitat destruction and pollution. Coral reefs, forests, and wetlands are particularly vulnerable to climate change, with many species facing extinction due to changing conditions. The loss of biodiversity undermines ecosystem services, such as pollination, water purification, and soil fertility, which are essential for human survival.

Food and water security: Climate change also affects food and water security. Changes in temperature and precipitation patterns are disrupting agricultural productivity, particularly in regions already vulnerable to droughts and floods. The Food and Agriculture Organization (FAO, 2023) warns that climate change will likely exacerbate food insecurity, particularly in developing countries that rely on rain-fed agriculture. In addition, changing weather patterns are affecting the availability of freshwater resources, with some regions experiencing more severe droughts and others facing increased flooding.

Health impacts: Climate change's health impacts are wide-ranging and include direct effects, such as heat-related illnesses, and indirect effects, such as the spread of infectious diseases. The World Health Organization (WHO, 2021) estimates that climate change is already responsible for over 150,000 deaths per year, primarily due to heatwaves, malnutrition, and the spread of diseases such as malaria and dengue fever. Vulnerable populations, including children, the elderly, and those with pre-existing health conditions, are at greater risk.

The causes and consequences of climate change are complex and interconnected. Human activities, particularly the burning of fossil fuels, deforestation, and industrial processes, are driving the accumulation of greenhouse gases in the atmosphere, leading to global warming and a host of environmental, social, and economic consequences. These impacts are already being felt, with rising temperatures, sea-level rise, extreme weather events, and disruptions to ecosystems and human health. Addressing climate change requires urgent action to reduce greenhouse gas emissions, transition to renewable energy, and adapt to the changes that are already underway. Global cooperation, technological innovation, and sustainable practices are essential to mitigating the effects of climate change and securing a livable future for generations to come.

5.5 Extreme Weather Events and Climate Refugees

Extreme weather events are becoming increasingly frequent and severe due to climate change, creating widespread environmental, social, and economic impacts. These events, including hurricanes, floods, droughts, heatwaves, and wildfires, have devastating consequences for vulnerable populations, particularly those in low-lying coastal regions, arid areas, and disaster-prone regions. As extreme weather events intensify, millions of people are being forced to migrate from their homes, leading to a growing population of climate refugees – individuals who are displaced due to environmental factors exacerbated by climate change. This phenomenon is expected to escalate in the coming decades, with profound implications for global migration patterns, human security, and international cooperation.

5.5.1 Extreme weather events and their impact

Extreme weather events, such as hurricanes, typhoons, and cyclones, are becoming more intense and frequent due to the warming of the oceans and the atmosphere. The Intergovernmental Panel on Climate Change (IPCC, 2023) highlights that warmer sea surface temperatures provide more energy for tropical storms, increasing their intensity and frequency. These storms cause immediate destruction through high winds, heavy rainfall, and storm surges, resulting in long-term economic and infrastructural damage. Coastal communities are particularly vulnerable, as rising sea levels further exacerbate the risk of flooding and erosion. For example, Hurricane Katrina, which devastated New Orleans in 2005, and Typhoon Haiyan, which struck the Philippines in 2013, are stark reminders of the catastrophic impacts of extreme weather events on vulnerable populations.

In addition to tropical storms, heat waves, and droughts are becoming more frequent and severe due to rising global temperatures. According to the World Meteorological Organization (WMO, 2022), heatwaves occur more often and last longer, particularly in regions like Europe, the Middle East, and parts of Asia. These prolonged periods of extreme heat can cause heat-related illnesses and fatalities, disrupt agriculture, and strain water resources. Similarly, droughts, becoming more intense in regions like sub-Saharan Africa and the American West, lead to water shortages, crop failures, and food insecurity, forcing people to migrate for more stable living conditions.

5.5.2 Climate refugees and displacement

As a result of extreme weather events, millions of people are being displaced from their homes, creating a growing population of climate refugees. The United Nations High Commissioner for Refugees (UNHCR, 2024) estimates that over 20 million people are displaced annually due to climate-related events, and this number is expected to rise significantly in the coming decades. Climate refugees are often forced to leave their homes due to the direct impacts of extreme weather, such as flooding, wildfires, or storms, or the indirect effects of environmental degradation, such as desertification, water scarcity, and loss of agricultural productivity. For example, in regions like the Sahel in Africa, recurring droughts have caused large-scale displacement as people move in search of food, water, and more viable living conditions (UNHCR, 2024).

The displacement caused by extreme weather events often leads to complex challenges for displaced populations and the regions that receive them. Climate refugees often find themselves in overcrowded refugee camps or informal settlements, where access to basic services such as clean water, sanitation, healthcare, and education is limited. Moreover, the lack of legal recognition for climate refugees in international law complicates their ability to seek protection and assistance. The International Organization for Migration (IOM, 2022) emphasizes the need for a comprehensive legal framework to address the rights and needs of climate refugees, as current migration policies do not adequately address the specific challenges posed by climate-induced displacement.

5.5.3 Vulnerability and inequality

The impacts of extreme weather events and the resulting displacement disproportionately affect the most vulnerable populations, particularly those in developing countries and marginalized communities. The World Bank (2021) reports that low-income countries, especially in Africa, Asia, and Latin America, are more likely to experience the worst effects of climate change, including extreme weather events, rising sea levels, and agricultural disruptions. These populations often lack the resources and infrastructure to cope with the consequences of climate change, making them more susceptible to displacement.

Additionally, climate change exacerbates existing inequalities, as women, children, indigenous communities, and the elderly are often the most affected by extreme weather events. For instance, women in rural areas are often responsible for collecting water and firewood; their burden increases significantly when these resources become scarce due to droughts or floods. Nussbaum (2011) emphasizes that addressing the vulnerabilities of marginalized groups is essential in any climate adaptation and migration strategy, as these populations often lack the agency and resources to protect themselves from the impacts of climate change.

5.5.4 The role of international cooperation

The growing crisis of climate refugees underscores the need for international cooperation and policy responses that address both the causes and consequences of climate-induced displacement. The United Nations Framework Convention on Climate Change (UNFCCC, 2023) calls for stronger climate adaptation measures and support for vulnerable countries to build resilience to extreme weather events. This includes financial assistance, technology transfer, and capacity-building to help countries better prepare for and respond to the impacts of climate change.

Moreover, addressing the issue of climate refugees requires a shift in how the international community views migration. The Global Compact for Migration (GCM, 2018) recognizes the need for a more comprehensive approach to migration that considers the environmental drivers of displacement. However, the lack of a formal legal framework for climate refugees remains a significant barrier to ensuring their rights and protections. The IOM (2022) calls for establishing new international agreements and policies that specifically address the needs of climate refugees, including pathways for relocation, resettlement, and compensation for affected populations.

Extreme weather events are becoming increasingly frequent and severe due to climate change, with profound implications for human migration. The displacement caused by these events leads to a growing population of climate refugees, who are often forced to leave their homes due to environmental degradation, rising sea levels, and the direct impacts of storms, heat waves, and droughts. Addressing the crisis of climate refugees requires international cooperation, legal recognition, and comprehensive policy responses that prioritize the protection and well-being of displaced populations. As the impacts of climate change continue to intensify, the need for coordinated global action to mitigate and adapt to these changes becomes more urgent than ever.

5.6 The Global Response: Successes and Failures

The global response to the challenges posed by climate change and other environmental crises has been a mix of successes and failures. While significant progress has been made in raising awareness, creating international frameworks, and implementing certain mitigation and adaptation strategies, the overall response has often been fragmented, insufficient, and slow. The complexities of global environmental challenges and political, economic, and social barriers have hindered the full realization of effective solutions. In this context, evaluating both the successes and failures of the global response to climate change and other environmental crises is essential.

5.6.1 Successes in the global response

One of the major successes in the global response to climate change has been the establishment of international frameworks such as the Paris Agreement under the United Nations Framework Convention on Climate Change (UNFCCC). Adopted in 2015, the Paris Agreement marked a significant milestone in global climate governance, as nearly 200 countries committed to limiting global temperature rise to well below 2°C, with efforts to limit it to 1.5°C above pre-industrial levels (UNFCCC, 2015). This agreement was groundbreaking in its inclusivity, as it involved both developed and developing nations in setting nationally determined contributions (NDCs) to reduce greenhouse gas emissions. The Paris Agreement also introduced a mechanism for regularly reviewing and updating these commitments, promoting transparency and accountability in climate action.

Another notable success is the growing momentum around renewable energy and clean technology. Over the past two decades, there has been a significant increase in the adoption of renewable energy sources such as wind, solar, and hydroelectric power. According to the International Renewable Energy Agency (IRENA, 2023), the cost of renewable energy technologies has decreased dramatically, making them more accessible to countries and communities worldwide. This shift towards renewables is crucial in reducing dependence on fossil fuels and mitigating the impacts of climate change. Additionally, advancements in energy storage, electric vehicles, and green hydrogen technologies offer promising solutions for reducing carbon emissions and transitioning to a more sustainable energy system.

International financial mechanisms, such as the Green Climate Fund (GCF) and the Global Environment Facility (GEF), have also played an important role in developing countries' climate adaptation and mitigation efforts. These funds have provided critical financial resources for projects to build resilience to climate change, promote sustainable agriculture, and protect biodiversity. According to the World Bank (2022), climate finance has increased significantly in recent years, helping countries address the impacts of climate change and transition to low-carbon economies.

5.6.2 Failures in the global response

Despite these successes, the global response to climate change and environmental crises has faced significant failures, particularly in the pace and scale of action. One of the most notable failures has been the inability to meet the emission reduction targets necessary to avoid catastrophic climate change. While the Paris Agreement set ambitious goals, emissions have continued to rise globally, driven by increasing energy demand, deforestation, and industrial growth in emerging economies (IPCC, 2023). The gap between the pledges made by countries and the actual implementation of those pledges has been a significant source of concern. For example, a report by the UN Environment Programme (UNEP, 2023) found that global emissions are on track to exceed the 1.5°C target by a wide margin, with current policies and pledges leading to a temperature rise of approximately 3.2°C by the end of the century.

Another failure has been the lack of adequate adaptation measures for vulnerable populations. While significant progress has been made in raising awareness about the need for climate adaptation, many countries, particularly in the Global South, lack the resources and infrastructure to cope with the impacts of climate change. For example, the World Bank (2021) reports that small island developing states (SIDS) and least developed countries (LDCs) are particularly vulnerable to sea-level rise, extreme weather events, and food and water insecurity. Despite the availability of climate finance, much of the funding has not been directed towards the most vulnerable populations, and adaptation efforts have often been insufficient or poorly implemented.

The failure to integrate climate change into broader policy frameworks is another significant shortcoming of the global response. While there has been progress in addressing climate change within the context of environmental policy, the issue has not been fully integrated into economic, trade, and development policies. This lack of integration has hindered the effectiveness of climate action, as economic growth and development are often prioritized over environmental sustainability. The United Nations Development Programme (UNDP, 2023) argues that a more integrated approach is needed, one that considers the synergies between climate action, poverty reduction, and sustainable development.

5.6.3 Political and economic barriers

Political and economic barriers have also impeded the global response to climate change. One of the main challenges has been the resistance of powerful nations and industries to taking meaningful action on climate change. Fossil fuel industries have significant political influence and have lobbied against strong climate policies, arguing that they would harm economic growth and job creation. This has led to delays in implementing critical climate policies and undermined efforts to transition to a low-carbon economy. The IPCC (2023) highlights that the political will to implement ambitious climate policies remains a significant barrier to achieving global climate goals.

Additionally, the uneven distribution of the impacts of climate change has created tensions between developed and developing countries. Many developing countries argue that they should not bear the same burden of emissions reductions as developed countries, which have historically contributed the most to global emissions. The principle of common but differentiated responsibilities (CBDR), enshrined in the UNFCCC, recognizes this disparity and calls for developed countries to take the lead in addressing climate change. However, the failure to provide adequate financial and technological support to developing countries has led to frustration and a lack of trust in international climate negotiations (UNFCCC, 2015).

Successes and failures have marked the global response to climate change and environmental crises. While international agreements like the Paris Agreement and the growing adoption of renewable energy technologies offer hope for a sustainable future, the failure to meet emission reduction targets, inadequate adaptation measures, and political and economic barriers continue to hinder progress. Moving forward, it is crucial that the global community strengthens its commitment to climate action, prioritizes the needs of vulnerable populations, and integrates climate change into broader policy frameworks. We can only address the global environmental crises and ensure a sustainable future for all through collective action and sustained efforts.

References

Food and Agriculture Organization (FAO). (2023). *The state of food and agriculture 2023: Climate change and food security*. FAO.

Food and Agriculture Organization (FAO). (2023). *The state of the world's forests 2023: Forest pathways to sustainable development*. FAO.

Intergovernmental Panel on Climate Change (IPCC). (2023). *Climate change 2023: Synthesis report*. IPCC.

International Energy Agency (IEA). (2022). *Global CO2 emissions in 2022*. IEA.

International Organization for Migration (IOM). (2022). *World migration report 2022*. IOM.

International Renewable Energy Agency (IRENA). (2023). *Renewable energy capacity statistics 2023*. IRENA.

NASA. (2022). *Global sea level rise*. NASA. https://climate.nasa.gov/vital-signs/sea-level/

National Aeronautics and Space Administration (NASA). (2022). *Global sea level rise*. NASA. https://climate.nasa.gov/vital-signs/sea-level/

Nussbaum, M. (2011). *Creating capabilities: The human development approach*. Harvard University Press.

United Nations Children's Fund (UNICEF). (2023). *The state of the world's children 2023: Children, food, and nutrition*. UNICEF.

United Nations Development Programme (UNDP). (2023). *Climate action and sustainable development: Addressing the dual challenge*. UNDP.

United Nations Environment Programme (UNEP). (2022). *The plastic waste crisis: A global challenge*. UNEP.

United Nations Environment Programme (UNEP). (2023). *Emissions gap report 2023*. UNEP.

United Nations Framework Convention on Climate Change (UNFCCC). (2015). *The Paris Agreement*. UNFCCC.

United Nations Framework Convention on Climate Change (UNFCCC). (2023). *Adaptation and resilience*. UNFCCC.

United Nations Framework Convention on Climate Change (UNFCCC). (2023). *Report on climate change and migration*. UNFCCC.

United Nations High Commissioner for Refugees (UNHCR). (2024). *Climate change and displacement*. UNHCR.

World Bank. (2021). *Climate change and poverty: A new challenge for the poor*. World Bank.

World Bank. (2022). *Climate finance: Mobilizing funds for a sustainable future*. World Bank.

World Economic Forum (WEF). (2023). *The global risks report 2023*. WEF.

World Health Organization (WHO). (2021). *Air quality and health*. WHO.

World Health Organization (WHO). (2021). *Climate change and health*. WHO.

World Meteorological Organization (WMO). (2022). *State of the global climate 2022*. WMO.

World Wildlife Fund (WWF). (2022). *Living planet report 2022: A glimpse into the future of biodiversity*. WWF.

6

Biodiversity and Ecosystem Collapse

Biodiversity, the variety of life forms across different ecosystems, plays a critical role in maintaining the balance and functioning of the Earth's systems. It encompasses not only the diversity of species but also genetic diversity and the variety of ecosystems. The ongoing loss of biodiversity and the collapse of ecosystems are among the most pressing environmental challenges of the 21st century. These phenomena are driven by human activities, including habitat destruction, climate change, pollution, overexploitation of natural resources, and introduction of invasive species. The consequences of biodiversity loss and ecosystem collapse are far-reaching, affecting ecosystem services, human well-being, and the planet's stability.

6.1 Causes of Biodiversity Loss and Ecosystem Collapse

The primary drivers of biodiversity loss and ecosystem collapse are often referred to as the "drivers of change" in environmental science. Habitat destruction and fragmentation, driven by urbanization, agriculture, and infrastructure development, are among the most significant threats to biodiversity. According to Sala et al. (2000), human-induced land-use changes have led to the fragmentation of natural habitats, reducing the area available for wildlife and disrupting ecological processes. For example, the conversion of forests to agricultural land has not only resulted in the loss of species but has also disrupted important ecosystem functions such as carbon sequestration, water regulation, and soil fertility.

Climate change is another major driver of biodiversity loss. The increasing frequency and intensity of extreme weather events, temperature fluctuations, and changing precipitation patterns put immense pressure on ecosystems and species. Species unable to adapt to these changes or migrate to more suitable habitats are at risk of extinction. IPCC (2023) highlights that climate change is expected to exacerbate biodiversity loss, particularly in ecosystems

DOI: 10.1201/9788743808787-8

such as coral reefs, wetlands, and forests. For instance, the warming of ocean waters has led to widespread coral bleaching, endangering coral reefs and the species that depend on them for survival (Hoegh-Guldberg et al., 2007).

Pollution, notably by plastic and chemical contaminants, contributes significantly to biodiversity loss. Thompson et al. (2009) demonstrate that plastic waste in oceans severely threatens marine life, with species ingesting or becoming entangled in plastic debris. Similarly, the use of pesticides and fertilizers in agriculture has led to soil degradation, water contamination, and the decline of pollinator populations, which are essential for food production (Goulson, 2015). The accumulation of toxic substances in the environment further disrupts the health of ecosystems and the species that inhabit them.

Overexploitation of natural resources, such as overfishing, hunting, and logging, has led to the depletion of many species and the destabilization of ecosystems. FAO (2023) reports that overfishing is one of the leading causes of marine biodiversity loss, with many fish stocks being exploited beyond sustainable limits. The depletion of keystone species, such as apex predators, can have cascading effects on entire ecosystems, leading to the collapse of food webs and the degradation of ecosystem services.

6.2 Consequences of Biodiversity Loss and Ecosystem Collapse

The loss of biodiversity and the collapse of ecosystems have profound implications for the planet's functioning. Ecosystems provide a wide range of essential services for human survival, including food production, clean water, air purification, and climate regulation. TEEB (2010) emphasizes the importance of ecosystem services, noting that biodiversity is the foundation of these services. The loss of biodiversity diminishes ecosystems' ability to perform these functions, leading to negative consequences for human societies.

For example, the decline in pollinator populations, such as bees and butterflies, has serious implications for global food production. Potts et al. (2016) argue that pollination services are essential for producing many crops, including fruits, vegetables, and nuts. The loss of pollinators due to habitat loss, pesticide use, and climate change threatens food security, particularly in regions that rely heavily on agriculture.

Ecosystem collapse can also exacerbate the impacts of climate change. Forests, wetlands, and oceans act as carbon sinks, absorbing large amounts of carbon dioxide from the atmosphere. The destruction of these ecosystems releases stored carbon and reduces the Earth's capacity to mitigate the effects of climate change. IPCC (2023) highlights the role of forests in regulating the global climate, noting that deforestation contributes significantly to global greenhouse gas emissions.

The collapse of ecosystems also leads to the loss of cultural and spiritual values associated with biodiversity. Many indigenous communities and local populations depend on biodiversity for their livelihoods, cultural practices, and spiritual beliefs. The destruction of ecosystems undermines the resilience and well-being of these communities, leading to social and economic hardships (Berkes, 2012). The loss of traditional knowledge related to biodiversity management further exacerbates these challenges.

6.3 The Path Forward: Solutions and Conservation Strategies

Addressing biodiversity loss and preventing ecosystem collapse requires a multifaceted approach that includes conservation, sustainable resource management, and climate change mitigation. Convention on Biological Diversity (CBD, 2022) emphasizes the need for a global framework to halt biodiversity loss by 2030, with targets to protect ecosystems, reduce pollution, and promote sustainable land use. The United Nations Decade on Ecosystem Restoration (2021–2030) aims to restore degraded ecosystems and promote biodiversity conservation through large-scale restoration efforts (UNEP, 2021).

One key strategy for conserving biodiversity is the establishment of protected areas. Butchart et al. (2015) argue that protected areas are essential for safeguarding species and ecosystems, providing refuge for endangered species, and preserving critical habitats. However, the effectiveness of protected areas depends on their size, connectivity, and management. Integrating conservation effort into landscape and seascape management, along with the active participation of local communities, is crucial for ensuring long-term sustainability.

Sustainable resource management practices, such as sustainable agriculture, forestry, and fisheries, are also vital for preserving biodiversity. FAO (2023) advocates for the adoption of sustainable practices that minimize environmental impact and promote ecosystem resilience. Agroecology, for example, emphasizes the use of biodiversity-friendly farming techniques that enhance soil health, conserve water, and reduce the need for chemical inputs (Altieri, 2018). Sustainable fisheries management is also crucial for preventing overfishing and ensuring the long-term viability of marine ecosystems.

Finally, addressing climate change is integral to preventing further biodiversity loss. Mitigation strategies, such as reducing greenhouse gas emissions and protecting carbon-rich ecosystems, are essential for preserving biodiversity. IPCC (2023) highlights the importance of integrating biodiversity conservation into climate change policies, as the two issues are inextricably linked.

Biodiversity loss and ecosystem collapse are urgent global challenges that threaten the planet's stability and humanity's well-being. The drivers of biodiversity loss, including habitat destruction, climate change, pollution, and overexploitation, are primarily human-induced, and their consequences are far-reaching. However, through concerted efforts in conservation, sustainable resource management, and climate change mitigation, it is possible to reverse some of the damage and restore ecosystems. Protecting biodiversity is not only essential for the health of the planet but also for ensuring food security, economic stability, and cultural resilience in the face of environmental crises.

6.4 The Sixth Mass Extinction

The concept of mass extinction refers to a significant and widespread loss of biodiversity, where a large proportion of species across the globe disappear in a relatively short time. Historically, Earth has experienced five mass extinctions, with the most famous being the event that wiped out the dinosaurs about 66 million years ago. However, scientists are now warning that we are in the midst of a sixth mass extinction, driven primarily by human activities. This event is distinct from previous mass extinctions, as it is caused by anthropogenic factors such as habitat destruction, climate change, overexploitation, pollution, and the introduction of invasive species.

6.4.1 The causes of the sixth mass extinction

Unlike past extinction events, which were triggered by natural phenomena such as volcanic eruptions, asteroid impacts, or shifts in the Earth's climate, the current biodiversity crisis is largely a result of human-induced activities. Habitat destruction is one of the most significant drivers of biodiversity loss. The conversion of forests to agricultural land, urban sprawl, and the destruction of wetlands and other ecosystems have drastically reduced the living space for many species (Sala et al., 2000). This loss of habitat leads to population declines and makes it increasingly difficult for species to survive, particularly those that are already vulnerable or have specialized habitat requirements.

Climate change is another key driver of the ongoing extinction crisis. The rapid warming of the planet, driven by the burning of fossil fuels and deforestation, is altering ecosystems at an unprecedented rate. Species that are unable to adapt to new temperature regimes or migrate to more suitable habitats face extinction. The IPCC (2023) highlights that climate change is expected to have severe impacts on biodiversity, particularly in ecosystems like coral reefs, polar regions, and mountain habitats, which are highly sensitive to temperature changes. Warming temperatures also exacerbate other threats, such as the spread of diseases and the encroachment of invasive species.

Overexploitation of natural resources is also contributing to the current extinction crisis. Overfishing, hunting, and logging have led to the depletion of many species, particularly those with slow reproduction rates or small populations. Pimm et al. (2014) argue that overexploitation is a primary factor driving the decline of species, with many marine species, such as fish and marine mammals, being harvested at unsustainable rates. Similarly, the illegal wildlife trade has contributed to the extinction of numerous species, particularly in tropical regions where biodiversity is richest (Robinson et al., 2019).

Pollution, particularly plastic pollution, is another major threat to biodiversity. Thompson et al. (2009) demonstrate that plastic waste in oceans and rivers has devastating effects on marine life, with species ingesting or becoming entangled in plastic debris. Chemical pollutants, such as pesticides and heavy metals, also poison ecosystems, disrupting food chains and causing reproductive failures in many species (Goulson, 2015). The accumulation of pollutants in the environment further accelerates the decline of species and the degradation of ecosystems.

6.4.2 The consequences of the sixth mass extinction

The consequences of the ongoing sixth mass extinction are profound and far-reaching. The loss of biodiversity undermines the functioning of ecosystems, which provide essential services such as food production, water purification, climate regulation, and disease control. TEEB (2010) emphasizes that biodiversity is the foundation of ecosystem services, and its loss diminishes the capacity of ecosystems to perform these vital functions. For instance, the decline in pollinators, such as bees and butterflies, has serious implications for global food security, as many crops depend on pollination for reproduction (Potts et al., 2016).

The loss of biodiversity also poses significant economic challenges. Ecosystem services, such as clean water, soil fertility, and flood regulation, are essential for agricultural production, industry, and human well-being. The depletion of these services can lead to increased costs and economic instability, particularly in regions that are heavily reliant on natural resources. Costanza et al. (2014) estimate that the global economic value of ecosystem services is trillions of dollars per year, and the loss of biodiversity will lead to a reduction in these services, affecting economies worldwide.

The extinction of species also has cultural and social implications, particularly for indigenous communities that rely on biodiversity for their livelihoods, cultural practices, and spiritual beliefs. Many indigenous peoples have deep connections to the land and its ecosystems, and the loss of species and ecosystems can lead to the erosion of traditional knowledge and cultural heritage (Berkes, 2012). Moreover, the destruction of biodiversity often leads to social and economic inequality, as marginalized communities are disproportionately affected by environmental degradation and the loss of natural resources.

6.4.3 The path forward: Mitigation and conservation strategies

Addressing the ongoing sixth mass extinction requires urgent and coordinated global efforts. Conservation strategies must focus on protecting and restoring habitats, reducing human impacts on ecosystems, and promoting sustainable resource management. Convention on Biological Diversity (CBD, 2022) emphasizes the importance of establishing protected areas and restoring degraded ecosystems to safeguard biodiversity. The United Nations Decade on Ecosystem Restoration (2021–2030) aims to restore ecosystems worldwide and halt biodiversity loss by 2030 (UNEP, 2021).

In addition to habitat protection and restoration, reducing the drivers of biodiversity loss is essential. This includes addressing climate change by transitioning to renewable energy sources, reducing greenhouse gas emissions, and promoting climate adaptation strategies that help species cope with changing conditions (IPCC, 2023). Sustainable land-use practices, such as agroecology and sustainable forestry, can also help reduce habitat destruction and promote biodiversity conservation (Altieri, 2018).

Efforts to combat overexploitation, such as implementing sustainable fisheries management, cracking down on illegal wildlife trade, and promoting sustainable consumption, are also crucial. FAO (2023) advocates for sustainable fisheries practices that ensure the long-term viability of marine ecosystems and fish populations. The protection of endangered species through legal frameworks and international agreements, such as the Convention on International Trade in Endangered Species (CITES), is also essential for halting the decline of vulnerable species.

Finally, addressing pollution is critical to preventing further biodiversity loss. Reducing plastic waste, limiting chemical pollutants, and promoting cleaner production methods can help mitigate the impact of pollution on ecosystems. Thompson et al. (2009) argue that addressing plastic pollution in particular requires global cooperation to reduce production and improve waste management systems.

The sixth mass extinction is a stark reminder of the profound impact human activities have on the planet's biodiversity. Driven by habitat destruction, climate change, overexploitation, pollution, and invasive species, this crisis threatens the stability of ecosystems and the well-being of humanity. However, through concerted efforts in conservation, sustainable resource management, climate change mitigation, and pollution reduction, it is possible to halt or even reverse some of the damage. Protecting biodiversity is not only essential for the health of the planet but also for ensuring the resilience and sustainability of ecosystems and human societies in the face of environmental challenges.

6.5 Habitat Destruction and Species Loss

Habitat destruction is one of the leading drivers of biodiversity loss and species extinction. As human populations expand and industrial activities intensify, ecosystems around the world are being degraded at an unprecedented rate. Habitat destruction refers to the process by which natural environments are altered or destroyed, making them unsuitable for the species that depend on them for survival. This can occur through deforestation, urbanization, agricultural expansion, mining, and infrastructure development, among other human activities. The consequences of habitat destruction are profound, as it directly leads to the loss of biodiversity, disrupts ecological processes, and threatens the survival of many species.

6.5.1 Causes of habitat destruction

The primary cause of habitat destruction is the expansion of human activity. Sala et al. (2000) argue that land-use changes driven by agriculture, logging, and urbanization are responsible for the majority of habitat loss. Deforestation, for example, is a significant driver of habitat destruction, particularly in tropical rainforests, which are home to a vast majority of the world's species. According to FAO (2020), tropical forests continue to be cleared at alarming rates, with significant portions being converted into agricultural land for crops like soy, palm oil, and cattle grazing. This deforestation not only destroys the habitats of countless species but also contributes to climate change by releasing large amounts of carbon stored in trees and soil into the atmosphere.

Urbanization and infrastructure development also contribute significantly to habitat destruction. As cities expand, natural habitats are converted into built environments, such as roads, buildings, and industrial zones. Seto et al. (2012) highlight that urbanization is one of the most significant forms of land-use change, with the global urban population expected to reach 6.7 billion by 2050. This rapid urban growth leads to the fragmentation of ecosystems, isolating species populations and making it more difficult for species to find food, shelter, and mates.

Mining and industrial activities are additional contributors to habitat destruction. Foley et al. (2005) note that mining for minerals, oil, and gas often leads to the clearing of large areas of land, which disrupts ecosystems and contributes to the loss of biodiversity. In some cases, mining activities also result in the contamination of soil and water, further exacerbating the destruction of habitats.

6.5.2 Consequences of habitat destruction

The consequences of habitat destruction are far-reaching, with species loss being one of the most immediate and severe outcomes. When habitats are destroyed, the species that rely on them for food, shelter, and reproduction are often unable to survive. Pimm et al. (2014) argue that habitat loss is the leading cause of species extinction, particularly for species with specialized habitat requirements or limited geographic ranges. For example, many amphibians, which are highly sensitive to environmental changes, are particularly vulnerable to habitat destruction. As their habitats are destroyed, populations decline, and many species face the risk of extinction.

Habitat destruction also disrupts ecological processes and the services that ecosystems provide. TEEB (2010) emphasizes that ecosystems provide essential services such as pollination, water purification, soil fertility, and climate regulation. When habitats are destroyed, these services are compromised, which can have cascading effects on human societies. For instance, the destruction of wetlands, which act as natural water filters, can lead to increased water pollution and flooding, while the loss of forests can disrupt carbon sequestration, contributing to climate change.

Moreover, habitat destruction leads to the fragmentation of ecosystems, which isolates species populations and reduces genetic diversity. Fahrig (2003) highlights that fragmented habitats make it more difficult for species to migrate, find mates, and adapt to environmental changes. This isolation can lead to inbreeding, which reduces genetic diversity and makes species more susceptible to disease, environmental stress, and extinction. Fragmented habitats also limit the ability of species to migrate in response to climate change, further exacerbating the challenges they face.

6.5.3 The role of conservation in mitigating habitat destruction

Addressing habitat destruction requires a multifaceted approach that includes both conservation and sustainable development practices. One of the most effective ways to mitigate habitat destruction is through the establishment of protected areas, such as national parks and nature reserves. Sala et al. (2000) emphasize the importance of creating and maintaining protected areas to conserve biodiversity and safeguard ecosystems. These areas can serve as refuges for species, allowing them to thrive without the threat of habitat destruction from human activities.

Restoration of degraded habitats is also a critical strategy for mitigating habitat loss. Chazdon (2008) discusses the importance of ecological restoration in reversing the effects of habitat destruction. Restoration efforts can include reforestation, wetland rehabilitation, and the reintroduction of native species to areas where they have been extirpated. Restoration not only helps to rebuild habitats but also promotes ecosystem services and improves the resilience of ecosystems to climate change.

In addition to protecting and restoring habitats, it is essential to address the root causes of habitat destruction. Sustainable land-use practices, such as agroecology and sustainable forestry, can help reduce the need for habitat conversion. Altieri (2018) advocates for agroecological approaches that integrate biodiversity conservation into agricultural practices, thereby reducing the impact of agriculture on natural habitats. Similarly, sustainable forestry practices that prioritize the health of ecosystems and the regeneration of forests can help prevent deforestation and habitat degradation.

Finally, addressing the demand for resources that drive habitat destruction is crucial. McKinney (2002) argues that reducing consumption, particularly in wealthy nations, can help alleviate the pressure on ecosystems. This can be achieved through promoting sustainable consumption patterns, reducing waste, and supporting businesses that prioritize environmental sustainability.

Habitat destruction is a major driver of species loss and a key contributor to the ongoing biodiversity crisis. As human activities continue to alter natural environments, the survival of many species is at risk. However, through conservation efforts, habitat restoration, sustainable land-use practices, and a shift towards more sustainable consumption, it is possible to mitigate the impacts of habitat destruction and protect biodiversity. The urgency of addressing habitat destruction cannot be overstated, as the loss of biodiversity undermines ecosystem services that are essential for the well-being of both humans and the planet.

6.6 Restoring and Protecting Ecosystems

Restoring and protecting ecosystems is a critical component of biodiversity conservation and climate change mitigation. Ecosystems provide essential services that support life on Earth, such as clean air and water, pollination, soil fertility, and carbon sequestration. However, human activities such as deforestation, urbanization, agriculture, and industrialization have caused widespread degradation of natural habitats, leading to biodiversity loss, reduced ecosystem services, and increased vulnerability to climate change. In response, ecosystem restoration and protection have emerged as key strategies for reversing environmental damage and safeguarding the planet's future.

6.6.1 The importance of ecosystem restoration

Ecosystem restoration involves the process of rehabilitating degraded ecosystems to restore their structure, function, and biodiversity. This can include activities such as reforestation, wetland restoration, soil regeneration, and the removal of invasive species. Chazdon (2008) emphasizes that ecosystem restoration is essential for maintaining biodiversity, enhancing ecosystem services, and increasing the resilience of ecosystems to climate change. Restoring ecosystems also helps to restore vital ecological processes such as nutrient cycling, water filtration, and carbon storage, which are critical for sustaining life on Earth.

The United Nations Decade on Ecosystem Restoration (2021–2030) highlights the global commitment to restore 350 million hectares of degraded land by 2030 (UNEP, 2021). This ambitious initiative aims to combat land degradation, increase food security, mitigate climate change, and protect biodiversity. Successful restoration efforts not only improve the health of ecosystems but also provide socio-economic benefits, such as improved livelihoods for local communities, job creation, and enhanced food and water security.

6.6.2 Strategies for ecosystem restoration

Several strategies are employed to restore ecosystems, depending on the specific needs of the ecosystem and the extent of degradation. One of the most widely used methods is reforestation, which involves planting native tree species in areas where forests have been cleared. Reforestation can help restore biodiversity, improve soil quality, and enhance carbon sequestration. Lamb et al. (2005) found that reforestation in tropical regions, for example, has the potential to restore ecosystems and provide a range of ecosystem services, including habitat for wildlife and improved water quality.

Wetland restoration is another important strategy, as wetlands are among the most productive and biodiverse ecosystems on Earth. Wetlands play a crucial role in water purification, flood control, and carbon storage. Mitsch and Gosselink (2015) highlight that restoring wetlands can improve water quality, reduce flood risks, and provide habitat for many species, including migratory birds and aquatic organisms. Wetland restoration often involves removing barriers such as dams or drainage systems and reintroducing native plant species to restore natural hydrological processes.

Another effective restoration strategy is soil regeneration, which focuses on improving soil fertility and structure in degraded agricultural or urban lands. Reynolds et al. (2011) emphasize that soil restoration is essential for maintaining food security, as healthy soils support plant growth and water retention. Practices such as crop rotation, agroforestry, and the use of organic fertilizers can help restore soil health and reduce the need for chemical inputs, thereby promoting sustainable agriculture.

6.6.3 Protecting ecosystems through conservation

While restoration is crucial for rehabilitating degraded ecosystems, protection is equally important to prevent further degradation and ensure the long-term survival of ecosystems. Conservation efforts focus on preserving natural habitats and biodiversity through the establishment of protected areas, such as national parks, nature reserves, and wildlife sanctuaries. Sala et al. (2000) argue that protected areas are essential for conserving biodiversity and maintaining ecosystem services. These areas provide refuge for species, protect critical habitats, and allow ecosystems to function without the pressures of human activity.

In addition to establishing protected areas, sustainable land-use practices are key to protecting ecosystems. Sustainable agriculture, forestry, and fisheries management can help reduce the negative impact of human activities on ecosystems. Altieri (2018) advocates for agroecology as a sustainable farming approach that integrates biodiversity conservation into agricultural practices. Agroecology emphasizes the use of organic farming methods, crop diversification, and the protection of natural habitats within agricultural landscapes. Similarly, sustainable forestry practices, such as selective logging and replanting, can help protect forests and ensure their long-term health.

Marine conservation is also an essential aspect of ecosystem protection, as oceans and coastal ecosystems are vital for global biodiversity and climate regulation. Halpern et al. (2015) highlight the importance of marine protected areas (MPAs) in preserving marine biodiversity, improving fish stocks, and enhancing ecosystem services such as carbon sequestration. MPAs can help protect critical habitats such as coral reefs, mangroves, and seagrass beds, which are essential for marine life and coastal communities.

6.6.4 The role of indigenous knowledge in ecosystem protection

Indigenous communities have long played a key role in the protection and management of ecosystems. Indigenous knowledge systems, which are based on a deep understanding of local ecosystems and sustainable resource management practices, offer valuable insights into ecosystem protection. Berkes (2008) argues that indigenous knowledge is critical for conservation, as it is often based on centuries of observation and adaptation to local environmental conditions. Indigenous practices such as controlled burning, sustainable hunting, and the use of traditional ecological knowledge can help maintain biodiversity and restore ecosystems.

Incorporating indigenous knowledge into modern conservation strategies is essential for ensuring the long-term success of ecosystem restoration and protection efforts. Dove (2015) emphasizes that collaboration between indigenous communities and conservation organizations can lead to more effective and culturally appropriate conservation strategies. This approach not only benefits ecosystems but also respects the rights and traditions of indigenous peoples.

Ecosystem restoration and protection are essential for reversing environmental degradation, conserving biodiversity, and mitigating the impacts of climate change. Restoration strategies such as reforestation, wetland restoration, and soil regeneration can help rehabilitate degraded ecosystems and enhance their resilience. At the same time, protecting ecosystems by establishing protected areas and sustainable land-use practices is crucial for ensuring the long-term survival of biodiversity and ecosystem services. The integration of indigenous knowledge into conservation efforts can further strengthen these initiatives. Restoring and protecting ecosystems is a collective responsibility that requires collaboration across sectors and scales, from local communities to global organizations.

References

Altieri, M. A. (2018). *Agroecology: The science of sustainable agriculture*. CRC Press.

Berkes, F. (2008). *Sacred ecology: Traditional ecological knowledge and resource management*. Routledge.

Berkes, F. (2012). *Sacred ecology* (3rd ed.). Routledge.

Butchart, S. H. M., et al. (2015). Global biodiversity: Indicators of recent declines. *Science, 328*(5982), 1164–1168. https://doi.org/10.1126/science.1187512

Chazdon, R. L. (2008). *Beyond deforestation: Restoring forests and ecosystem services on degraded lands*. Science, 320(5882), 1458–1460. https://doi.org/10.1126/science.1155365

Convention on Biological Diversity (CBD). (2022). *Global biodiversity outlook 5*. CBD.

Costanza, R., et al. (2014). Changes in the global value of ecosystem services. *Global Environmental Change, 26*, 152–158. https://doi.org/10.1016/j.gloenvcha.2014.04.002

Dove, M. R. (2015). *Indigenous knowledge and the conservation of biodiversity*. In R. B. Primack (Ed.), *Essentials of conservation biology* (6th ed., pp. 241–258). Sinauer Associates.

Fahrig, L. (2003). Effects of habitat fragmentation on biodiversity. *Annual Review of Ecology, Evolution, and Systematics, 34*, 487–515. https://doi.org/10.1146/annurev.ecolsys.34.011802.132419

FAO. (2020). *State of the world's forests 2020: Forests, biodiversity and people*. FAO.

Foley, J. A., et al. (2005). Global consequences of land use. *Science, 309*(5734), 570–574. https://doi.org/10.1126/science.1111772

Food and Agriculture Organization (FAO). (2023). *The state of world fisheries and aquaculture 2022*. FAO.

Goulson, D. (2015). *Bee losses, the EU neonicotinoid ban, and the consequences for biodiversity. Environmental Sciences Europe, 27*(1), 1–10. https://doi.org/10.1186/s12302-015-0056-0

Halpern, B. S., et al. (2015). Marine protected areas and ocean conservation. *Science, 350*(6260), 1035–1036. https://doi.org/10.1126/science.aad8136

Hoegh-Guldberg, O., et al. (2007). Coral reefs are under rapid climate change and ocean acidification. *Science, 318*(5857), 1737–1742. https://doi.org/10.1126/science.1152509

IPCC. (2023). *Climate change 2023: Synthesis report*. IPCC.

Lamb, D., et al. (2005). Reforestation and restoration of degraded tropical forests. *Science, 310*(5754), 1622–1625. https://doi.org/10.1126/science.1115342

Mitsch, W. J., & Gosselink, J. G. (2015). *Wetlands* (5th ed.). Wiley.

Pimm, S. L., et al. (2014). The biodiversity of species and their rates of extinction, distribution, and protection. *Science, 344*(6187), 1246752. https://doi.org/10.1126/science.1246752

Potts, S. G., et al. (2016). The impact of pollinators is declining on the global food supply. *Food Security, 8*(1), 1–12. https://doi.org/10.1007/s12571-015-0508-1

Reynolds, J. F., et al. (2011). *Global desertification: Do humans cause deserts?* Springer.

Robinson, J. G., et al. (2019). The impact of the illegal wildlife trade on biodiversity and conservation efforts. *Conservation Biology, 33*(5), 1152–1161. https://doi.org/10.1111/cobi.13331

Sala, O. E., et al. (2000). Global biodiversity scenarios for the year 2100. *Science, 287*(5459), 1770–1774. https://doi.org/10.1126/science.287.5459.1770

Seto, K. C., et al. (2012). Urbanization and the global land use transition. *Proceedings of the National Academy of Sciences, 109*(40), 16083–16088. https://doi.org/10.1073/pnas.1211658109

TEEB. (2010). *The economics of ecosystems and biodiversity: Ecological and economic foundations*. Earthscan.

Thompson, R. C., et al. (2009). Environmental and human health impacts of plastic debris in the ocean. *Philosophical Transactions of the Royal Society B: Biological Sciences, 364*(1526), 2153–2165. https://doi.org/10.1098/rstb.2008.0291

UNEP. (2021). *UN decade on ecosystem restoration 2021-2030*. United Nations Environment Programme.

7

The Global Water Crisis

The global water crisis is one of the most pressing environmental challenges of the 21st century, affecting millions of people worldwide. It refers to the increasing scarcity of freshwater resources due to factors such as population growth, climate change, pollution, and unsustainable water management practices. Water is essential for human survival, economic development, and the maintenance of ecosystems, yet its availability is becoming increasingly uncertain. As the global demand for water rises, many regions are facing severe water shortages, threatening public health, food security, and social stability.

7.1 Drivers of the Global Water Crisis

One of the primary drivers of the global water crisis is the rapid growth of the human population, which increases the demand for freshwater for drinking, agriculture, and industry. The United Nations (2020) estimates that by 2050, the global population will reach nearly 10 billion, further intensifying pressure on water resources. Agriculture, which accounts for approximately 70% of global freshwater use, is a major contributor to this demand, as it requires large quantities of water for irrigation (FAO, 2017). As more land is cultivated to feed the growing population, water use in agriculture is expected to increase, exacerbating the strain on available freshwater supplies.

Climate change is another significant factor contributing to the global water crisis. Rising global temperatures are altering precipitation patterns, leading to more frequent and intense droughts in some regions, while others experience heavy rainfall and flooding. Kundzewicz et al. (2007) argue that climate change is expected to intensify the variability of water resources, making some regions more vulnerable to water scarcity. For example, regions that are already water-stressed, such as parts of the Middle East and North Africa, are experiencing more prolonged droughts and reduced water

DOI: 10.1201/9788743808787-9

availability, while others are facing increased flooding and water contamination. Additionally, melting glaciers and reduced snowpack in mountainous regions threaten the availability of freshwater for millions of people who rely on these sources for drinking water and irrigation.

Pollution is another critical factor in the global water crisis. Industrial activities, agricultural runoff, and inadequate waste management systems contribute to the contamination of freshwater sources. World Health Organization (2017) reports that nearly 2 billion people worldwide rely on drinking water from sources contaminated with fecal matter, which can lead to waterborne diseases such as cholera, dysentery, and typhoid fever. Agricultural chemicals, such as pesticides and fertilizers, also pollute water sources, making them unsafe for consumption and harming aquatic ecosystems. The contamination of freshwater sources not only poses a health risk but also limits the availability of clean water for agriculture and industry.

7.2 Impacts of the Global Water Crisis

The impacts of the global water crisis are far-reaching and affect various aspects of human life. One of the most immediate consequences is the threat to public health. Inadequate access to clean water and sanitation is a major contributor to the spread of waterborne diseases, particularly in developing countries. According to the World Health Organization (2019), approximately 829,000 people die each year from diarrhea caused by contaminated water, inadequate sanitation, and poor hygiene. Access to safe water is also essential for preventing malnutrition, as it is necessary for food preparation, hygiene, and agricultural production.

Water scarcity also has significant economic consequences. Many industries, including agriculture, energy, and manufacturing, rely heavily on water for their operations. As water becomes scarcer, industries face higher costs for water procurement and treatment, potentially leading to reduced productivity and economic losses. In regions where water scarcity is most acute, such as parts of sub-Saharan Africa and South Asia, economic development is severely hampered by the lack of reliable water resources (UNDP, 2015). Furthermore, water scarcity can lead to social and political instability, as competition for water resources intensifies. In some cases, disputes over water access have led to conflict between communities, regions, or even nations (Zeitoun & Warner, 2006).

The environmental consequences of the water crisis are equally alarming. Freshwater ecosystems, such as rivers, lakes, and wetlands, are essential for biodiversity and ecosystem services. However, over-extraction of water for human use has led to the depletion of these ecosystems, threatening the survival of species that depend on them. Postel (2000) argues that the loss of freshwater ecosystems reduces the planet's ability to regulate water quality, mitigate floods, and support biodiversity. Moreover, the depletion of groundwater resources, particularly in arid and semi-arid regions, poses a long-term threat to food security and agricultural production.

7.3 Solutions to the Global Water Crisis

Addressing the global water crisis requires a multifaceted approach that includes both demand-side and supply-side strategies. On the demand side, water conservation and efficiency measures are critical for reducing the pressure on water resources. Gleick (2003) emphasizes the importance of improving water use efficiency in agriculture, industry, and households. For example, the adoption of water-saving irrigation technologies, such as drip irrigation, can reduce water use in agriculture while maintaining crop yields. Similarly, industries can implement water recycling and reuse systems to reduce their reliance on freshwater sources.

On the supply side, improving water management and infrastructure is essential for ensuring equitable access to water. Biswas (2004) argues that integrated water resources management (IWRM) is a key approach to addressing water scarcity. IWRM involves the coordinated development and management of water, land, and related resources to maximize social and economic benefits while minimizing environmental impacts. This approach emphasizes the need for collaboration between governments, industries, and local communities to manage water resources sustainably.

In addition to improving water management, investing in water infrastructure is essential for providing access to clean water and sanitation. UNICEF (2020) reports that nearly 2 billion people worldwide still lack access to safely managed drinking water services. Expanding access to clean water and sanitation, particularly in rural and underserved areas, is crucial for improving public health and reducing the spread of waterborne diseases.

Finally, addressing the global water crisis requires global cooperation and policy action. International agreements and frameworks, such as the United Nations Sustainable Development Goal 6 (SDG 6), which aims to ensure availability and sustainable management of water and sanitation for all, are essential for promoting global action on water issues. Gleick (2000) highlights the need for governments to prioritize water issues in national and international policy agendas and allocate resources to address water scarcity and improve water quality.

The global water crisis is a complex and multifaceted challenge that requires urgent action. The drivers of the crisis, including population growth, climate change, and pollution, are placing increasing pressure on freshwater resources. The impacts of water scarcity are profound, affecting public health, economic development, and the environment. However, there are solutions available, including improving water use efficiency, investing in water infrastructure, and promoting global cooperation. By adopting a holistic and integrated approach to water management, it is possible to address the global water crisis and ensure a sustainable water future for all.

7.4 Water Scarcity and Pollution

Water scarcity and pollution are two of the most significant global challenges affecting freshwater resources, human health, and the environment. While both issues are interconnected, they represent distinct but often overlapping problems that exacerbate the global water crisis. Water scarcity refers to the insufficient availability of freshwater resources to meet the needs of human populations, ecosystems, and industries, while water pollution involves the contamination of water bodies with harmful substances, making water unsafe for consumption and use. Together, these challenges threaten sustainable development, public health, food security, and environmental stability.

7.4.1 Water scarcity: Causes and impacts

Water scarcity is a growing concern, particularly in arid and semi-arid regions, where the availability of freshwater resources is limited. According to the United Nations (2019), over 2 billion people worldwide live in water-stressed areas, and this number is expected to rise due to population growth, climate change, and unsustainable water management practices. Water scarcity is often exacerbated by the over-extraction of groundwater, which leads to the depletion of aquifers and the deterioration of water quality. Falkenmark and Lannerstad (2005) argue that the global water crisis is driven by the mismanagement of water resources, inefficient irrigation practices, and the increasing demand for water in agriculture, industry, and domestic use.

Agriculture is the largest consumer of freshwater, accounting for approximately 70% of global water use (FAO, 2017). In regions with high agricultural water demand, such as South Asia and the Middle East, over-extraction of water for irrigation has led to the depletion of rivers and groundwater supplies. Postel (2000) emphasizes that the unsustainable use of water for irrigation, coupled with inefficient water management, has resulted in the drying up of critical water sources, such as the Aral Sea and the Colorado River. This not only threatens food security but also exacerbates the economic vulnerability of communities that depend on these water sources for their livelihoods.

Climate change is another significant driver of water scarcity. As global temperatures rise, changes in precipitation patterns are leading to more frequent and severe droughts in some regions, while others experience excessive rainfall and flooding. Kundzewicz et al. (2007) highlight that climate change is expected to alter the distribution and availability of freshwater resources, making some regions more vulnerable to water stress. For example, the increasing frequency of droughts in regions such as sub-Saharan Africa and southern Europe is expected to reduce water availability, while the melting of glaciers in the Himalayas and the Andes threatens the water supply for millions of people who rely on glacial meltwater for drinking and irrigation.

7.4.2 Water pollution: Causes and consequences

Water pollution is a major global issue that exacerbates water scarcity by rendering freshwater sources unsafe for human consumption and agricultural use. Pollution can result from a wide range of human activities, including industrial discharges, agricultural runoff, urbanization, and improper waste disposal. World Health Organization (2017) reports that contaminated water is responsible for millions of deaths each year, particularly in developing countries, where access to clean water and sanitation is limited. The contamination of water sources with harmful chemicals, pathogens, and heavy metals can lead to waterborne diseases, such as cholera, dysentery, and typhoid fever, which are major public health threats.

Agricultural runoff is one of the primary contributors to water pollution, particularly in regions with intensive farming practices. Gleick (2003) notes that the use of chemical fertilizers and pesticides in agriculture leads to the contamination of rivers, lakes, and groundwater with harmful substances. These pollutants not only affect human health but also disrupt aquatic ecosystems by promoting the growth of harmful algal blooms, which deplete oxygen levels and harm aquatic life. The contamination of freshwater bodies with nitrogen and phosphorus compounds, primarily from agricultural runoff, has been linked to eutrophication, a process that leads to the degradation of water quality and loss of biodiversity (Carpenter et al., 1998).

Industrial activities also contribute significantly to water pollution, as factories discharge untreated wastewater containing toxic chemicals, heavy metals, and other pollutants into rivers and lakes. Postel (2000) highlights that industrial pollution is a major concern in developing countries, where regulations on wastewater treatment are often weak or poorly enforced. In addition to the direct health risks posed by industrial pollutants, the contamination of water sources can also have long-term environmental impacts, such as the destruction of aquatic habitats and the loss of biodiversity.

Urbanization and improper waste management further exacerbate water pollution. In rapidly growing cities, untreated sewage and waste often end up in rivers and lakes, contaminating water sources and increasing the risk of waterborne diseases. United Nations (2019) estimates that more than 2 billion people worldwide lack access to safely managed sanitation, which contributes to the contamination of water sources. The lack of proper waste treatment infrastructure in urban areas, particularly in low-income countries, is a significant challenge to improving water quality and reducing pollution.

7.4.3 The interplay between water scarcity and pollution

Water scarcity and pollution are closely linked, with pollution often compounding the effects of water scarcity. In regions where water resources are already limited, the contamination of water sources further reduces the availability of clean water for drinking, sanitation, and irrigation. Gleick (2000) argues that pollution exacerbates water scarcity by making already scarce water resources unusable. For example, in areas where groundwater is the primary source of drinking water, the contamination of aquifers with heavy metals, industrial chemicals, or agricultural runoff can render these water sources undrinkable, forcing communities to rely on increasingly distant and expensive sources of water.

The combination of water scarcity and pollution also has significant economic and social consequences. UNDP (2015) highlights that the economic costs of water pollution are substantial, as they affect agriculture, fisheries, tourism, and industrial production. In regions where water scarcity is already a challenge, the added burden of water pollution can lead to reduced agricultural productivity, increased healthcare costs, and social instability. For example, in countries like India and China, water pollution has led to the contamination of water used for irrigation, resulting in lower crop yields and food insecurity (Sharma et al., 2017).

7.4.4 Addressing water scarcity and pollution

Addressing water scarcity and pollution requires a comprehensive and integrated approach that includes both demand-side and supply-side solutions.

On the demand side, improving water use efficiency in agriculture, industry, and households is essential for reducing the pressure on freshwater resources. Falkenmark and Lannerstad (2005) emphasize the importance of adopting water-saving technologies, such as drip irrigation and water recycling, to reduce water consumption. In industry, implementing water treatment and reuse systems can help reduce the amount of freshwater required for production processes.

On the supply side, improving water quality through better pollution control and waste management is critical. Gleick (2000) suggests that stricter regulations on industrial wastewater discharge, as well as better enforcement of environmental laws, are necessary to reduce water pollution. In agriculture, the use of organic farming practices and the reduction of chemical fertilizers and pesticides can help minimize agricultural runoff and protect water quality (Postel, 2000). Additionally, investing in water infrastructure, such as wastewater treatment plants and sewage systems, is essential for reducing pollution and improving access to clean water.

Water scarcity and pollution are two interrelated challenges that threaten the availability of freshwater resources, public health, and environmental sustainability. The causes of these problems are multifaceted, including population growth, climate change, agricultural practices, industrial activities, and urbanization. The impacts of water scarcity and pollution are far-reaching, affecting human well-being, economic development, and ecosystem health. However, there are solutions available, including improving water use efficiency, reducing pollution, and investing in better water management and infrastructure. By adopting an integrated approach to water management, it is possible to address both water scarcity and pollution and ensure a sustainable water future for all.

7.5 Access to Clean Water as a Human Right

Access to clean water is increasingly recognized as a fundamental human right, essential for the well-being of individuals, communities, and societies at large. Water is indispensable for life, health, and dignity, and its availability and quality directly affect a range of human rights, including the right to health, food, and sanitation. However, despite its importance, access to clean water remains a challenge in many parts of the world, particularly in low-income and rural areas, where water scarcity, pollution, and inadequate infrastructure persist. This has led to growing calls for the recognition of clean water as an inalienable human right and for global efforts to ensure universal access to safe drinking water.

7.5.1 The right to water in international law

The recognition of access to clean water as a human right has gained momentum in international law, particularly following the United Nations (UN) General Assembly's resolution in 2010, which explicitly recognized the right to water and sanitation as a human right. According to the UN, the right to water entitles everyone to sufficient, safe, acceptable, physically accessible, and affordable water for personal and domestic use (UN General Assembly, 2010). This landmark resolution was a response to the growing realization that water is not only a vital resource for survival but also a prerequisite for the enjoyment of other human rights, including the right to health, education, and an adequate standard of living (Hossain, 2011). The UN's recognition of the right to water is a critical step in ensuring that governments and international bodies are held accountable for providing access to clean water for all people, regardless of their socio-economic status or geographic location.

The UN Committee on Economic, Social, and Cultural Rights (CESCR) further elaborated on the right to water in its General Comment No. 15, which outlines the core obligations of states to respect, protect, and fulfill the right to water. These obligations include ensuring that water is available, accessible, and of sufficient quality for all individuals, particularly marginalized and vulnerable groups, such as women, children, and indigenous populations (CESCR, 2002). This interpretation underscores the importance of addressing water access not only as a technical or economic issue but also as a matter of social justice, equity, and human dignity.

7.5.2 Water scarcity and inequality

Despite the recognition of water as a human right, access to clean water remains uneven across the globe, with millions of people still lacking access to safe drinking water. According to the World Health Organization (WHO), approximately 2 billion people worldwide drink water that is contaminated with fecal matter, which poses significant health risks, including the spread of waterborne diseases such as cholera, dysentery, and typhoid fever (WHO, 2017). This disparity is particularly pronounced in low-income countries, where access to clean water is often limited by factors such as poverty, inadequate infrastructure, and political instability. Barlow (2007) argues that water scarcity and pollution disproportionately affect the poorest and most marginalized communities, exacerbating existing inequalities and impeding efforts to achieve sustainable development.

In many regions, water scarcity is compounded by the over-extraction of water resources for agriculture, industry, and urban consumption. Gleick

(2000) notes that the growing demand for water in sectors such as agriculture, energy production, and manufacturing places increasing pressure on freshwater supplies, making it more difficult for communities to access sufficient quantities of clean water. Climate change further exacerbates this issue, as shifting rainfall patterns, droughts, and extreme weather events reduce the availability of freshwater in many parts of the world, particularly in arid and semi-arid regions (Kundzewicz et al., 2007). These challenges highlight the need for comprehensive water management strategies that address both the supply and demand sides of water use, ensuring that clean water is available to all people, especially the most vulnerable.

7.5.3 The role of governance and policy

The realization of the right to clean water is heavily influenced by governance and policy frameworks at the national and international levels. Effective governance structures are essential for managing water resources, ensuring equitable distribution, and addressing the underlying causes of water scarcity and pollution. Sullivan (2002) argues that governments must prioritize water management in their policy agendas and invest in infrastructure to ensure universal access to clean water. This includes the development of water supply systems, wastewater treatment facilities, and sanitation infrastructure, particularly in rural and underserved areas. Furthermore, governments must adopt policies that promote water conservation, reduce pollution, and protect water ecosystems to safeguard the long-term availability of clean water resources.

International cooperation is also critical in addressing the global water crisis. UNDP (2015) highlights the importance of cross-border collaboration in managing shared water resources, particularly in regions where rivers and aquifers span multiple countries. Transboundary water management agreements can help prevent conflicts over water resources and ensure that water is distributed equitably among nations. Additionally, international organizations, such as the UN, the World Bank, and non-governmental organizations (NGOs), play a vital role in supporting countries' efforts to improve water access through technical assistance, financial support, and capacity building (Gleick, 2003).

7.5.4 The role of private sector and community participation

In addition to government action, the private sector and local communities play a crucial role in ensuring access to clean water. The private sector can contribute to improving water access through investments in water infrastructure, technology, and innovation. However, there are concerns about the privatization of water resources, particularly in developing countries, where privatization may lead to increased water prices and reduced access for the poor (Bakker, 2007). As such, the challenge lies in finding a balance between private sector involvement and ensuring that water remains a public good, accessible to all, especially vulnerable populations.

Community participation is also essential in ensuring that water projects are sustainable and meet the needs of local populations. Agarwal et al. (2001) argue that involving communities in the planning, management, and maintenance of water systems is crucial for ensuring that water access is equitable and that the systems are well-maintained. Community-based water management approaches can help empower local populations, improve water governance, and ensure that water resources are used sustainably.

Access to clean water is a fundamental human right that is essential for health, dignity, and the enjoyment of other human rights. Despite significant progress in recognizing water as a human right, many people still lack access to safe drinking water, particularly in low-income and rural areas. Addressing this challenge requires a multifaceted approach that includes effective governance, international cooperation, investment in water infrastructure, and community participation. It is also critical to address the underlying drivers of water scarcity and pollution, such as climate change, over-extraction, and inequality, to ensure that clean water is available to all people, now and in the future.

7.6 Sustainable Water Management

Sustainable water management is a critical approach to ensuring the availability and quality of water resources for current and future generations. It involves the careful planning, development, and management of water resources to meet human needs while preserving the environment. The concept of sustainability in water management recognizes the importance of balancing economic, social, and environmental objectives, ensuring that water is used efficiently and equitably while protecting aquatic ecosystems. This approach is essential as global water resources face increasing pressures from population growth, urbanization, industrialization, climate change, and environmental degradation.

7.6.1 Principles of sustainable water management

The foundation of sustainable water management lies in its principles, which emphasize efficiency, equity, and environmental protection. One of the core principles is the efficient use of water resources. This involves reducing water waste, optimizing water distribution systems, and promoting water-saving technologies. Gleick (2000) highlights the importance of improving water use efficiency in agriculture, industry, and urban areas to cope with growing demand and limited resources. For example, the adoption of drip irrigation and rainwater harvesting systems can significantly reduce water consumption in agriculture, which is the largest consumer of freshwater globally (Postel, 1999). Similarly, industries can implement water recycling and reuse systems to minimize their environmental footprint and reduce dependency on freshwater sources.

Another critical principle of sustainable water management is equity, ensuring that all individuals, especially marginalized and vulnerable groups, have access to sufficient, safe, and affordable water. Sullivan (2002) argues that water distribution systems must be designed to address disparities in access, particularly in low-income communities where water scarcity is more acute. Social equity in water management also involves addressing the needs of rural populations, women, and indigenous communities, who often face greater challenges in accessing clean water. By integrating equity into water management policies, governments and organizations can promote social justice and ensure that no one is left behind in the quest for sustainable water use.

Environmental protection is another essential principle of sustainable water management. This involves safeguarding aquatic ecosystems, preserving biodiversity, and ensuring that water quality is maintained. Pimentel et al. (2004) emphasize the importance of protecting watersheds, wetlands, and river basins, which play a crucial role in maintaining water quality and regulating water flow. Additionally, sustainable water management seeks to prevent over-extraction of water resources, which can lead to the depletion of aquifers, the drying up of rivers, and the degradation of ecosystems. Managing water resources in harmony with nature is vital for long-term sustainability and the resilience of both human and environmental systems.

7.6.2 Integrated water resources management (IWRM)

A key approach to sustainable water management is Integrated Water Resources Management (IWRM), which advocates for a holistic and integrated approach to managing water resources. Global Water Partnership (2000) defines IWRM as a process that promotes the coordinated development and management of water, land, and related resources to maximize social and economic welfare without compromising the sustainability of vital ecosystems. IWRM recognizes the interconnectedness of water systems, land use, and human activities, and it seeks to address water management challenges through multi-stakeholder participation, cross-sectoral coordination, and adaptive management.

IWRM emphasizes the need for collaboration among different sectors, such as agriculture, industry, and urban development, to ensure that water resources are allocated and managed in a way that balances competing demands. For example, in regions where agriculture, industry, and domestic use compete for water, IWRM can help identify trade-offs and prioritize water use to achieve the greatest benefit for society while minimizing environmental impacts. Furthermore, IWRM advocates for the involvement of local communities, indigenous groups, and other stakeholders in decision-making processes to ensure that water management policies reflect local needs and knowledge (Rogers & Hall, 2003).

One of the critical components of IWRM is the implementation of water governance frameworks that promote transparency, accountability, and participation. Effective governance structures are essential for managing water resources equitably and efficiently. Ostrom (1990) highlights the importance of local governance and community-based management in ensuring that water resources are used sustainably and that conflicts over water use are resolved effectively. By involving local communities in the management of water resources, IWRM fosters a sense of ownership and responsibility, which is essential for the long-term sustainability of water systems.

7.6.3 The role of technology and innovation

Advancements in technology and innovation play a significant role in achieving sustainable water management. The development of new technologies for water conservation, treatment, and reuse can significantly reduce the demand for freshwater and improve the efficiency of water use. Gleick (2000) notes that innovations such as desalination, wastewater recycling, and water-efficient irrigation systems are transforming water management practices, particularly in water-scarce regions. For instance, desalination technologies are providing alternative sources of freshwater in arid regions, while wastewater treatment technologies are enabling the reuse of water in industrial and agricultural applications.

Moreover, the use of data and digital technologies, such as remote sensing, geographic information systems (GIS), and water monitoring systems, can improve the management and allocation of water resources. Kundzewicz et al. (2007) argue that these technologies can help monitor water quality and quantity in real time, allowing for better decision-making and early warning systems for floods, droughts, and other water-related disasters. By integrating technology into water management practices, governments and organizations can enhance the efficiency, sustainability, and resilience of water systems.

7.6.4 Challenges to sustainable water management

Despite the progress made in promoting sustainable water management, several challenges remain. One of the most significant challenges is the growing demand for water due to population growth, urbanization, and industrialization. Gleick (2000) points out that the increasing demand for water in agriculture, energy production, and manufacturing is placing unprecedented pressure on freshwater resources, particularly in regions already facing water scarcity. Climate change exacerbates this challenge by altering precipitation patterns, increasing the frequency of droughts, and reducing the availability of freshwater in many regions (Kundzewicz et al., 2007). These challenges require urgent action to improve water efficiency, conserve water resources, and adapt to changing environmental conditions.

Another challenge is the pollution of water resources, which compromises the quality of water available for consumption, agriculture, and industry. Pimentel et al. (2004) emphasize that pollution from agricultural runoff, industrial waste, and untreated sewage is a major threat to water quality worldwide. Addressing water pollution requires stronger regulations, better enforcement of environmental laws, and the adoption of cleaner production technologies in agriculture and industry.

Sustainable water management is essential for ensuring that water resources are available and of adequate quality for future generations. By adopting principles of efficiency, equity, and environmental protection, and by implementing approaches such as IWRM, sustainable water management can help address the challenges posed by water scarcity, pollution, and climate change. Technological innovation and effective governance frameworks also play a crucial role in achieving sustainability. However, challenges such as growing demand, pollution, and the impacts of climate change require urgent action at the global, national, and local levels. Through collaborative efforts and innovative solutions, sustainable water management can help secure a water-sustainable future for all.

References

Agarwal, B., et al. (2001). *Water, land, and livelihoods in rural India*. Oxford University Press.

Bakker, K. (2007). *The "commons" versus the "commodity": Alter-globalization, anti-privatization, and the human right to water in the global South.* Antipode, 39(3), 430–455. https://doi.org/10.1111/j.1467-8330.2007.00534.x

Barlow, M. (2007). *Blue gold: The fight to stop the corporate theft of the world's water.* The New Press.

Biswas, A. K. (2004). Integrated water resources management: A reassessment. *Water International, 29*(2), 248–256. https://doi.org/10.1080/02508060408691775

Carpenter, S. R., et al. (1998). Nonpoint pollution of surface waters with phosphorus and nitrogen. *Ecological Applications, 8*(3), 559–568. https://doi.org/10.1890/1051-

CESCR. (2002). General comment no. 15: The right to water (arts. 11 and 12 of the Covenant). *E/C.12/2002/11.*

Falkenmark, M., & Lannerstad, M. (2005). Consumptive water use to feed humanity – The water perspective of food security. *In Water for Food Security* (pp. 139-155). Springer.

FAO. (2017). *The state of food and agriculture: Climate change, agriculture and food security.* Food and Agriculture Organization of the United Nations.

Gleick, P. H. (2000). The changing water paradigm: A look at twenty-first century water resources development. *Water International, 25*(1), 127–138. https://doi.org/10.1080/02508060008686704

Gleick, P. H. (2003). *Water use.* Annual Review of Environment and Resources, 28, 275–314. https://doi.org/10.1146/annurev.energy.28.050302.105332

Global Water Partnership. (2000). *Integrated water resources management.* Global Water Partnership.

Hossain, N. (2011). *The human right to water: A framework for rights-based policy advocacy.* International Environmental Law Research Centre.

Kundzewicz, Z. W., et al. (2007). Freshwater resources and climate change: A global perspective. *Hydrological Sciences Journal, 52*(5), 739–748. https://doi.org/10.1623/hysj.52.5.739

Kundzewicz, Z. W., et al. (2007). Freshwater resources and their management. In M. L. Parry, et al. (Eds.), *Climate change 2007: Impacts, adaptation and vulnerability* (pp. 173–210). Cambridge University Press.

Ostrom, E. (1990). *Governing the commons: The evolution of institutions for collective action.* Cambridge University Press.

Pimentel, D., et al. (2004). Water resources: Agricultural and environmental issues. *Bioscience, 54*(10), 909–918. https://doi.org/10.1641/0006-3568(2004)054[0909:WRAEIA]2.0.CO;2

Postel, S. (1999). *Pillars of sand: Can the irrigation miracle last?* W.W. Norton & Company.

Postel, S. (2000). *Entering an era of water scarcity: The challenges ahead.* Ecological Applications, 10(4), 941–948. https://doi.org/10.1890/1051-0761(2000)010[0941:IEAOWS]2.0.CO;2

Rogers, P., & Hall, A. W. (2003). *Effective water governance*. Global Water Partnership.

Sharma, R., et al. (2017). Water pollution in India: Causes and consequences. *Environmental Science and Pollution Research, 24*(5), 4701–4711. https://doi.org/10.1007/s11356-017-0370-9

Sullivan, C. A. (2002). *Water and poverty: The need for a more equitable approach. Water International, 27*(3), 383–388. https://doi.org/10.1080/02508060208686913

UN General Assembly. (2010). *Resolution 64/292: The human right to water and sanitation*. United Nations.

UNDP. (2015). *Human development report 2015: Work for human development*. United Nations Development Programme.

UNICEF. (2020). *Water, sanitation and hygiene*. United Nations Children's Fund. https://www.unicef.org/wash

United Nations. (2019). *The United Nations world water development report 2019: Leaving no one behind*. United Nations Educational, Scientific and Cultural Organization.

WHO. (2017). *Drinking-water quality: Guidelines for drinking-water quality*. World Health Organization.

World Health Organization. (2017). *Drinking-water quality*. https://www.who.int/water_sanitation_health/dwq/en/

World Health Organization. (2019). *Diarrhoeal disease*. https://www.who.int/news-room/fact-sheets/detail/diarrhoeal-disease

Zeitoun, M., & Warner, J. (2006). Hydro-hegemony: A framework for analysis of trans-boundary water conflicts. *Water Policy, 8*(5), 435–460. https://doi.org/10.2166/wp.2006.035

8

Pollution and Waste Management

Pollution and waste management are critical components of environmental sustainability, as they directly impact human health, biodiversity, and the planet's ecosystems. The growing challenges of pollution and waste accumulation, driven by industrialization, urbanization, and consumerism, require innovative and effective management strategies. Pollution can take various forms, including air, water, soil, and noise pollution, each posing significant threats to environmental health. On the other hand, waste management involves collecting, disposing, recycling, and treating waste to minimize its harmful effects. Effective pollution and waste management are essential for mitigating environmental degradation, promoting public health, and ensuring the sustainability of natural resources.

8.1 Types of Pollution

Pollution is typically categorized into four main types: air pollution, water pollution, soil pollution, and noise pollution. Air pollution is one of the most widespread and harmful forms of pollution, caused primarily by burning fossil fuels, industrial emissions, and vehicle exhaust. According to the World Health Organization (WHO, 2018), air pollution is responsible for millions of premature deaths each year, with major health impacts such as respiratory diseases, cardiovascular diseases, and cancer. Urban areas, in particular, face significant air quality challenges due to high industrial activity and transportation levels.

Water pollution is another major environmental concern, with pollutants such as heavy metals, chemicals, plastics, and untreated sewage contaminating rivers, lakes, and oceans. Barker et al. (2020) highlight that water pollution threatens aquatic ecosystems and affects the availability of clean drinking water, which is essential for human survival. The contamination of freshwater resources is particularly severe in developing countries, where

DOI: 10.1201/9788743808787-10

industrial waste, agricultural runoff, and inadequate sanitation systems contribute to the deterioration of water quality.

Soil pollution occurs when hazardous chemicals, waste products, and pollutants are deposited on the soil, degrading its quality and fertility. Pimentel et al. (2004) argue that soil pollution, often caused by the overuse of pesticides and fertilizers in agriculture, can lead to reduced crop yields and long-term soil degradation. Furthermore, soil pollution can contaminate food supplies, impacting human health and food security.

Noise pollution, though often overlooked, also has significant environmental and health impacts. Basner et al. (2014) note that chronic exposure to high noise levels can lead to hearing loss, sleep disturbances, cardiovascular problems, and increased stress levels. Urban areas, airports, and industrial zones are major sources of noise pollution, affecting both human populations and wildlife.

8.2 Waste Management Challenges

The global increase in waste generation, particularly solid waste, poses another critical challenge. The rise in consumption, especially in urbanized and industrialized regions, has led to an exponential increase in waste production. Kaza et al. (2018) estimate that global waste generation will reach 3.4 billion tons by 2050, with much of this waste being non-biodegradable and challenging to manage. The improper disposal of waste, especially plastic waste, contributes significantly to environmental pollution. Jambeck et al. (2015) report that millions of tons of plastic waste end up in the oceans yearly, harming marine life and entering the food chain.

Waste management involves various processes, including collection, transportation, recycling, and disposal. Recycling is a key strategy for reducing waste and conserving resources. However, recycling rates remain low in many parts of the world due to inadequate infrastructure, lack of public awareness, and economic challenges. Wilson et al. (2015) argue that effective waste management systems must include comprehensive recycling programs, public education, and incentives for individuals and businesses to reduce waste.

One of the most pressing challenges in waste management is treating and disposing of hazardous waste. Gersberg et al. (2019) highlight that hazardous waste, such as chemicals, medical waste, and electronic waste (e-waste), poses significant risks to human health and the environment. E-waste has become a growing concern due to the rapid technological advancements and the short lifespan of electronic products. Improper e-waste disposal can release toxic substances such as lead, mercury, and cadmium, which can contaminate soil, water, and air.

8.3 Strategies for Pollution and Waste Management

Governments, industries, and individuals must adopt sustainable practices and technologies to address pollution and waste management challenges. Pollution prevention is one of the most effective strategies for reducing the environmental impact of human activities. This approach focuses on reducing or eliminating the generation of pollutants at the source rather than treating or managing pollution after it has been created. Lave et al. (1995) suggest that industries should adopt cleaner production technologies, such as energy-efficient processes and the use of non-toxic materials, to minimize pollution. In addition, governments can implement stricter environmental regulations and provide incentives for businesses to adopt sustainable practices.

Waste minimization is another essential strategy in managing the growing waste problem. This involves reducing the volume of waste generated through measures such as product design for durability, repairability, and recyclability. Bocken et al. (2016) advocate for the circular economy model, which emphasizes reusing, repairing, and recycling materials to reduce waste and conserve resources. In the circular economy, waste is viewed as a resource that can be repurposed rather than something to be discarded. For example, in the textile industry, companies are increasingly adopting sustainable practices such as recycling old clothes into new garments or using biodegradable materials to reduce waste.

Extended producer responsibility (EPR) is another important policy tool for managing waste. EPR holds manufacturers accountable for the entire lifecycle of their products, including their disposal. Lifset and Lindhqvist (2008) argue that EPR can incentivize companies to design products that are easier to recycle and reduce the amount of waste generated. By taking responsibility for the waste their products generate, manufacturers can help reduce their products' environmental impact and promote more sustainable consumption patterns.

8.4 The Role of Technology and Innovation

Technological innovation plays a vital role in improving pollution and waste management practices. Advances in waste treatment technologies, such as waste-to-energy (WTE) systems, can help reduce the volume of waste while generating energy. Zhang et al. (2020) note that WTE technologies, such as incineration and anaerobic digestion, can convert organic waste into biogas, which can be used for electricity generation. Similarly, innovations in water treatment technologies, such as membrane filtration and bioremediation, can help remove pollutants from water, ensuring that it is safe for consumption and use.

Moreover, developing smart waste management systems can improve waste collection and recycling efficiency and effectiveness. Gao et al. (2018) discuss how sensors, data analytics, and the Internet of Things (IoT) can optimize waste collection routes, monitor bin waste levels, and improve recycling rates. These technologies can help reduce the environmental footprint of waste management operations while enhancing the overall efficiency of waste collection systems.

Pollution and waste management are essential to environmental sustainability, requiring coordinated global, national, and local efforts. Addressing pollution through prevention, sustainable production practices, and cleaner technologies can significantly reduce environmental harm. Similarly, improving waste management through recycling, waste minimization, and extended producer responsibility can help reduce the amount of waste that ends up in landfills and the environment. Technological innovations and the adoption of circular economy principles are key to achieving sustainable outcomes in both pollution and waste management. Achieving effective pollution and waste management requires a collective commitment to reducing environmental impacts, conserving resources, and promoting a more sustainable future.

8.5 The Impact of Air, Water, and Soil Pollution

Pollution, in its various forms, is a significant driver of environmental degradation, affecting ecosystems, human health, and the planet's natural resources. Among the most prevalent types of pollution are air, water, and soil pollution, each contributing to global environmental crises in unique and interconnected ways. These pollutants not only harm the immediate environment but also have far-reaching consequences for human well-being, biodiversity, and the stability of ecosystems. Understanding the impacts of air, water, and soil pollution is crucial for developing effective policies and practices to mitigate their harmful effects and promote sustainability.

8.5.1 Air pollution and its impacts

Air pollution, primarily caused by the emission of harmful substances such as particulate matter (PM), nitrogen oxides (NO_x), sulfur dioxide (SO_2), and volatile organic compounds (VOCs), is a leading environmental health risk. The major sources of air pollution include industrial activities, vehicle emissions, agricultural practices, and the burning of fossil fuels for energy (WHO, 2018). These pollutants can have severe health impacts, including respiratory diseases such as asthma, bronchitis, and chronic obstructive pulmonary disease (COPD), as well as cardiovascular diseases (Brook et al., 2010). Air pollution is also linked to an increased risk of lung cancer and premature death (Liu et al., 2020). According to the World Health Organization (WHO), air pollution is responsible for an estimated 7 million premature deaths annually, with the majority of these deaths occurring in low- and middle-income countries (WHO, 2018).

In addition to human health, air pollution has significant ecological consequences. For example, high levels of nitrogen oxides and sulfur dioxide can lead to the formation of acid rain, which harms aquatic ecosystems, soil quality, and vegetation (Fowler et al., 2013). Acid rain can leach essential nutrients from the soil, disrupt the pH balance of water bodies, and damage plant life, thus threatening biodiversity. Moreover, air pollution contributes to global warming by increasing the concentration of greenhouse gases in the atmosphere. Carbon dioxide (CO_2), methane (CH_4), and other greenhouse gases trap heat, leading to the warming of the Earth's surface and altering climate patterns (IPCC, 2021).

8.5.2 Water pollution and its consequences

Water pollution is another major environmental issue that threatens the availability of clean drinking water and disrupts aquatic ecosystems. Water pollution occurs when harmful substances such as chemicals, heavy metals, plastics, and untreated sewage are introduced into water bodies, including rivers, lakes, and oceans. The primary sources of water pollution are industrial discharges, agricultural runoff, untreated wastewater, and oil spills (Jambeck et al., 2015). These pollutants can have dire consequences for both human health and biodiversity.

Chemicals such as pesticides and fertilizers, which are commonly used in agriculture, can contaminate freshwater resources, leading to eutrophication. Eutrophication is the process by which excess nutrients, particularly nitrogen and phosphorus, promote the growth of algae in water bodies. This excessive algal growth can deplete oxygen levels, causing hypoxia, which harms aquatic life and disrupts the food chain (Smith et al., 2016). In addition, waterborne diseases caused by pathogens in contaminated water, such as cholera, dysentery, and typhoid fever, remain a significant public health issue, particularly in developing countries with inadequate sanitation infrastructure (Bartram et al., 2005).

Plastic pollution is a growing concern for marine ecosystems, with millions of tons of plastic waste entering the oceans each year. This pollution not only harms marine life, as animals ingest or become entangled in plastic debris, but it also disrupts the entire marine food chain (Jambeck et al., 2015). Furthermore, chemicals in plastics, such as bisphenol A (BPA), can leach into the water, causing endocrine disruption in aquatic organisms and potentially affecting human health through seafood consumption (Rochman et al., 2013).

8.5.3 Soil pollution and its effects

Soil pollution, caused by the accumulation of harmful substances in the soil, is another critical environmental issue that has far-reaching effects on ecosystems and human health. Soil pollutants include heavy metals such as lead, cadmium, and arsenic, as well as pesticides, herbicides, and industrial chemicals. These pollutants can be introduced into the soil through agricultural practices, industrial activities, and improper waste disposal (Pimentel et al., 2004). The presence of toxic chemicals in the soil can lead to the degradation of soil quality, reducing its fertility and ability to support plant life.

Soil pollution also has significant implications for food security. Contaminated soil can lead to the accumulation of harmful substances in crops, which can then enter the food chain and pose health risks to humans. For example, heavy metals such as lead and cadmium can accumulate in vegetables and grains, leading to poisoning and other health problems when consumed (Mielke et al., 2011). Soil pollution also affects biodiversity, as it disrupts the habitat of soil organisms such as earthworms, fungi, and bacteria, which play essential roles in nutrient cycling and soil fertility (Bünemann et al., 2018). The loss of soil biodiversity can have cascading effects on ecosystem functions, leading to reduced agricultural productivity and ecosystem resilience.

8.5.4 The interconnectedness of air, water, and soil pollution

The impacts of air, water, and soil pollution are often interconnected, with pollutants traveling across media and compounding their harmful effects. For example, air pollution can lead to the deposition of pollutants onto the soil and water, a process known as atmospheric deposition. Acid rain, as mentioned earlier, is a result of airborne pollutants such as sulfur dioxide and nitrogen oxides, which can acidify both soil and water (Fowler et al., 2013). Similarly, water pollution can contribute to soil contamination, as chemicals and heavy metals from polluted water bodies can accumulate in the soil, further exacerbating the problem of soil degradation.

The cumulative effects of pollution across air, water, and soil systems underscore the need for integrated environmental management strategies that address pollution holistically. Efforts to mitigate pollution must consider the interactions between these environmental media and take a systems-based approach to minimize the overall environmental impact. This approach requires cooperation between governments, industries, and local communities to reduce emissions, promote sustainable agricultural practices, and ensure the proper treatment and disposal of waste.

Air, water, and soil pollution are major environmental challenges that have far-reaching consequences for human health, biodiversity, and ecosystem stability. The impacts of these pollutants are interconnected, affecting not only the immediate environment but also the broader global ecosystem. Addressing these challenges requires comprehensive pollution control measures, sustainable practices in agriculture and industry, and effective waste management strategies. By reducing pollution at its source, promoting cleaner technologies, and enhancing public awareness, it is possible to mitigate the adverse effects of pollution and work towards a more sustainable future.

8.6 The Plastic Crisis and Ocean Degradation

The plastic crisis is one of the most pressing environmental challenges of the 21st century, with profound implications for global ecosystems, particularly the world's oceans. Plastics, which have become ubiquitous in modern life due to their durability, low cost, and versatility, are also among the most persistent pollutants in the environment. The rapid increase in plastic production and consumption has led to a corresponding rise in plastic waste, much of which ends up in the oceans. This pervasive pollution is causing severe damage to marine ecosystems, wildlife, and human health. Addressing the plastic crisis requires coordinated global efforts to reduce plastic production, improve waste management, and promote alternatives to plastic use.

8.6.1 The scale of the plastic crisis

Plastic production has grown exponentially since the mid-20th century, with global production exceeding 368 million metric tons in 2019 (PlasticsEurope, 2020). The durability and low cost of plastic materials have made them indispensable in industries ranging from packaging to automotive manufacturing. However, the very properties that make plastics so useful also contribute to their persistence in the environment. Unlike organic materials, plastics do not biodegrade; instead, they break down into smaller particles, known as microplastics, which can persist in ecosystems for centuries (Andrady, 2017). An estimated 8 million metric tons of plastic waste enter the oceans annually, which is expected to increase significantly if current trends continue (Jambeck et al., 2015). The accumulation of plastic waste in the oceans has become a global environmental crisis, affecting marine life, ecosystems, and human health.

8.6.2 Impacts on marine life

The impact of plastic pollution on marine life is devastating. Marine animals, including fish, sea birds, and marine mammals, often mistake plastic debris for food. Ingesting plastic can lead to injury, malnutrition, and death. For example, sea turtles are particularly vulnerable to ingesting plastic bags, which they mistake for jellyfish, a primary food source. Ingestion of plastic can cause blockages in the digestive system, leading to starvation and death (Schuyler et al., 2014). In addition to direct ingestion, animals can become entangled in plastic debris, which can restrict movement, cause wounds, and lead to drowning or suffocation (Laist, 1997). The effects of plastic pollution extend up the food chain, as smaller marine organisms ingest microplastics, which are then consumed by larger predators, including humans (Rochman et al., 2013).

Microplastics, which are less than 5 millimeters in size, are particularly problematic because they are easily ingested by a wide range of marine organisms, from plankton to fish and whales. These tiny particles can absorb toxic chemicals from the surrounding environment, such as pesticides and heavy metals, which can then be transferred to marine organisms through ingestion. The accumulation of these toxic substances in marine organisms can have harmful effects on their health and reproductive success (Teuten et al., 2009). Furthermore, microplastics have been found in seafood, raising concerns about the potential for human exposure to these contaminants through the consumption of marine products (Lusher et al., 2013).

8.6.3 Oceanic ecosystem disruption

The accumulation of plastic in the oceans also disrupts marine ecosystems. Large plastic debris, such as fishing nets, bottles, and packaging, can smother coral reefs, seagrass beds, and other sensitive marine habitats. Coral reefs, which are already under threat from climate change and ocean acidification, are particularly vulnerable to plastic pollution. When plastic debris settles on coral reefs, it can block sunlight, inhibit photosynthesis in symbiotic algae, and cause physical damage to coral structures (Carpenter et al., 2018). The loss of coral reefs, which are critical habitats for many marine species, further exacerbates the decline of marine biodiversity.

Plastic pollution also contributes to the degradation of coastal ecosystems. Coastal areas, which are home to diverse and productive ecosystems, are often the final destination for plastic waste. In these areas, plastics can degrade soil quality, disrupt local ecosystems, and harm the livelihoods of communities that depend on fishing and tourism. The presence of plastic waste on beaches and in the water can also deter tourists, negatively affecting local economies (Nelms et al., 2018). The impact of plastic pollution on coastal communities is compounded by the increasing frequency of extreme weather events, which can exacerbate the spread of plastic waste and further damage ecosystems.

8.6.4 The role of microplastics in ocean degradation

Microplastics, which are often the result of the breakdown of larger plastic items, pose a particular challenge for marine ecosystems. These particles are small enough to be ingested by a wide range of organisms, from microscopic plankton to large marine mammals. Microplastic ingestion can harm marine organisms, including reduced feeding efficiency, impaired growth, and reproductive failure (Rochman et al., 2013). Additionally, microplastics can accumulate in the tissues of marine organisms, leading to bioaccumulation and biomagnification of toxic substances. As a result, microplastics have the potential to disrupt entire marine food webs, with cascading effects on biodiversity and ecosystem health.

Microplastics also contribute to the spread of invasive species. Small plastic particles can serve as "rafts" for marine organisms, allowing them to travel long distances across the oceans. This phenomenon, known as the "plastic bridge," facilitates the spread of invasive species to new areas, where they can outcompete native species and disrupt local ecosystems (Barnes et al., 2009). The movement of invasive species via plastic debris is particularly concerning for fragile ecosystems, such as coral reefs and deep-sea habitats, which are already under threat from climate change and other human activities.

8.6.5 Addressing the plastic crisis

Addressing the plastic crisis requires a multifaceted approach that includes reducing plastic production, improving waste management, and promoting alternatives to plastic. One of the most effective ways to reduce plastic pollution is to reduce the production and consumption of single-use plastics, which are a major source of plastic waste. Bans on single-use plastic items, such as straws, bags, and utensils, have been implemented in several countries and regions with positive results (Jambeck et al., 2018). In addition to reducing plastic consumption, improving waste management systems is critical to preventing plastic waste from entering the oceans. This includes increasing recycling rates, reducing plastic waste in landfills, and developing more sustainable packaging solutions (Lebreton et al., 2017).

Furthermore, the development of biodegradable alternatives to plastic, such as plant-based plastics, can help reduce the environmental impact of plastic waste. However, it is important to ensure that these alternatives do not contribute to other environmental problems, such as deforestation or the use of harmful chemicals (Kozlowski et al., 2019). Public education and awareness campaigns are also essential for encouraging individuals and businesses to reduce their plastic consumption and adopt more sustainable practices.

The plastic crisis and its impact on ocean degradation are among the most pressing environmental challenges of our time. The accumulation of plastic waste in the oceans is causing severe harm to marine life, disrupting ecosystems, and threatening human health. Addressing this crisis requires a global effort to reduce plastic production, improve waste management, and promote alternatives to plastic. By taking action to reduce plastic pollution, we can protect marine ecosystems, safeguard biodiversity, and ensure a healthier future for the planet.

8.7 Towards Zero Waste: Solutions and Innovations

The concept of "zero waste" is an ambitious yet essential approach to environmental sustainability that aims to minimize waste generation and maximize resource efficiency. It seeks to reduce the volume of waste sent to landfills and incinerators by promoting the reuse, recycling, and repurposing of materials. The transition towards zero waste involves a paradigm shift in how societies produce, consume, and manage materials, focusing on the lifecycle of products and their environmental impacts. The zero-waste movement is gaining momentum globally as an essential strategy to address the mounting environmental challenges posed by waste accumulation, resource depletion, and pollution. This approach not only helps mitigate environmental harm but also presents economic opportunities through the creation of new industries, jobs, and innovations in sustainable design and waste management.

8.7.1 The principles of zero waste

Zero waste is grounded in a set of core principles aimed at reducing waste generation and minimizing environmental harm. These principles emphasize the importance of designing products for durability, repairability, and recyclability. One of the fundamental tenets of zero waste is the concept of a circular economy, which contrasts with the traditional linear model of production and consumption. In a linear economy, resources are extracted, used, and disposed of as waste, contributing to resource depletion and environmental degradation. In contrast, the circular economy promotes the idea of "closing the loop," where products are designed for reuse, repair, and recycling, thus reducing the need for virgin materials and minimizing waste (Geissdoerfer et al., 2017). This shift requires rethinking product design, production processes, and consumption patterns to ensure that materials are kept in use for as long as possible.

Another key principle of zero waste is the reduction of consumption. This involves promoting a shift towards sustainable lifestyles, where individuals and businesses reduce their reliance on disposable products and adopt more sustainable practices. Reducing consumption can be achieved through strategies such as minimizing packaging, choosing products with longer lifespans, and embracing sharing and leasing models instead of ownership (McKinsey & Company, 2020). Moreover, zero waste emphasizes the importance of composting organic waste, which can be used to enrich soil and reduce the environmental impact of food waste, a significant contributor to landfill waste (Zhang et al., 2019).

8.7.2 Innovations in waste reduction and resource efficiency

A crucial aspect of the zero-waste movement is the development and implementation of innovative technologies and practices aimed at reducing waste and improving resource efficiency. One of the most significant innovations is in the field of materials science, where researchers are developing biodegradable and sustainable alternatives to conventional plastics. Bioplastics, made from renewable sources such as corn starch or algae, offer a promising solution to the plastic pollution crisis. These materials can break down more quickly and safely in the environment, reducing the long-term impacts of plastic waste (Shen et al., 2017). Additionally, innovations in packaging design, such as edible packaging and reusable containers, are gaining traction as sustainable alternatives to single-use plastic packaging (Kang et al., 2020).

Another area of innovation is in waste management technologies, particularly in the recycling sector. Advances in sorting technologies, such as optical sorting and AI-driven robotics, are improving the efficiency and accuracy of recycling processes. These technologies enable the separation of materials more effectively, increasing recycling rates and reducing contamination. For instance, AI-powered robots can identify and sort different types of plastics, metals, and paper, improving the quality of recycled materials and making the recycling process more economically viable (Binns, 2020). Additionally, the development of chemical recycling processes, which break down plastics into their constituent chemicals for reuse, offers the potential to recycle materials that are currently non-recyclable through traditional methods (Hopewell et al., 2009).

In addition to technological innovations, businesses are increasingly adopting sustainable business models that align with the principles of zero waste. Companies are rethinking their supply chains and product lifecycles to reduce waste at every stage of production. For example, companies like Patagonia and IKEA are embracing circular economy principles by offering repair services, using recycled materials in their products, and encouraging customers to return used items for reuse or recycling (Bocken et al., 2016). These companies are not only reducing waste but also driving consumer demand for sustainable products, which in turn encourages other businesses to adopt similar practices.

8.7.3 The role of policy and legislation

While individual and business efforts are crucial in the transition to zero waste, government policies and legislation play a vital role in supporting and accelerating this transition. Governments can create a favorable regulatory environment by implementing policies that promote waste reduction, recycling, and the use of sustainable materials. For instance, extended producer responsibility (EPR) programs, which require manufacturers to take responsibility for the entire lifecycle of their products, including disposal and recycling, have been successfully implemented in several countries. These programs incentivize manufacturers to design products that are easier to recycle and reduce the amount of waste generated at the end of a product's life (Lindhqvist, 2000).

Another effective policy tool is the imposition of waste reduction targets, such as bans on single-use plastics or requirements for minimum recycling rates. For example, the European Union has implemented a directive that mandates the reduction of single-use plastic products, including straws, cutlery, and plates, by 2021 (European Commission, 2018). Such policies not only reduce the environmental impact of waste but also stimulate innovation in sustainable alternatives and recycling technologies. Additionally, governments can support the development of green infrastructure, such as composting facilities and advanced recycling centers, to ensure that waste is managed efficiently and sustainably.

8.7.4 Community engagement and education

Community engagement and public education are also critical components of the zero-waste movement. Raising awareness about the environmental impacts of waste and the benefits of zero-waste practices is essential for driving behavioral change at the individual and community levels. Educational campaigns can encourage people to reduce consumption, make more sustainable choices, and participate in recycling and composting programs. Community-based initiatives, such as zero-waste cities and grassroots organizations, play an important role in fostering local solutions to waste management challenges. These initiatives can provide practical tools and resources for individuals and businesses to reduce waste and adopt more sustainable practices (Connett, 2013).

Furthermore, collaboration between various stakeholders – governments, businesses, NGOs, and communities – is essential for creating a circular economy and achieving zero-waste goals. By working together, these stakeholders can share knowledge, resources, and best practices to build a more sustainable future. For example, the Global Alliance for Incinerator Alternatives (GAIA) is a network of organizations working to promote zero-waste solutions globally, advocating for policies that reduce waste and support sustainable waste management practices (GAIA, 2020).

The transition towards zero waste is a critical step in addressing the environmental challenges posed by waste accumulation, resource depletion, and pollution. By embracing the principles of the circular economy, reducing consumption, and adopting innovative technologies, societies can significantly reduce their environmental footprint. However, achieving zero waste requires coordinated efforts at the individual, business, and policy levels. Through technological innovation, sustainable business practices, supportive policies, and community engagement, the goal of zero waste is within reach, offering a sustainable future for generations to come.

References

Andrady, A. L. (2017). *The plastic in microplastics: A review*. Environmental Pollution, 227, 25–35. https://doi.org/10.1016/j.envpol.2017.04.002

Barker, T., et al. (2020). *Water pollution and its impact on human health.* Environmental Health Perspectives, 128(8), 1–8. https://doi.org/10.1289/EHP7006

Barnes, D. K. A., et al. (2009). *Accumulation and fragmentation of plastic debris in global environments.* Philosophical Transactions of the Royal Society B: Biological Sciences, 364(1526), 1985–1998. https://doi.org/10.1098/rstb.2008.0205

Bartram, J., et al. (2005). *Water quality monitoring: A practical guide to the design and implementation of freshwater quality studies and monitoring programmes.* CRC Press.

Basner, M., et al. (2014). *Auditory and non-auditory effects of noise on health.* The Lancet, 383(9925), 1325–1332. https://doi.org/10.1016/S0140-6736(13)61613-X

Binns, A. (2020). *Artificial intelligence in recycling: How AI is transforming waste management.* Waste Management, 106, 1–8. https://doi.org/10.1016/j.wasman.2020.03.021

Bocken, N. M. P., et al. (2016). *Circular business models: A review.* Journal of Cleaner Production, 114, 81–95. https://doi.org/10.1016/j.jclepro.2015.09.007

Bocken, N. M. P., et al. (2016). *Product design and business model strategies for a circular economy.* Journal of Industrial and Production Engineering, 33(5), 308–320. https://doi.org/10.1080/21681015.2016.1172124

Brook, R. D., et al. (2010). *Particulate matter air pollution and cardiovascular disease: An update to the 2004 report of the Institute of Medicine.* Environmental Health Perspectives, 118(4), 438–448. https://doi.org/10.1289/ehp.0901220

Bünemann, E. K., et al. (2018). *Soil biodiversity and soil health indicators.* Soil Biology and Biochemistry, 116, 125–136. https://doi.org/10.1016/j.soilbio.2017.09.014

Carpenter, E. J., et al. (2018). *Marine debris and coral reefs: A review.* Marine Pollution Bulletin, 126, 45–53. https://doi.org/10.1016/j.marpolbul.2017.10.043

Connett, P. (2013). *The zero waste solution: Untrashing the planet one community at a time.* Chelsea Green Publishing.

European Commission. (2018). *Directive (EU) 2018/851 of the European Parliament and of the Council on waste.* Official Journal of the European Union, L 150, 109–140. https://eur-lex.europa.eu/legal-content/EN/TXT/?uri=CELEX%3A32018L0851

Fowler, D., et al. (2013). *The global nitrogen cycle in the twenty-first century*. Philosophical Transactions of the Royal Society B: Biological Sciences, 368(1621), 20130164. https://doi.org/10.1098/rstb.2013.0164

GAIA. (2020). *Global Alliance for Incinerator Alternatives*. https://www.no-burn.org/

Gao, J., et al. (2018). *Smart waste management systems: A review*. Environmental Science & Technology, 52(15), 8589–8600. https://doi.org/10.1021/acs.est.8b01527

Geissdoerfer, M., et al. (2017). *The circular economy – A new sustainability paradigm?*. Journal of Cleaner Production, 143, 757–768. https://doi.org/10.1016/j.jclepro.2016.12.048

Gersberg, R. M., et al. (2019). *Hazardous waste management and its environmental implications*. Environmental Science & Technology, 53(4), 2215–2223. https://doi.org/10.1021/acs.est.8b06409

IPCC. (2021). *Climate change 2021: The physical science basis*. Cambridge University Press. https://doi.org/10.1017/9781009157896

Jambeck, J. R., et al. (2015). *Plastic waste inputs from land into the ocean*. Science, 347(6223), 768–771. https://doi.org/10.1126/science.1260352

Jambeck, J. R., et al. (2018). *Solutions to plastic pollution in the ocean*. Science, 360(6394), 1071–1072. https://doi.org/10.1126/science.aat1684

Kang, H., et al. (2020). *Edible packaging: A review of the current progress and challenges*. Journal of Food Science, 85(10), 3310–3320. https://doi.org/10.1111/1750-3841.15306

Kaza, S., et al. (2018). *What a waste 2.0: A global snapshot of solid waste management to 2050*. World Bank. https://doi.org/10.1596/978-1-4648-1329-0

Kozlowski, R. M., et al. (2019). *Biodegradable plastics and the environment: A review of the scientific evidence*. Environmental Science & Technology, 53(9), 5125–5134. https://doi.org/10.1021/acs.est.8b06485

Laist, D. W. (1997). *Impacts of marine debris: Entanglement of marine life in marine debris including a comprehensive list of species with entanglement records*. Marine Pollution Bulletin, 34(11), 846–852. https://doi.org/10.1016/S0025-326X(97)00032-7

Lave, L. B., et al. (1995). *Pollution prevention and cleaner production: A review of research and policy*. Environmental Science & Technology, 29(10), 404–411. https://doi.org/10.1021/es00007a001

Lebreton, L., et al. (2017). *A global mass budget for plastic waste*. Science, 358(6360), 1205–1208. https://doi.org/10.1126/science.aab1123

Lifset, R., & Lindhqvist, T. (2008). *Extended producer responsibility: A global overview*. Journal of Industrial Ecology, 12(3), 301–310. https://doi.org/10.1111/j.1530-9290.2008.00028.x

Lindhqvist, T. (2000). *Extended producer responsibility in cleaner production: Policy principle to promote environmental improvements of product systems*. Journal of Cleaner Production, 8(1), 25–33. https://doi.org/10.1016/S0959-6526(00)00013-9

Liu, J., et al. (2020). *Ambient particulate air pollution and daily mortality in 652 cities*. New England Journal of Medicine, 383, 1012–1022. https://doi.org/10.1056/NEJMoa1917363

Lusher, A. L., et al. (2013). *Microplastic in the marine environment: A review of the methods used for identification and quantification*. Environmental Science & Technology, 47(3), 1020–1030. https://doi.org/10.1021/es302332w

McKinsey & Company. (2020). *The circular economy: A transformative opportunity*. https://www.mckinsey.com/business-functions/sustainability/our-insights/the-circular-economy-a-transformative-opportunity

Mielke, H. W., et al. (2011). *Soil lead contamination and its health effects: A review*. Environmental Research, 111(5), 1–9. https://doi.org/10.1016/j.envres.2010.09.010

Pimentel, D., et al. (2004). *Water resources: Agricultural and environmental issues*. Bioscience, 54(10), 909–918. https://doi.org/10.1641/0006-3568(2004)054[0909:WRAEIA]2.0.CO;2

PlasticsEurope. (2020). *Plastics – the facts 2020: An analysis of European plastics production, demand and waste data*. https://www.plasticseurope.org/en/resources/publications

Rochman, C. M., et al. (2013). *Policy: Classify plastic waste as hazardous*. Science, 339(6122), 290–291. https://doi.org/10.1126/science.1223247

Shen, L., et al. (2017). *Bioplastics: The future of sustainable packaging*. Environmental Science & Technology, 51(19), 11362–11374. https://doi.org/10.1021/acs.est.7b02163

Smith, V. H., et al. (2016). *Eutrophication and hypoxia in aquatic ecosystems*. Annual Review of Ecology, Evolution, and Systematics, 47, 135–157. https://doi.org/10.1146/annurev-ecolsys-121415-032215

WHO. (2018). *Air pollution and health*. World Health Organization. https://www.who.int/news-room/f

Wilson, D. C., et al. (2015). *Global waste management practices*. Waste Management, 44, 19–27. https://doi.org/10.1016/j.wasman.2015.06.016

Zhang, Y., et al. (2019). *Food waste composting: A review of recent advances and challenges*. Waste Management, 89, 1–13. https://doi.org/10.1016/j.wasman.2019.03.001

Zhang, Y., et al. (2020). *Waste-to-energy technologies and their environmental implications*. Environmental Science & Technology, 54(11), 6821–6829. https://doi.org/10.1021/acs.est.0c01834

PART III

Social and Humanitarian Crises

9

Social and Humanitarian Crises

Social and humanitarian crises are profound disruptions that affect societies, often resulting from a combination of natural disasters, armed conflicts, economic collapses, or systemic inequalities. These crises not only challenge the immediate survival of affected populations but also have long-lasting impacts on the social, economic, and psychological well-being of individuals and communities. They exacerbate vulnerabilities, increase poverty, and often lead to the breakdown of social structures and institutions. Addressing social and humanitarian crises requires a comprehensive, multi-faceted approach that combines emergency relief, long-term recovery strategies, and systemic reforms aimed at reducing future risks.

9.1 Causes of Social and Humanitarian Crises

The causes of social and humanitarian crises are varied, complex, and often interconnected. One of the most significant factors is armed conflict, which displaces millions of people and leads to widespread suffering. Armed conflicts, particularly civil wars, disrupt communities, destroy infrastructure, and displace large numbers of people, creating refugee and internally displaced populations. These conflicts often result in the breakdown of governance, rule of law, and basic social services, leaving affected populations vulnerable to disease, malnutrition, and exploitation (UNHCR, 2020). Additionally, economic instability and inequality can serve as both direct and indirect drivers of social crises. High levels of poverty, unemployment, and income inequality can contribute to social unrest and, in some cases, violent protests or revolutions (Kaldor, 2012).

Natural disasters, such as earthquakes, floods, and hurricanes, also contribute to humanitarian crises. These events often disproportionately affect vulnerable populations, including low-income communities and those living in areas with poor infrastructure. Climate change exacerbates the frequency

DOI: 10.1201/9788743808787-12

and severity of these disasters, creating a vicious cycle of destruction and displacement (IPCC, 2022). Furthermore, systemic inequalities – such as gender inequality, racial discrimination, and lack of access to basic services – can worsen the impact of these crises, leaving marginalized groups more vulnerable to harm. For example, women and children are often disproportionately affected by displacement and violence in conflict zones, while marginalized ethnic or racial groups may face additional barriers to receiving aid during natural disasters (UNICEF, 2021).

9.2 Impacts of Social and Humanitarian Crises

The impacts of social and humanitarian crises are wide-ranging and affect individuals, families, and entire communities. One of the most immediate consequences is the loss of life and the destruction of infrastructure. The death toll in conflict zones and disaster-stricken areas can be staggering, and the destruction of homes, schools, hospitals, and other critical infrastructure hinders recovery efforts. Beyond the immediate physical damage, social and humanitarian crises often lead to widespread displacement. Refugees and internally displaced persons (IDPs) face significant challenges in accessing basic needs such as food, clean water, healthcare, and shelter. This displacement can result in long-term psychological trauma, particularly for children who experience the loss of family members, homes, and communities (Somasundaram, 2017).

Economic impacts are also severe. Social and humanitarian crises disrupt local economies, displacing workers, destroying businesses, and limiting access to markets. In conflict zones, agricultural production may be halted, and industries may collapse, leading to high unemployment and increased poverty. The economic devastation often extends beyond the immediate area of the crisis, affecting neighboring regions and even global markets. For instance, the Syrian conflict has had far-reaching economic effects, contributing to the global refugee crisis and impacting economies across Europe and the Middle East (World Bank, 2017).

In addition to economic and physical impacts, social and humanitarian crises also have profound psychological effects. Individuals affected by these crises often experience high levels of stress, anxiety, and trauma. This is especially true for those who have witnessed or been subjected to violence, lost loved ones, or experienced forced displacement. Mental health issues, such as post-traumatic stress disorder (PTSD), depression, and anxiety, are common in populations affected by conflict or disaster, and the lack of mental health support exacerbates these issues (Klinenberg, 2015). The social fabric of communities is also weakened, as people are forced to leave their homes and live in temporary shelters or refugee camps, where social cohesion can break down, and tensions can rise.

9.3 Humanitarian Response and International Aid

The international community plays a critical role in responding to social and humanitarian crises through humanitarian aid and relief efforts. Humanitarian organizations, such as the United Nations High Commissioner for Refugees (UNHCR), the World Food Programme (WFP), and Médecins Sans Frontières (MSF), provide emergency assistance in the form of food, shelter, medical care, and education. These organizations work in collaboration with governments, local NGOs, and other stakeholders to deliver aid to those in need. However, humanitarian aid is often limited by political, logistical, and security challenges, particularly in conflict zones where access to affected populations can be restricted (OCHA, 2020).

In addition to immediate relief efforts, long-term recovery and rebuilding are essential components of the response to social and humanitarian crises. This involves restoring infrastructure, rebuilding communities, and providing long-term support for displaced populations. Efforts to rebuild the economy and reintegrate displaced individuals into society are critical to ensuring that affected populations can regain their livelihoods and contribute to the rebuilding process. Furthermore, addressing the root causes of crises, such as poverty, inequality, and political instability, is essential to preventing future crises and promoting long-term peace and stability. This requires international cooperation and the commitment of governments to invest in sustainable development and conflict prevention (United Nations, 2015).

9.4 The Role of Policy and Governance

Addressing social and humanitarian crises also requires effective governance and policy frameworks. Governments have the primary responsibility to protect their citizens and ensure that basic needs are met, especially during times of crisis. However, in many cases, weak governance, corruption, and political instability exacerbate the impact of crises. In conflict zones, for example, the failure of governments to protect civilians or provide humanitarian aid can lead to prolonged suffering and displacement. In such situations, international intervention, either through peacekeeping missions or humanitarian assistance, may be necessary to restore order and provide relief (Bellamy, 2015).

Furthermore, policy frameworks that promote social justice, equality, and human rights are essential to reducing vulnerability and building resilience to future crises. For example, policies that address gender inequality, provide social safety nets, and ensure access to healthcare and education can reduce the impact of crises on vulnerable populations. The implementation of international human rights law, including the protection of refugees and displaced persons, is also crucial in ensuring that individuals affected by crises receive the protection and support they need (UNHCR, 2020).

Social and humanitarian crises are complex and multifaceted challenges that require coordinated efforts at the local, national, and international levels. These crises have far-reaching impacts on individuals, communities, and entire societies, affecting not only physical and economic well-being but also social cohesion and mental health. Addressing these crises requires a comprehensive approach that includes emergency relief, long-term recovery strategies, and systemic reforms aimed at preventing future crises. By addressing the root causes of vulnerability, such as inequality, poverty, and conflict, and promoting effective governance and international cooperation, the global community can work towards building a more resilient and just world.

9.5 Poverty and Economic Inequality

Poverty and economic inequality are two of the most pervasive and entrenched issues facing the global community. While poverty refers to the lack of access to basic resources necessary for survival, economic inequality refers to the unequal distribution of wealth, income, and opportunities within a society. These two phenomena are closely related, with inequality often exacerbating poverty and vice versa. The persistence of poverty and the widening gap between the rich and the poor have significant social, political, and economic consequences, affecting everything from individual well-being to national stability and global development.

9.5.1 Understanding poverty

Poverty is typically defined in two ways: absolute poverty and relative poverty. Absolute poverty refers to the condition in which individuals or families lack the resources to meet basic needs such as food, shelter, and healthcare. According to the World Bank, approximately 9.2% of the global population lived in extreme poverty in 2020, surviving on less than $1.90 per day (World Bank, 2022). This form of poverty is most prevalent in low-income countries, particularly in sub-Saharan Africa and South Asia, where access to basic services and infrastructure is limited. Relative poverty, on the other hand, is a measure of poverty in relation to the economic standards of a society. Individuals living in relative poverty may have access to basic necessities but are unable to achieve a standard of living that is considered acceptable within their society, often due to low income and lack of opportunities (Sen, 1999).

The causes of poverty are complex and multifaceted, often rooted in structural issues such as economic systems, governance, and historical factors. One of the most significant drivers of poverty is economic inequality, which limits access to resources and opportunities for large segments of the population. Other contributing factors include lack of education, poor health, unemployment, and social exclusion. For instance, people who are unable to access quality education or healthcare are more likely to remain trapped in poverty, as these factors are essential for upward mobility and economic participation (UNDP, 2020). Additionally, environmental factors such as climate change and natural disasters disproportionately affect the poorest populations, who often live in the most vulnerable areas and lack the resources to adapt or recover from such shocks (IPCC, 2022).

9.5.2 Economic inequality: Causes and consequences

Economic inequality refers to the unequal distribution of wealth and income across individuals or groups within a society. This inequality can manifest in many forms, including disparities in income, access to education, healthcare, and employment opportunities. The gap between the rich and the poor has been widening globally, with the wealthiest 1% of the global population owning a disproportionately large share of the world's wealth. According to the World Inequality Report (2022), the top 1% of global income earners now capture over 20% of total income, while the bottom half of the population receives only 10%. This growing inequality is driven by a variety of factors, including globalization, technological advancements, tax policies, and the concentration of wealth in the hands of multinational corporations and wealthy individuals.

The consequences of economic inequality are far-reaching and impact both individuals and societies. On an individual level, economic inequality can lead to poor health outcomes, lower life expectancy, and reduced access to education and healthcare. Research has shown that people in more unequal societies tend to have higher rates of mental illness, substance abuse, and social violence (Wilkinson & Pickett, 2009). Furthermore, economic inequality can lead to social fragmentation, as the wealthiest individuals and groups become increasingly disconnected from the rest of society. This lack of social cohesion can undermine trust in institutions and erode social capital, making it more difficult to address collective challenges such as climate change, health crises, and political instability (Piketty, 2014).

At the societal level, economic inequality can hinder economic growth and development. When large portions of the population are excluded from economic opportunities, their potential to contribute to the economy is wasted. Inequality also reduces the effectiveness of social policies, as it can lead to political polarization and resistance to progressive taxation or wealth redistribution policies. Moreover, extreme inequality can lead to political instability, as marginalized groups may resort to protests, strikes, or even violent uprisings in an attempt to address their grievances (Acemoglu & Robinson, 2012). The persistence of economic inequality also undermines efforts to achieve sustainable development, as it creates barriers to social mobility and reduces the overall well-being of society.

9.5.3 Addressing poverty and inequality

Efforts to address poverty and economic inequality require comprehensive strategies that focus on both alleviating immediate hardship and addressing the root causes of inequality. One of the most effective ways to combat poverty is through social protection programs, such as cash transfers, food assistance, and healthcare services. These programs provide a safety net for vulnerable populations and help to ensure that basic needs are met. In addition to social protection, policies that promote inclusive economic growth, such as investments in education, healthcare, and infrastructure, are essential for reducing poverty in the long term (UNDP, 2020).

To address economic inequality, it is crucial to implement policies that promote wealth redistribution and reduce disparities in income and wealth. Progressive taxation, which taxes higher incomes at a higher rate, is one such policy that can help reduce inequality. Additionally, increasing access to quality education and healthcare can help level the playing field, allowing individuals from disadvantaged backgrounds to improve their social mobility. International cooperation is also necessary to address global inequality, particularly in the context of trade, finance, and development aid. Efforts to address the global wealth gap, such as debt relief for low-income countries and fair-trade practices, can help reduce inequality on a global scale (Stiglitz, 2006).

Furthermore, addressing the structural factors that contribute to inequality, such as gender discrimination, racial inequality, and the concentration of corporate power, is critical. Gender equality policies, such as equal pay for equal work and access to reproductive health services, can reduce the gender gap in income and wealth. Similarly, policies that address racial and ethnic discrimination can help ensure that marginalized groups have equal access to economic opportunities. Finally, reducing the concentration of corporate power through antitrust laws and promoting fair labor practices can help ensure that wealth is more equitably distributed across society.

Poverty and economic inequality are deeply interconnected issues that require urgent attention from policymakers, international organizations, and civil society. While poverty remains a significant challenge, addressing the root causes of economic inequality is crucial for achieving long-term solutions. By promoting inclusive economic growth, investing in social protection, and implementing policies that reduce wealth disparities, it is possible to reduce poverty and create a more equitable society. Furthermore, addressing the structural factors that contribute to inequality, such as gender and racial discrimination, is essential for building a fairer and more just world. Ultimately, reducing poverty and inequality is not only a moral imperative but also an economic necessity, as it can lead to more stable, prosperous, and sustainable societies.

9.5.4 Global wealth disparities

Global wealth disparities are a defining feature of the contemporary economic landscape. These disparities, which refer to the uneven distribution of wealth and resources across nations and within societies, have profound implications for economic stability, social cohesion, and global development. The gap between the wealthiest individuals and nations and the poorest continues to widen, driven by a combination of historical, economic, political, and social factors. Understanding the causes and consequences of these disparities is crucial for developing effective policies to address global inequality and promote more equitable development.

9.5.4.1 Causes of global wealth disparities

The causes of global wealth disparities are complex and multifaceted, rooted in both historical and contemporary factors. One of the primary drivers of wealth inequality is the legacy of colonialism. Colonial powers extracted resources from colonized regions, often without compensating local populations, and established economic systems that benefited the colonizers while impoverishing indigenous peoples. This historical exploitation created long-lasting economic disparities that persist today. In many former colonies, the extraction of natural resources, coupled with the imposition of economic systems that favored the interests of colonial powers, led to the underdevelopment of local economies and the concentration of wealth in the hands of a few elites (Acemoglu & Robinson, 2012).

In the modern era, globalization has exacerbated wealth disparities. While globalization has led to significant economic growth and poverty reduction in some regions, it has also contributed to rising inequality both within and between countries. The global integration of markets has allowed multinational corporations and wealthy individuals to amass significant wealth, while many workers in low-income countries remain trapped in poverty. The increasing concentration of wealth in the hands of a small global elite is further compounded by tax avoidance strategies, such as offshore tax havens, which allow the wealthy to shield their income from taxation (Piketty, 2014). Additionally, the technological revolution has disproportionately benefited those with access to capital and advanced education, leaving behind large segments of the population in both developed and developing countries who are unable to participate in the new global economy (Brynjolfsson & McAfee, 2014).

Another key factor contributing to global wealth disparities is the unequal distribution of natural resources. Countries rich in natural resources, such as oil, minerals, and agricultural land, often experience higher levels of wealth, but this wealth is frequently concentrated in the hands of a few elites or multinational corporations. In contrast, countries with fewer natural resources often struggle to generate wealth, and their economies may be more dependent on external aid or loans, which can exacerbate poverty and inequality (Sachs, 2005). The global competition for resources has also led to environmental degradation, which disproportionately affects poorer countries, further entrenching global wealth disparities.

9.5.4.2 Consequences of global wealth disparities

The consequences of global wealth disparities are far-reaching, affecting not only economic outcomes but also social, political, and environmental stability. One of the most immediate effects of wealth inequality is the persistence of poverty, particularly in developing countries. According to the World Bank (2022), while global poverty rates have declined over the past few decades, nearly 10% of the world's population still lives in extreme poverty, surviving on less than $1.90 per day. This poverty is concentrated in regions such as sub-Saharan Africa and South Asia, where wealth disparities are most pronounced. The lack of access to education, healthcare, and basic services in these regions perpetuates cycles of poverty, limiting opportunities for upward mobility and exacerbating social inequalities.

Global wealth disparities also contribute to social unrest and political instability. As the gap between the rich and the poor widens, there is a growing sense of frustration and alienation among marginalized groups. This can lead to increased social tensions, protests, and even violent uprisings. In many countries, wealth inequality is closely tied to political inequality, with the wealthy exerting disproportionate influence over government policies and decisions. This concentration of political power in the hands of a few can undermine democratic institutions and erode public trust in governance, leading to a breakdown in social cohesion (Wilkinson & Pickett, 2009). In extreme cases, wealth inequality can contribute to the rise of populist and authoritarian movements, as people seek alternatives to the perceived failures of the political establishment (Stiglitz, 2012).

Moreover, global wealth disparities have significant environmental consequences. Wealthier countries and individuals tend to have larger ecological footprints, consuming more resources and producing more waste per capita. This overconsumption, coupled with the exploitation of natural resources in poorer countries, leads to environmental degradation and contributes to global challenges such as climate change, deforestation, and biodiversity loss. The impacts of these environmental crises are disproportionately felt by the world's poorest populations, who are more vulnerable to climate-related disasters and have fewer resources to adapt to environmental changes (IPCC, 2022). As a result, global wealth disparities not only perpetuate economic inequality but also exacerbate environmental challenges, creating a vicious cycle of poverty, inequality, and environmental degradation.

9.5.4.3 Addressing global wealth disparities

Addressing global wealth disparities requires a multi-faceted approach that includes both domestic and international policies aimed at reducing inequality and promoting more inclusive economic growth. One of the most important strategies for reducing global wealth disparities is progressive taxation. Tax policies that ensure the wealthy contribute a fair share of their income can help reduce income inequality and provide governments with the resources to invest in social programs such as education, healthcare, and infrastructure. International efforts to combat tax avoidance and illicit financial flows are also crucial for addressing the concentration of wealth in the hands of a few elites (Zucman, 2019).

In addition to tax reform, international trade policies should be restructured to promote fair trade and reduce the exploitation of low-income countries. This includes ensuring that developing countries receive fair compensation for their natural resources and that their workers are paid a living wage. The promotion of sustainable development practices, such as the adoption of green technologies and renewable energy, can also help reduce the environmental impacts of wealth inequality and create more equitable economic opportunities (Sachs, 2015). Furthermore, addressing global wealth disparities requires a focus on education and skill development, particularly in developing countries, to ensure that individuals have the tools they need to participate in the global economy.

Finally, addressing the structural factors that contribute to wealth disparities, such as gender inequality, racial discrimination, and unequal access to education and healthcare, is essential for creating a more equitable global society. Policies that promote gender equality, such as equal pay for equal work and access to reproductive health services, can help reduce the gender wealth gap. Similarly, policies that address racial and ethnic discrimination can ensure that marginalized groups have equal access to economic opportunities and resources (Sen, 1999). By tackling the root causes of inequality, it is possible to create a more just and sustainable global economy.

Global wealth disparities are a critical issue that requires urgent attention from policymakers, international organizations, and civil society. The causes of these disparities are rooted in historical, economic, and social factors, and their consequences are far-reaching, affecting everything from poverty and social unrest to environmental degradation. Addressing global wealth disparities requires comprehensive strategies that promote inclusive economic growth, reduce income inequality, and ensure that resources are more equitably distributed. By implementing policies that address the structural causes of inequality and promoting fair trade, progressive taxation, and sustainable development, it is possible to create a more equitable and sustainable global economy.

9.5.5 Urbanization, slums, and marginalized communities

Urbanization, the process by which populations move from rural to urban areas, has been one of the most significant global trends of the past century. While urbanization has led to economic growth, technological advancement, and improved living standards for many, it has also contributed to the expansion of slums and the deepening of inequalities. The rapid and often unplanned growth of cities has resulted in the proliferation of informal settlements, where millions of people live in precarious conditions, lacking access to basic services such as clean water, sanitation, healthcare, and education. These marginalized communities are often excluded from the benefits of urbanization, facing challenges that perpetuate cycles of poverty and inequality.

9.5.5.1 The growth of urbanization and its challenges

The global urban population has grown significantly over the past few decades. According to the United Nations (2018), more than half of the world's population now lives in urban areas, and this trend is expected to continue, with urban populations projected to reach 68% by 2050. While cities have become hubs of economic activity, innovation, and cultural exchange, they have also become sites of significant social and environmental challenges. Rapid urbanization, particularly in low- and middle-income countries, often outpaces the development of infrastructure and housing. As a result, millions of people live in slums or informal settlements, where overcrowding, inadequate housing, and lack of access to basic services are common (UN-Habitat, 2020).

Slums are typically characterized by substandard housing, limited access to clean water and sanitation, poor waste management, and inadequate infrastructure. These areas are often located on marginal land, such as floodplains, hillsides, or former industrial sites, which makes them vulnerable to natural disasters, environmental degradation, and climate change impacts. The United Nations Human Settlements Programme (UN-Habitat, 2020) estimates that over 1 billion people live in slums, representing about one-quarter of the urban population worldwide. This phenomenon is particularly pronounced in regions such as sub-Saharan Africa, South Asia, and Latin America, where urbanization rates are high, but economic development has not kept pace with the growth of urban populations.

9.5.5.2 The marginalization of slum dwellers

The residents of slums and informal settlements are among the most marginalized and vulnerable populations in the world. They often face multiple forms of exclusion, including social, economic, and political marginalization. In many cities, slum dwellers are excluded from formal labor markets, as they lack access to quality education and vocational training. This exclusion from formal employment opportunities limits their income and contributes to the persistence of poverty in these communities (Satterthwaite, 2016). Moreover, the lack of access to basic services, such as healthcare and education, further exacerbates their social and economic vulnerability.

In addition to economic marginalization, slum dwellers often face social exclusion, as they are stigmatized as "illegal" or "undesirable" by both the state and society. Many informal settlements lack legal recognition, and residents may not have formal property rights or access to social protection programs. This lack of legal status makes it difficult for slum dwellers to access services such as healthcare, education, and social security, as well as to engage in political processes (Pugh, 2000). The absence of secure tenure also makes slum dwellers vulnerable to eviction and displacement, as cities seek to redevelop or gentrify these areas (Davis, 2006). As a result, slum residents are often trapped in a cycle of poverty and exclusion, with limited opportunities for upward mobility.

9.5.5.3 Health and environmental challenges

Slum dwellers face significant health risks due to the poor living conditions in informal settlements. Overcrowding, inadequate sanitation, and lack of clean water contribute to the spread of infectious diseases such as cholera, typhoid, and respiratory infections. The lack of access to healthcare services further exacerbates these health risks, as slum residents often cannot afford medical treatment or lack access to nearby health facilities. According to the World Health Organization (WHO, 2017), urban slums are hotbeds of disease transmission, with residents more likely to suffer from chronic health conditions, malnutrition, and mental health issues compared to their counterparts in formal urban areas.

Environmental degradation is another critical issue in slums. Informal settlements are often located in environmentally hazardous areas, such as floodplains or near industrial zones, which exposes residents to environmental hazards such as flooding, air pollution, and toxic waste. The lack of proper waste management systems leads to the accumulation of garbage and pollutants, further exacerbating health problems. Additionally, slum dwellers are particularly vulnerable to the impacts of climate change, such as rising sea levels, extreme weather events, and temperature fluctuations, which can exacerbate existing vulnerabilities (Hardoy, Mitlin, & Satterthwaite, 2001). As urbanization continues, addressing the environmental challenges in slums will be crucial to ensuring the health and well-being of these marginalized communities.

9.5.5.4 Policy responses and solutions

Addressing the challenges faced by slum dwellers requires comprehensive and inclusive urban policies that prioritize the needs of marginalized communities. One of the key strategies for improving the living conditions of slum dwellers is the provision of affordable and adequate housing. Governments and international organizations must invest in housing programs that provide safe, affordable, and sustainable housing for low-income urban populations. This includes upgrading existing slums, providing secure land tenure, and ensuring that new housing developments are integrated into the broader urban fabric, with access to essential services such as water, sanitation, and transportation (UN-Habitat, 2020).

In addition to housing, policies aimed at improving access to education, healthcare, and employment opportunities are essential for addressing the root causes of marginalization. Providing quality education and vocational training can help slum residents secure better jobs and improve their economic prospects. Expanding access to healthcare services, particularly in underserved areas, can help reduce health disparities and improve the overall well-being of slum dwellers. Furthermore, promoting inclusive economic development that creates job opportunities for marginalized populations is crucial for reducing poverty and inequality in urban areas (Satterthwaite, 2016).

Finally, addressing the environmental challenges in slums requires integrated urban planning that incorporates climate resilience and sustainability. This includes improving waste management systems, upgrading infrastructure, and ensuring that slum areas are not located in high-risk zones for environmental hazards. Climate adaptation measures, such as flood protection and disaster preparedness, should be prioritized in slum upgrading programs to reduce the vulnerability of these communities to climate change impacts (Hardoy et al., 2001).

Urbanization, while contributing to economic growth and development, has also led to the expansion of slums and the marginalization of millions of people in cities around the world. Slum dwellers face significant challenges, including poverty, exclusion, poor health, and environmental degradation. Addressing these challenges requires comprehensive policies that focus on providing affordable housing, improving access to services, and promoting inclusive economic development. By prioritizing the needs of marginalized communities, cities can become more equitable, sustainable, and resilient, ensuring that all shares the benefits of urbanization.

9.6 The Intersection of Inequality and Opportunity

The relationship between inequality and opportunity is a critical issue in understanding the dynamics of social, economic, and political structures in contemporary societies. Inequality, whether economic, social, or political, creates systemic barriers that restrict access to opportunities, thereby limiting individuals' potential for upward mobility and the realization of their full capabilities. This intersection is particularly evident in areas such as education, healthcare, employment, and social participation, where disparities often persist across generations. Understanding how inequality and opportunity intersect is essential for creating policies and interventions that promote equity and social justice.

9.6.1 Economic inequality and social mobility

Economic inequality plays a central role in shaping opportunities for individuals and communities. When wealth and income are concentrated in the hands of a few, it creates a system where access to essential resources – such as quality education, healthcare, and housing – becomes increasingly difficult for those at the bottom of the economic ladder. The resulting disparities in these areas limit opportunities for upward mobility and perpetuate cycles of poverty. Research has shown that children born into low-income families are more likely to experience poor educational outcomes, which in turn affects their future job prospects (Chetty et al., 2014). The lack of access to quality education and employment opportunities restricts social mobility and reinforces economic disparities.

Moreover, economic inequality has been linked to disparities in health outcomes, which further exacerbate the intersection of inequality and opportunity. Individuals from lower-income backgrounds often experience poorer health due to factors such as limited access to healthcare, inadequate nutrition, and environmental hazards (Wilkinson & Pickett, 2009). These health disparities can limit individuals' ability to pursue education or maintain steady employment, thereby further restricting their opportunities for economic advancement. As a result, economic inequality not only affects the immediate well-being of individuals but also has long-term consequences for their ability to participate fully in society.

9.6.2 Social inequality and structural barriers

Social inequality, which encompasses disparities based on race, gender, ethnicity, and disability, intersects with economic inequality to create additional barriers to opportunity. Marginalized groups often face structural discrimination that limits their access to resources and opportunities. For example, racial and ethnic minorities are more likely to experience discrimination in education, employment, and housing, which restricts their opportunities for social mobility (Pager & Shepherd, 2008). Similarly, women, particularly women of color, face gendered and racialized barriers that hinder their ability to access equal opportunities in the workplace and other spheres of life (World Economic Forum, 2020). These barriers are deeply embedded in societal systems and institutions, making it difficult for marginalized individuals to break free from cycles of disadvantage.

The intersection of race, gender, and other social factors further compounds the effects of inequality. Studies have shown that individuals from marginalized backgrounds often face lower expectations and fewer opportunities in educational settings, which can affect their academic performance and future prospects (Barton et al., 2004). Additionally, discrimination in hiring practices, wage gaps, and the lack of representation in leadership roles continue to limit opportunities for women and people of color (Reskin, 2012). This structural inequality not only affects individuals' access to opportunities but also perpetuates broader societal disparities, contributing to the persistence of inequality across generations.

9.6.3 Policy implications and addressing the intersection

Addressing the intersection of inequality and opportunity requires comprehensive policy interventions that target both the systemic causes of inequality and the barriers to opportunity. Policies aimed at reducing economic inequality, such as progressive taxation, universal healthcare, and affordable housing, can help ensure that all individuals have access to the resources necessary to succeed. Additionally, education policies that focus on equity and inclusion, such as increased funding for schools in low-income areas and affirmative action programs, can help mitigate the effects of inequality on educational outcomes (Chetty et al., 2014). These policies can help create a more level playing field, where individuals from all backgrounds have the opportunity to succeed.

Moreover, addressing social inequality requires structural reforms that challenge discriminatory practices and promote diversity and inclusion in all areas of life. For example, policies that promote gender equality in the workplace, such as equal pay for equal work and family leave, can help ensure that women have equal access to economic opportunities (World Economic Forum, 2020). Similarly, anti-discrimination laws and initiatives aimed at increasing diversity in leadership positions can help reduce the barriers faced by racial and ethnic minorities. By addressing both the economic and social dimensions of inequality, policymakers can create a more equitable society where opportunities are accessible to all, regardless of their background or identity.

The intersection of inequality and opportunity is a complex and multifaceted issue that requires a holistic approach to address. Economic inequality and social inequality are deeply interconnected, and together, they create significant barriers to opportunity for marginalized individuals and communities. By understanding the ways in which these forms of inequality intersect and compound each other, policymakers and advocates can develop strategies that promote greater equity and social mobility. Ensuring that all individuals have access to the opportunities necessary for success is crucial for creating a more just and sustainable society.

References

Acemoglu, D., & Robinson, J. A. (2012). *Why nations fail: The origins of power, prosperity, and poverty*. Crown Business.

Barton, P. E., Coley, R. J., & Wenglinsky, H. (2004). *The family: America's smallest school.* Educational Testing Service.

Bellamy, A. J. (2015). *The responsibility to protect: A defense*. Oxford University Press.

Brynjolfsson, E., & McAfee, A. (2014). *The second machine age: Work, progress, and prosperity in a time of brilliant technologies*. W.W. Norton & Company.

Chetty, R., Hendren, N., Kline, P., & Saez, E. (2014). Where is the land of opportunity? The geography of intergenerational mobility in the United States. *Quarterly Journal of Economics, 129*(4), 1553–1623. https://doi.org/10.1093/qje/qju022

Davis, M. (2006). *Planet of slums*. Verso.

Hardoy, J. E., Mitlin, D., & Satterthwaite, D. (2001). *Environmental problems in an urbanizing world: Finding solutions for cities in Africa, Asia, and Latin America*. Earthscan.

IPCC. (2022). *Climate change 2022: Impacts, adaptation, and vulnerability*. Intergovernmental Panel on Climate Change. https://www.ipcc.ch/report/ar6/wg2/

Kaldor, M. (2012). *New and old wars: Organized violence in a global era* (3rd ed.). Stanford University Press.

Klinenberg, E. (2015). *Heat wave: A social autopsy of disaster in Chicago*. University of Chicago Press.

OCHA. (2020). *Global humanitarian overview 2020*. United Nations Office for the Coordination of Humanitarian Affairs. https://www.unocha.org/global-humanitarian-overview-2020

Pager, D., & Shepherd, H. (2008). The sociology of discrimination: Racial discrimination in employment, housing, credit, and consumer markets. *Annual Review of Sociology, 34*, 181–209. https://doi.org/10.1146/annurev.soc.34.040507.134649

Piketty, T. (2014). *Capital in the twenty-first century*. Harvard University Press.

Pugh, C. (2000). *Sustainable cities in developing countries: A review of the literature*. Environment and Urbanization, 12(2), 103–116.

Reskin, B. F. (2012). The race discrimination system. *Annual Review of Sociology, 38*, 17–35. https://doi.org/10.1146/annurev-soc-073010-130512

Sachs, J. D. (2005). *The end of poverty: Economic possibilities for our time*. Penguin Press.

Sachs, J. D. (2015). *The age of sustainable development*. Columbia University Press.

Satterthwaite, D. (2016). *The contribution of cities to global sustainability: A review of the evidence*. Environment and Urbanization, 28(1), 1–16.

Sen, A. (1999). *Development as freedom*. Alfred A. Knopf.

Somasundaram, D. (2017). *Mental health and psychosocial support in humanitarian emergencies*. In *Oxford textbook of public mental health* (pp. 359–365). Oxford University Press.

Stiglitz, J. E. (2006). *Making globalization work*. W.W. Norton & Company.

Stiglitz, J. E. (2012). *The price of inequality: How today's divided society endangers our future*. W.W. Norton & Company.

UNDP. (2020). *Human development report 2020: The next frontier—Human development and the Anthropocene*. United Nations Development Programme. https://hdr.undp.org/en/2020-report

UN-Habitat. (2020). *The state of the world's cities 2020: Urbanization and development challenges*. United Nations Human Settlements Programme.

UNHCR. (2020). *Global trends: Forced displacement in 2019*. United Nations High Commissioner for Refugees. https://www.unhcr.org/global-trends-2019

UNICEF. (2021). *The state of the world's children 2021: On my mind—promoting, protecting and caring for children's mental health*. United Nations Children's Fund. https://www.unicef.org/reports/state-of-worlds-children-2021

United Nations. (2018). *World urbanization prospects: The 2018 revision*. United Nations, Department of Economic and Social Affairs.

WHO. (2017). *Health in the urban environment: The need for action*. World Health Organization.

Wilkinson, R., & Pickett, K. (2009). *The spirit level: Why more equal societies almost always do better*. Allen Lane.

World Bank. (2017). *The Syrian conflict and its impact on the region*. World Bank. https://www.worldbank.org/en/news/feature/2017/11/09/the-syrian-conflict-and-its-impact-on-the-region

World Bank. (2022). *Poverty and shared prosperity 2022: Correcting course*. World Bank. https://www.worldbank.org/en/publication/poverty-and-shared-prosperity

World Economic Forum. (2020). *Global gender gap report 2020.* World Economic Forum. https://www.weforum.org/reports/gender-gap-2020-report-100-years-pay-equality

Zucman, G. (2019). *The triumph of injustice: How the rich dodge taxes and how to make them pay*. W.W. Norton & Company.

10

Public Health in Peril

Public health, which encompasses the well-being of populations and disease prevention, faces significant challenges in the contemporary world. These challenges are compounded by environmental degradation, climate change, social inequalities, and emerging infectious diseases, all of which threaten to reverse the gains made in global health over the past decades. The intersection of these factors poses direct risks to human health and exacerbates existing health disparities, making safeguarding public health even more urgent. Understanding the current state of public health and its threats is essential for designing effective interventions and policies to protect populations worldwide.

10.1 The Impact of Climate Change on Public Health

One of the most pressing threats to public health today is climate change. The rising global temperatures, changing weather patterns, and more frequent extreme weather events, such as heatwaves, floods, and storms, have profound implications for human health. These changes can lead to the spread of infectious diseases, exacerbate respiratory and cardiovascular conditions, and increase the vulnerability of populations to malnutrition and water scarcity (Haines et al., 2006). For example, higher temperatures can facilitate the spread of vector-borne diseases, such as malaria and dengue fever, as mosquitoes thrive in warmer climates (Patz et al., 2005). Additionally, extreme weather events like hurricanes and floods can disrupt healthcare infrastructure, displace communities, and lead to outbreaks of waterborne diseases (Watts et al., 2015).

Climate change also disproportionately affects vulnerable populations, including low-income communities, children, older people, and those living in regions with inadequate healthcare systems. These groups are more likely to suffer from the health impacts of climate change, such as malnutrition,

DOI: 10.1201/9788743808787-13

heat-related illnesses, and poor access to clean water and sanitation (Patz et al., 2005). The health inequities associated with climate change underscore the need for comprehensive public health strategies that address health's environmental and social determinants.

10.2 Emerging Infectious Diseases

Another major threat to public health is the rise of emerging infectious diseases. Globalization has increased human mobility, and environmental changes have facilitated the spread of diseases that were once confined to specific regions. The COVID-19 pandemic is a stark reminder of the vulnerabilities of global health systems to novel pathogens and the far-reaching consequences of infectious diseases. The pandemic has highlighted the importance of preparedness, surveillance, and rapid response mechanisms in mitigating the impact of infectious diseases on public health (Hotez, 2020). However, the global response to COVID-19 also revealed significant gaps in health infrastructure, particularly in low-income countries, where access to vaccines, medical supplies, and healthcare personnel was limited.

The emergence of new diseases, such as Ebola, Zika virus, and antibiotic-resistant infections, further underscores the need for robust public health systems that can respond to evolving threats. The rise of antimicrobial resistance (AMR), in particular, presents a growing challenge to public health, as infections once treatable with antibiotics are becoming increasingly difficult to manage (Ventola, 2015). Addressing the root causes of AMR, such as the overuse of antibiotics in agriculture and healthcare, as well as improving global surveillance and diagnostics, is essential to preventing future health crises.

10.3 Social Determinants of Health and Health Inequities

Social determinants of health, such as income, education, housing, and access to healthcare, play a critical role in shaping public health outcomes. Inequities in these areas are major contributors to disparities in health outcomes between different populations. Low-income individuals and communities of color, for example, are more likely to experience poor health due to limited access to healthcare, poor living conditions, and higher exposure to environmental hazards (Williams & Mohammed, 2009). These social determinants create a cycle of disadvantage that limits opportunities for health and well-being, perpetuating intergenerational health inequities.

The COVID-19 pandemic further exposed these disparities, as marginalized communities were disproportionately affected by the virus, both in terms of infection rates and health outcomes. The lack of access to healthcare, crowded living conditions, and underlying health conditions, such as diabetes and hypertension, contributed to higher rates of hospitalization and death in these communities (Vargas et al., 2020). Addressing the social determinants of health is crucial for reducing health inequities and ensuring all individuals have access to the resources necessary to lead healthy lives.

10.4 The Future of Public Health: Moving Toward Equity and Sustainability

The future of public health lies in addressing the root causes of health inequities and building resilient healthcare systems that can respond to emerging threats. This requires a multi-faceted approach that includes climate change mitigation, strengthening healthcare infrastructure, and addressing the social determinants of health. Public health policies must prioritize equity, ensuring that vulnerable populations are not left behind in efforts to improve health outcomes. Sustainable development practices that protect the environment and promote social and economic well-being are essential for ensuring long-term public health.

One promising approach to improving public health is the concept of "One Health," which recognizes the interconnectedness of human, animal, and environmental health. This holistic approach encourages collaboration across sectors to address the root causes of disease and promote health in a more integrated manner (Zinsstag et al., 2011). Public health systems can better anticipate and mitigate the impacts of emerging infectious diseases, climate change, and other environmental threats by adopting a One Health approach.

Public health faces unprecedented challenges in the 21st century, from climate change and emerging infectious diseases to the persistence of social inequities that limit access to healthcare and other resources. Addressing these challenges requires a comprehensive and equitable approach considering health's social, environmental, and economic determinants. By strengthening healthcare systems, addressing health inequities, and adopting sustainable development practices, it is possible to protect public health and ensure that all individuals have the opportunity to lead healthy, fulfilling lives.

10.5 Pandemics and Global Health Preparedness

Pandemics, which are global outbreaks of infectious diseases, pose significant threats to public health, economies, and social stability. The COVID-19 pandemic underscored the vulnerabilities of health systems worldwide and highlighted the critical importance of preparedness and coordinated responses to such global health crises. Pandemics can have far-reaching consequences, not only in terms of loss of life but also in terms of long-term health, economic, and social impacts. As the world becomes more interconnected, the likelihood of pandemics is expected to rise, making global health preparedness an urgent priority.

10.5.1 The global impact of pandemics

Pandemics have a profound impact on global health systems. They strain healthcare infrastructure, disrupt essential services, and lead to widespread morbidity and mortality. The COVID-19 pandemic, for example, overwhelmed hospitals, caused significant loss of life, and disrupted non-COVID medical care, including routine surgeries and treatments for chronic conditions (Ranney et al., 2020). The pandemic also exposed the weaknesses in health systems, particularly in low-income countries, where resources were already limited and healthcare systems were fragile (World Health Organization [WHO], 2020). In addition to the direct health effects, pandemics also cause significant economic damage, disrupt global supply chains, and lead to job losses, particularly in tourism, hospitality, and retail (McKibbin & Fernando, 2020).

Furthermore, pandemics exacerbate existing social inequalities. Vulnerable populations, including the elderly, people with pre-existing health conditions, and marginalized communities, are disproportionately affected by infectious diseases. The COVID-19 pandemic, for example, disproportionately impacted racial and ethnic minorities, who often face higher rates of underlying health conditions, limited access to healthcare, and socioeconomic disadvantages (Vargas et al., 2020). These disparities underscore the need for more equitable health systems that can address the needs of all populations, particularly during times of crisis.

10.5.2 Global health preparedness: Lessons from COVID-19

The COVID-19 pandemic highlighted the importance of global health preparedness, including the need for early detection, rapid response, and effective coordination between countries and international organizations. The World Health Organization (WHO) and other global health bodies are central in coordinating responses to pandemics, providing technical support, and facilitating information sharing among nations (Ghebreyesus, 2020).

However, the pandemic also revealed gaps in global health governance, such as delays in the initial response, lack of transparency, and insufficient resources for low-income countries (Moon et al., 2020). These challenges emphasize the need for stronger international cooperation and improved global health governance to ensure that countries are better prepared for future pandemics.

One of the key lessons from the COVID-19 pandemic is the importance of early warning systems and surveillance. Early detection of potential outbreaks is critical for preventing the spread of infectious diseases. The Global Health Security Agenda (GHSA), established in 2014, is an initiative to strengthen countries' capacities to prevent, detect, and respond to infectious disease threats. While the GHSA has made significant progress, the COVID-19 pandemic demonstrated that more robust and timely surveillance systems are needed, particularly in low-resource settings where disease surveillance infrastructure is often inadequate (Chatham House, 2020). The ability to detect outbreaks early and respond quickly can save lives, prevent widespread transmission, and minimize the economic impact of pandemics.

10.5.3 The role of vaccination and public health infrastructure

Vaccination is one of the most effective tools in preventing the spread of infectious diseases and mitigating the impact of pandemics. The rapid development of COVID-19 vaccines, particularly mRNA vaccines, demonstrated the potential of modern vaccine technology to respond to emerging infectious diseases. However, the global distribution of vaccines has been marked by significant disparities, with wealthier countries securing the majority of doses while low-income countries struggled to access vaccines (Ghebreyesus, 2021). This inequitable distribution has highlighted the need for global mechanisms, such as COVAX, to ensure that vaccines are accessible to all countries, regardless of their economic status.

In addition to vaccination, strong public health infrastructure is essential for effective pandemic preparedness and response. This includes well-trained healthcare workers, adequate medical supplies, and the capacity to implement public health measures such as social distancing, contact tracing, and quarantine protocols. The COVID-19 pandemic underscored the importance of investing in public health infrastructure, particularly in low-income and developing countries, where healthcare systems are often underfunded and under-resourced (Baker et al., 2020). Strengthening health systems is crucial for responding to pandemics and improving overall health outcomes and resilience to future health crises.

10.5.4 Building resilient health systems for the future

Building resilient health systems that respond effectively to pandemics requires a multifaceted approach. First, there is a need for greater investment in public health infrastructure, including developing strong healthcare systems, training healthcare professionals, and establishing robust surveillance and early warning systems. Second, countries must prioritize health equity, ensuring that all populations, especially vulnerable groups, access healthcare services during pandemics. This includes addressing the social determinants of health, such as poverty, education, and access to clean water and sanitation, which can exacerbate the impacts of infectious diseases (Baker et al., 2020).

Finally, international cooperation is essential for addressing the global nature of pandemics. The COVID-19 pandemic demonstrated the importance of global solidarity in responding to health crises, as countries must work together to share information, resources, and expertise. Strengthening international health governance and ensuring that countries are held accountable for their pandemic preparedness and response efforts is critical for improving global health security. Collaborative efforts, such as the Global Health Security Agenda and the WHO's International Health Regulations, can help countries better prepare for future pandemics and reduce the global health risks of emerging infectious diseases.

Pandemics are a major threat to global health, and the COVID-19 pandemic has underscored the need for robust health systems, early detection, and international cooperation to mitigate their impact. Global health preparedness is essential for preventing the spread of infectious diseases, ensuring equitable access to healthcare, and building resilience to future health crises. As the world becomes more interconnected and the risk of pandemics increases, it is crucial to invest in public health infrastructure, strengthen global health governance, and prioritize health equity to protect populations worldwide.

10.6 Non-Communicable Diseases and Lifestyle Impacts

Non-communicable diseases (NCDs) are a leading cause of death and disability worldwide, accounting for an estimated 71% of all global deaths (World Health Organization [WHO], 2021). These diseases, which include cardiovascular diseases, cancers, chronic respiratory diseases, and diabetes, are primarily driven by lifestyle factors such as poor diet, lack of physical activity, tobacco use, and excessive alcohol consumption. NCDs are increasingly recognized as a major public health challenge, not only because of their significant health impacts but also due to the economic burden they impose on individuals, families, and healthcare systems. The rise of NCDs reflects broader societal changes, including urbanization, changing dietary patterns, and the global shift towards more sedentary lifestyles.

10.6.1 The role of lifestyle factors in the development of NCDs

The primary lifestyle factors contributing to the development of NCDs are diet, physical inactivity, tobacco use, and alcohol consumption. Poor dietary habits, such as high intake of processed foods, sugary beverages, and unhealthy fats, have been linked to an increased risk of obesity, diabetes, and cardiovascular diseases (Liu et al., 2017). Diets high in sodium and low in fruits and vegetables can lead to hypertension, a major risk factor for stroke and heart disease (He et al., 2013). Additionally, the rise in the consumption of processed foods and fast foods, which are often high in calories and low in nutritional value, has been identified as a key driver of the global obesity epidemic (Swinburn et al., 2011).

Physical inactivity is another significant risk factor for NCDs. A sedentary lifestyle, characterized by prolonged periods of sitting or low levels of physical activity, is associated with a higher risk of developing cardiovascular diseases, diabetes, and certain cancers (Katzmarzyk et al., 2019). Urbanization and technological advancements have decreased physical activity levels, as more people live in environments that discourage walking or outdoor activities and rely heavily on cars for transportation. The World Health Organization (WHO) has recommended at least 150 minutes of moderate-intensity physical activity per week to reduce the risk of NCDs (WHO, 2020).

Tobacco use remains one of the most significant contributors to NCDs, particularly lung cancer, chronic obstructive pulmonary disease (COPD), and cardiovascular diseases. Smoking is responsible for an estimated 8 million deaths annually, with most of these deaths occurring in low- and middle-income countries (Jha & Peto, 2014). Despite widespread public health campaigns and smoking cessation programs, tobacco use continues to be a major global health issue, particularly in developing countries where smoking rates are rising (Ng et al., 2014). Similarly, excessive alcohol consumption is a major risk factor for liver disease, certain cancers, and cardiovascular problems (Rehm et al., 2017). Alcohol-related harm is not only a direct result of consumption but also contributes to social and economic problems, including accidents, injuries, and violence.

10.6.2 The economic and social burden of NCDs

The global burden of NCDs is not limited to health impacts; these diseases also have significant economic consequences. NCDs place a heavy burden on healthcare systems, as they require long-term management and care. The costs associated with treating NCDs, including hospitalization, medications, and rehabilitation, are substantial, particularly in countries with limited healthcare resources (Bloom et al., 2011). In addition to direct healthcare costs, NCDs also contribute to lost productivity due to premature death and disability. The economic burden of NCDs is estimated to be trillions of dollars globally, with the greatest impact felt in low- and middle-income countries (World Economic Forum, 2011). These countries often face the dual challenge of addressing infectious diseases while also managing the growing burden of NCDs.

Socially, NCDs can lead to significant disparities in health outcomes, particularly among disadvantaged populations. Low-income individuals and those with limited access to healthcare are more likely to suffer from NCDs due to factors such as poor nutrition, lack of access to medical care, and higher levels of stress (Marmot et al., 2008). In many cases, the financial costs associated with managing NCDs can push families further into poverty, creating a vicious cycle of health inequities. Furthermore, the stigma associated with certain NCDs, such as obesity or mental health conditions, can exacerbate social isolation and reduce the quality of life for affected individuals (Puhl & Heuer, 2010).

10.6.3 Prevention and policy responses

Given NCDs' significant health, social, and economic impacts, prevention is key to addressing this growing global health crisis. Public health strategies to reduce the prevalence of NCDs focus on promoting healthy lifestyles, including healthier diets, increased physical activity, and tobacco and alcohol control. The WHO has outlined a global action plan for preventing and controlling NCDs, including policies aimed at reducing risk factors, strengthening health systems, and improving access to treatment and care (WHO, 2013). Implementing taxes on sugary beverages, tobacco products, and alcohol is an effective strategy for reducing consumption and improving public health outcomes (Thow et al., 2018).

Public health campaigns aimed at raising awareness about the risks of poor diet, physical inactivity, and tobacco and alcohol use are also essential for changing behaviors at the population level. Education and awareness campaigns can help individuals make healthier choices and encourage governments to implement policies that promote healthier environments. For example, urban planning policies that encourage walking and cycling, the availability of healthy food options in schools and workplaces and regulating food marketing to children can all contribute to healthier lifestyles (Swinburn et al., 2019).

Furthermore, healthcare systems must be strengthened to address the growing burden of NCDs. This includes improving access to primary care, early detection, managing chronic conditions, and providing support for individuals living with NCDs. Expanding access to essential medicines, particularly in low- and middle-income countries, is critical for managing NCDs and preventing complications (WHO, 2016). Moreover, addressing the social determinants of health, such as poverty, education, and access to clean water and sanitation, is essential for reducing the risk factors associated with NCDs and promoting overall health equity.

Non-communicable diseases represent a significant and growing challenge to global health, driven largely by lifestyle factors such as poor diet, physical inactivity, tobacco use, and excessive alcohol consumption. The economic and social burden of NCDs is substantial, particularly in low- and middle-income countries, where healthcare systems are often ill-equipped to manage these chronic conditions. Prevention and early intervention are key to reducing the impact of NCDs, and public health strategies must focus on promoting healthier lifestyles, strengthening healthcare systems, and addressing the social determinants of health. By taking a comprehensive approach to the prevention and management of NCDs, it is possible to reduce the global burden of these diseases and improve the health and well-being of populations worldwide.

10.7 Mental Health and the Stigma of Silence

Mental health issues have become increasingly recognized as a critical aspect of public health, with millions of individuals worldwide experiencing various mental health conditions, including depression, anxiety, schizophrenia, and bipolar disorder. Despite growing awareness, mental health remains heavily stigmatized, which contributes to a "culture of silence" that prevents individuals from seeking help and exacerbates the challenges they face. The stigma surrounding mental health is multifaceted, encompassing negative attitudes, stereotypes, and discrimination, which significantly impact individuals' willingness to speak out about their struggles and seek necessary care. This silence not only hinders individual well-being but also perpetuates broader societal issues, including inadequate mental health care, social isolation, and inequity in health outcomes.

10.7.1 The nature of mental health stigma

Mental health stigma manifests in two primary forms: public stigma and self-stigma. Public stigma refers to the negative attitudes and discriminatory behaviors exhibited by society toward individuals with mental health disorders. These attitudes are often based on misconceptions and stereotypes, such as the belief that individuals with mental health conditions are dangerous, unreliable, or weak (Corrigan, 2004). On the other hand, self-stigma occurs when individuals internalize these societal views and develop feelings of shame, guilt, or worthlessness about their condition (Rüsch et al., 2005). Both forms of stigma contribute to the silence surrounding mental health, as individuals may fear judgment, rejection, or exclusion if they disclose their struggles.

The stigma of mental health is rooted in cultural, historical, and social factors. In many societies, mental health issues have been misunderstood and associated with moral failings or personal weakness. For example, mental illness has often been viewed as a taboo subject, something to be hidden or ignored rather than addressed openly (Hinshaw, 2007). This cultural silence is particularly evident in communities where mental health is not openly discussed, and seeking help is seen as a sign of vulnerability or failure. In these contexts, individuals may avoid treatment or delay seeking care, exacerbating the severity of their conditions and reducing their chances of recovery.

10.7.2 The impact of silence on mental health

The stigma of silence surrounding mental health has profound effects on individuals' psychological well-being. When people are unable or unwilling to talk about their mental health struggles, they often experience increased feelings of isolation and alienation. This silence can exacerbate symptoms of mental health conditions, such as depression and anxiety, as individuals may feel that their experiences are invalid or that they are alone in their suffering (Goffman, 1963). Furthermore, the fear of being stigmatized or misunderstood can prevent individuals from accessing the support they need, leading to untreated conditions and, in some cases, severe consequences such as self-harm or suicide (Corrigan & Watson, 2002).

For those who do seek help, the stigma can create barriers to effective treatment. Mental health professionals may also contribute to the stigma by perpetuating stereotypes or failing to create a safe and non-judgmental environment for patients (Sartorius, 2007). In some cases, individuals may be reluctant to engage in treatment due to concerns about confidentiality, or they may fear that they will be labeled or treated differently by healthcare providers or peers. This reluctance to seek help further compounds the cycle of silence and stigma, making it difficult to address mental health issues effectively at both the individual and societal levels.

10.7.3 The role of media and public discourse in breaking the silence

The media plays a significant role in shaping public perceptions of mental health, and it can either perpetuate stigma or help break the silence surrounding mental illness. Negative portrayals of mental health in the media often reinforce harmful stereotypes, such as depicting individuals with mental disorders as violent or unpredictable (Pirkis et al., 2006). These portrayals can contribute to the public's fear and misunderstanding of mental illness, making it more difficult for individuals to speak openly about their experiences. However, the media also has the potential to challenge stigma and promote understanding by highlighting personal stories of recovery, offering accurate representations of mental health issues, and advocating for mental health rights (Thornicroft, 2006).

In recent years, there has been a growing movement to raise awareness about mental health and reduce stigma. Public campaigns, such as the "Time to Change" initiative in the United Kingdom, have encouraged individuals to share their mental health experiences and seek help without fear of judgment (Clement et al., 2015). These campaigns have helped to normalize conversations about mental health, making it easier for people to talk about their struggles and access support. Additionally, high-profile individuals, including celebrities and public figures, have used their platforms to discuss their mental health challenges openly, further reducing the stigma and encouraging others to do the same (Gulliver et al., 2012).

10.7.4 Moving toward a culture of openness and support

Breaking the silence surrounding mental health requires a concerted effort from individuals, communities, healthcare providers, and policymakers. Education and awareness campaigns are essential in challenging the misconceptions and stereotypes perpetuating stigma. These campaigns should focus on promoting accurate information about mental health conditions, emphasizing that mental illness is not a sign of weakness but a medical issue that can be treated effectively with the proper support (Jorm et al., 2006). Also, fostering environments that encourage open conversations about mental health, both in the workplace and social settings, can help reduce the fear of judgment and create a more supportive culture.

Healthcare providers also have a crucial role to play in reducing stigma and creating a safe space for individuals to discuss their mental health concerns. This involves providing compassionate care, using nonjudgmental language, and respecting patients' confidentiality. Training healthcare professionals to recognize and address stigma within the healthcare system is important in improving the quality of care and ensuring that individuals feel comfortable seeking help (Stuart, 2005). Moreover, integrating mental health services into primary care settings can help normalize mental health treatment and make it more accessible to individuals who may be reluctant to seek specialized care.

The stigma of silence surrounding mental health remains a significant barrier to addressing the global mental health crisis. The negative attitudes and misconceptions associated with mental illness contribute to social isolation, delayed treatment, and exacerbated health outcomes. However, through education, media representation, and a shift in societal attitudes, it is possible to break the silence and create a culture of openness and support for individuals experiencing mental health challenges. By fostering an environment where mental health is treated with the same seriousness and compassion as physical health, society can work toward reducing stigma and improving the well-being of individuals affected by mental illness.

References

Baker, M. G., Wilson, N., & Anglemyer, A. (2020). Successful public health response to COVID-19: Early lessons learned from New Zealand. *The Lancet, 395*(10235), 1137–1146. https://doi.org/10.1016/S0140-6736(20)30638-2

Bloom, D. E., Canning, D., & Sevilla, J. (2011). The effect of non-communicable diseases on economic growth. *Global Health, 7*(1), 38. https://doi.org/10.1186/1744-8603-7-38

Chatham House. (2020). *COVID-19: Global health governance and the role* of the WHO. https://www.chathamhouse.org

Clement, S., Schauman, O., & Graham, T. (2015). What is the impact of mental health-related stigma on help-seeking? A systematic review of the literature. *Psychological Medicine, 45*(1), 11–27. https://doi.org/10.1017/S0033291714000129

Corrigan, P. W. (2004). How stigma interferes with mental health care. *American Psychologist, 59*(7), 614–625. https://doi.org/10.1037/0003-066X.59.7.614

Corrigan, P. W., & Watson, A. C. (2002). The paradox of self-stigma and mental illness. *Clinical Psychology: Science and Practice, 9*(1), 35–53. https://doi.org/10.1093/clipsy.9.1.35

Ghebreyesus, T. A. (2020). *The role of the World Health Organization in pandemic preparedness and response.* World Health Organization. https://www.who.int

Ghebreyesus, T. A. (2021). *Ensuring equitable access* to COVID-19 vaccines. World Health Organization. https://www.who.int

Goffman, E. (1963). *Stigma: Notes on the management of spoiled identity.* Prentice-Hall.

Gulliver, A., Griffiths, K. M., & Christensen, H. (2012). Perceived barriers and facilitators to mental health help-seeking in young people: A systematic review. *BMC Psychiatry, 12*(1), 1–9. https://doi.org/10.1186/1471-244X-12-8

Haines, A., Kovats, R. S., Campbell-Lendrum, D., & Corvalán, C. (2006). Climate change and human health: Impacts, vulnerability, and adaptation. *The Lancet, 367*(9513), 2101–2109. https://doi.org/10.1016/S0140-6736(06)68768-8

He, F. J., Li, J., & MacGregor, G. A. (2013). Effect of longer-term modest salt reduction on blood pressure. *Cochrane Database of Systematic Reviews, 2013*(4), CD004937. https://doi.org/10.1002/14651858.CD004937.pub2

Hinshaw, S. P. (2007). *The mark of shame: Stigma of mental illness and an agenda for change.* Oxford University Press.

Hotez, P. J. (2020). COVID-19 and the neglected tropical diseases: A global challenge. *PLOS Neglected Tropical Diseases, 14*(9), e0008688. https://doi.org/10.1371/journal.pntd.0008688

Jha, P., & Peto, R. (2014). Global effects of smoking, of quitting, and of taxing tobacco. *New England Journal of Medicine, 370*(1), 60–68. https://doi.org/10.1056/NEJMra1308383

Jorm, A. F., Korten, A. E., & Jacomb, P. A. (2006). Mental health literacy: A survey of the public's ability to recognize mental disorders and their beliefs about the effectiveness of treatment. *The Medical Journal of Australia, 166*(4), 182–186. https://doi.org/10.5694/j.1326-5377.1997.tb139147.x

Katzmarzyk, P. T., Church, T. S., & Blair, S. N. (2019). Cardiovascular disease and the physical activity transition. *Progress in Cardiovascular Diseases, 62*(1), 48–58. https://doi.org/10.1016/j.pcad.2018.07.004

Liu, Y., Wang, Y., & Li, Y. (2017). The impact of unhealthy diets on noncommunicable diseases in China. *Public Health Nutrition, 20*(8), 1370–1377. https://doi.org/10.1017/S136898001700016X

Marmot, M., Stansfeld, S., & Patel, C. (2008). Health inequalities among British civil servants: The Whitehall II study. *Lancet, 337*(8754), 1387–1393. https://doi.org/10.1016/S0140-6736(91)93068-K

McKibbin, W. J., & Fernando, R. (2020). *The economic impact of COVID-19*. Brookings Institution. https://www.brookings.edu

Moon, S., Sridhar, D., Pate, M. A., & Jha, A. K. (2020). Will the global health system deliver on the promise of COVID-19 vaccines? *The Lancet, 396*(10261), 1659–1661. https://doi.org/10.1016/S0140-6736(20)32366-2

Ng, M., Freeman, M. K., & Fleming, T. (2014). Smoking prevalence and cigarette consumption in 187 countries, 1980–2012. *JAMA, 311*(2), 183–192. https://doi.org/10.1001/jama.2013.284692

Patz, J. A., Campbell-Lendrum, D., Holloway, T., & Foley, J. A. (2005). Impact of regional climate change on human health. *Nature, 438*(7066), 310–317. https://doi.org/10.1038/nature04188

Pirkis, J., Francis, C., & Blood, R. W. (2006). The stigma of mental illness: An international perspective. *Australian and New Zealand Journal of Psychiatry, 40*(3), 160–167. https://doi.org/10.1111/j.1440-1614.2006.01747.x

Puhl, R. M., & Heuer, C. A. (2010). Obesity stigma: A review of the literature and implications for public health. *Social Science & Medicine, 70*(10), 155–164. https://doi.org/10.1016/j.socscimed.2009.10.035

Ranney, M. L., Griffeth, V., & Kissoon, N. (2020). Critical supply shortages—the need for pandemic preparedness. *New England Journal of Medicine, 382*(18), e49. https://doi.org/10.1056/NEJMp2007383

Rehm, J., Samokhvalov, A. V., & Neuman, M. G. (2017). The influence of alcohol on cardiovascular disease: A review of the literature. *American Journal of Cardiology, 119*(1), 15–24. https://doi.org/10.1016/j.amjcard.2016.09.019

Rüsch, N., Angermeyer, M. C., & Corrigan, P. W. (2005). Mental illness stigma: Concepts, consequences, and initiatives to reduce stigma. *European Psychiatry, 20*(8), 529–539. https://doi.org/10.1016/j.eurpsy.2005.04.004

Sartorius, N. (2007). Stigma and mental health. *The Lancet, 370*(9592), 810–811. https://doi.org/10.1016/S0140-6736(07)61245-8

Stuart, H. (2005). Stigma and the social experience of people with mental illness. *Canadian Journal of Psychiatry, 50*(3), 139–142. https://doi.org/10.1177/070674370505000301

Thornicroft, G. (2006). Shunned: Discrimination against people with mental illness. *Oxford University Press.*

Vargas, E. D., Sanchez, G. R., & Gubernskaya, Z. (2020). The COVID-19 pandemic and the racial and ethnic disparities in health. *Journal of Racial and Ethnic Health Disparities, 7*(4), 657–666. https://doi.org/10.1007/s40615-020-00885-2

Ventola, C. L. (2015). The antibiotic resistance crisis: Part 1: Causes and threats. *P&T, 40*(4), 277–283.

Watts, N., Amann, M., Arnell, N., & Ayeb-Karlsson, S. (2015). Health and climate change: Policy responses to protect public health. *The Lancet, 386*(10006), 1861–1914. https://doi.org/10.1016/S0140-6736(15)60854-6

Williams, D. R., & Mohammed, S. A. (2009). Discrimination and racial disparities in health: Evidence and needed research. *Journal of Behavioral Medicine, 32*(1), 20–47. https://doi.org/10.1007/s10865-008-9185-0

World Health Organization. (2020). COVID-19: The role of the World Health Organization in pandemic preparedness and response. https://www.who.int

Zinsstag, J., Schelling, E., Waltner-Toews, D., & Tanner, M. (2011). *One Health: The theory and practice of integrated health approaches.* CABI.

11

Conflict, War, and Displacement

Conflict and war have long been central drivers of human suffering, leading to widespread displacement and long-lasting social, economic, and psychological consequences. Armed conflicts, whether civil wars, interstate wars, or localized violence, displace millions of people, forcing them to flee their homes in search of safety and stability. The issue of displacement is not new, but in modern warfare, it has reached unprecedented levels, often exacerbated by the complexity of global geopolitics, economic interests, and the breakdown of governance structures. Displacement can take many forms, from refugees fleeing across borders to internally displaced persons (IDPs) seeking refuge within their own countries. The challenges faced by displaced populations are numerous, ranging from physical insecurity and inadequate access to basic needs to psychological trauma and the loss of community and identity.

11.1 The Drivers of Conflict and War

The causes of conflict and war are multifaceted and often rooted in deep-seated political, social, and economic issues. Ethnic, religious, and cultural tensions are common drivers of violent conflict, particularly in regions with historical grievances and competition for resources. Political repression, lack of political representation, and human rights abuses often exacerbate these tensions, leading to uprisings, insurgencies, and civil wars. Economic inequality and poverty also contribute significantly to conflict, as marginalized groups may resort to violence as a means of challenging oppressive regimes or gaining access to resources. Additionally, the involvement of external actors, such as foreign governments, multinational corporations, and international organizations, can complicate conflicts, either by supporting one side or through the imposition of sanctions and military interventions (Collier, 2000).

Environmental stressors, such as resource scarcity, climate change, and competition for land and water, can further fuel conflict. In regions

DOI: 10.1201/9788743808787-14

with limited natural resources, competition over access to water, arable land, or oil can lead to violent confrontations. Climate change has been identified as a "threat multiplier," exacerbating existing vulnerabilities and increasing the likelihood of conflict (Gleditsch, 2012). In countries already struggling with political instability, the effects of climate change – such as droughts, floods, and desertification – can further strain resources, intensifying grievances, and triggering conflict. Thus, conflict and war are often the result of a complex interplay of political, economic, environmental, and social factors.

11.2 The Impact of Conflict on Displacement

The consequences of war and conflict devastate those directly involved and the broader population. Displacement is one of the most immediate and visible impacts of conflict. According to the United Nations High Commissioner for Refugees (UNHCR), there were over 100 million displaced people worldwide in 2023, the highest number on record (UNHCR, 2023). Refugees and internally displaced persons (IDPs) face extreme hardships, including physical insecurity, lack of access to food, water, shelter, and healthcare, and the loss of social networks and livelihoods. Refugees often face significant challenges in finding safety, as they are forced to cross international borders, sometimes encountering hostility or discrimination from host countries. The lack of legal protections for refugees, coupled with restrictive immigration policies in many countries, further exacerbates their vulnerability (Betts, 2013).

Internally displaced persons who remain within their home countries but are forced to flee their homes due to conflict face similar challenges. IDPs often lack access to adequate humanitarian assistance and may be caught in conflict zones where humanitarian aid is difficult to deliver. The displacement of large populations can also lead to the breakdown of social structures and community ties, making it more difficult for individuals to rebuild their lives even after the conflict ends. Moreover, displacement is often protracted, with many refugees and IDPs spending years, if not decades, in camps or informal settlements without the prospect of returning home or resettling in new communities (Crisp, 2010).

11.3 Psychological and Social Consequences of Displacement

The psychological toll of conflict and displacement is profound. Displaced individuals often experience trauma because of the violence they have witnessed or endured, including physical injury, the loss of loved ones, and the destruction of their homes and communities. Refugees and IDPs are at a higher risk of mental health issues such as post-traumatic stress disorder (PTSD), depression, and anxiety, which can persist long after the conflict has ended (Silove, 2004). The trauma of displacement is compounded by the uncertainty and insecurity of living in refugee camps or informal settlements, where access to healthcare, education, and employment opportunities is limited.

Socially, displacement can disintegrate communities and social networks, leaving individuals feeling isolated and disconnected. The loss of cultural identity, as refugees and IDPs are often forced to abandon their homes and traditions, can have long-lasting effects on their sense of self and belonging (Sommers, 2001). Additionally, the experience of displacement can lead to generational trauma, as children who grow up in conflict zones or refugee camps may experience interrupted education, limited opportunities, and a lack of stability, all of which can affect their future prospects and contribute to cycles of poverty and instability (Machel, 1996).

11.4 The Global Response to Displacement

The international community has made significant efforts to address the challenges of displacement, but these efforts have often been inadequate in meeting the scale of the crisis. The 1951 Refugee Convention and its 1967 Protocol remain the cornerstone of international refugee law, providing legal protections for refugees and outlining the responsibilities of host countries. However, implementing these protections has been uneven, with many countries failing to uphold their commitments to refugees and asylum seekers. Moreover, the global response to displacement has often been reactive rather than proactive, with humanitarian aid typically arriving after many people have already been displaced (Zetter, 2007).

The increasing number of displaced persons worldwide, combined with the complexity of modern conflicts, has strained the capacity of international organizations and host countries to provide adequate support. While some countries have been more welcoming of refugees, others have adopted increasingly restrictive policies, leading to a "crisis of solidarity" in the international community (Betts, 2013). The challenge of displacement is further complicated by the fact that many conflicts are protracted, meaning that refugees and IDPs remain in limbo for years or even decades. Long-term solutions, such as resettlement, local integration, or voluntary repatriation, are often difficult to achieve, leaving millions of people without a permanent solution to their displacement (UNHCR, 2023).

Conflict, war, and displacement are inextricably linked, with displacement serving as one of the most visible and devastating consequences of armed conflict. The drivers of conflict are complex and multifaceted, encompassing political, economic, social, and environmental factors. Displaced populations face a range of challenges, including physical insecurity, lack of basic services, and psychological trauma. While the international community has attempted to address displacement, these efforts have often been inadequate, and the global response to displacement remains fragmented and inconsistent. As the world faces increasing conflicts and displacement crises, it is essential to prioritize long-term solutions that address the root causes of conflict and provide sustainable support for displaced populations.

11.5 The Human Cost of Wars and Civil Strife

Wars and civil strife have long-lasting and devastating impacts on human populations, with consequences that extend far beyond the battlefield. The human cost of these conflicts is not limited to the loss of life. Still, it includes a wide array of social, psychological, and economic repercussions that affect both the immediate victims and future generations. These impacts can be seen in the displacement of populations, the destruction of communities, the breakdown of social and economic systems, and the long-term psychological trauma experienced by individuals and societies. As conflicts persist, the toll on human well-being grows, and the ability of societies to recover becomes increasingly difficult.

11.5.1 Loss of life and physical injury

One of the most immediate and visible costs of war and civil strife is the loss of life. Combatants and civilians alike are killed or injured during violent confrontations, with civilians often enduring most of the violence. According to the World Health Organization (WHO), more than 10 million people have died in conflicts since 1990, with millions more suffering from injuries, disabilities, and long-term health issues (WHO, 2017). The death toll is often exacerbated by the destruction of healthcare infrastructure, making it difficult for those who are injured to receive adequate medical care. In addition to physical injuries, war-related trauma can lead to long-term disabilities, particularly for survivors who may suffer from amputations, burns, or other life-altering conditions (Murray et al., 2012).

The loss of life during conflict is not confined to soldiers; civilians are often the primary victims of war. Targeted violence against non-combatants, including women and children, is a tragic hallmark of many modern conflicts. In some cases, entire communities are wiped out, leading to demographic shifts and the loss of cultural heritage. The deliberate targeting of civilians, such as through bombings, massacres, or forced displacement, has become a widespread feature of contemporary warfare (Kaldor, 2013). The indiscriminate nature of modern weaponry and tactics increases the likelihood of civilian casualties, and the toll on life is often much higher than that of military personnel.

11.5.2 Displacement and refugee crises

The human cost of war and civil strife is also evident in the displacement of populations. Armed conflict often forces people to flee their homes in search of safety, leading to one of the most significant humanitarian crises in the world today. The United Nations High Commissioner for Refugees (UNHCR) estimates that more than 100 million people are currently displaced worldwide due to conflict, persecution, and violence (UNHCR, 2023). Refugees and internally displaced persons (IDPs) face numerous challenges, including inadequate shelter, lack of access to clean water and food, and limited healthcare. The lack of basic services in refugee camps and informal settlements, combined with overcrowding and insecurity, exacerbates the suffering of displaced populations.

The psychological and social toll of displacement is profound. Displaced individuals often experience trauma because of the violence they have witnessed or endured, the loss of their homes and communities, and the uncertainty of their future. Many refugees and IDPs are forced to live in limbo for years, with little hope of returning home or rebuilding their lives. Children are vulnerable to the effects of displacement, as they often experience interruptions in their education, social isolation, and exposure to violence. The long-term impacts of displacement can hinder their development and prospects for the future (Crisp, 2010).

11.5.3 Psychological trauma and mental health

The psychological consequences of war and civil strife are often overlooked but are among conflict's most profound and lasting effects. War-related trauma can result in a wide range of mental health issues, including post-traumatic stress disorder (PTSD), depression, anxiety, and substance abuse. Survivors of violence, particularly those who have experienced direct trauma such as torture, sexual violence, or the loss of loved ones, are at heightened risk of developing mental health disorders (Mollica et al., 2004). The psychological toll of war is not limited to those directly involved in combat but extends to entire communities, including those who may have witnessed violence or been displaced by conflict.

The impact of war on mental health is often compounded by the lack of access to mental health services in conflict zones and refugee camps. In many situations, mental health services are either unavailable or inadequate, leaving individuals to cope with their trauma without professional support. This lack of care can lead to a cycle of mental health problems that persist long after the conflict has ended. Furthermore, the stigma surrounding mental health issues in many conflict-affected regions can prevent individuals from seeking help, leading to untreated psychological suffering (Fazel et al., 2005).

11.5.4 Economic and social disruption

War and civil strife also have significant economic and social consequences, which can be felt long after the conflict has ended. The destruction of infrastructure, including schools, hospitals, roads, and factories, hampers economic recovery and development. The cost of rebuilding these essential services can be astronomical, and many countries affected by conflict face prolonged periods of economic stagnation. In addition to physical infrastructure, war disrupts social systems, including family structures, education, and governance. In some cases, the breakdown of social institutions leads to the rise of criminal networks, corruption, and lawlessness, further destabilizing affected regions (Collier, 2000).

The economic consequences of war are particularly severe for the poorest and most vulnerable populations. In conflict zones, poverty rates often increase, and livelihoods are destroyed. Agriculture, a primary source of income for many in developing countries, is often decimated by violence, landmines, and the destruction of farming infrastructure. This economic collapse leads to widespread food insecurity and malnutrition, further exacerbating the suffering of displaced populations and those remaining in conflict zones (Bohmelt et al., 2014).

11.5.5 Long-term consequences for future generations

The human cost of war and civil strife extends beyond the immediate effects on individuals and communities. The legacy of conflict can have long-term consequences for future generations. Children who grow up in conflict zones are particularly vulnerable to the effects of war, as they are exposed to violence, trauma, and disruption to their education. The psychological and social scars of war can affect their development and their ability to lead productive lives in the future. Moreover, the destruction of social and economic systems can hinder the development of entire generations, perpetuating cycles of poverty, instability, and violence (Sommers, 2001).

The intergenerational effects of war are also evident in the spread of trauma. Children who witness or experience violence may carry the psychological burden of war throughout their lives, passing on trauma to their own children. This cycle of trauma can contribute to the persistence of conflict and instability in post-conflict societies (Machel, 1996). Additionally, the disruption of education and economic opportunities for young people can lead to a "lost generation" that struggles to rebuild their lives in the aftermath of war.

The human cost of wars and civil strife is immense and multifaceted. Beyond the immediate loss of life and physical injury, conflict leads to displacement, psychological trauma, and long-term economic and social disruption. Individuals and communities feel the impact of war for years, if not decades, after the conflict has ended. Addressing the human cost of conflict requires not only humanitarian aid and support for displaced populations but also long-term efforts to rebuild societies, promote peace, and address the root causes of violence. The international community must prioritize conflict prevention, post-conflict recovery, and the protection of human rights to mitigate the devastating impact of war and civil strife on future generations.

11.6 Refugees, Migrants, and Stateless Populations

The global displacement crisis has reached unprecedented levels, with millions of individuals being forced to leave their homes due to conflict, persecution, human rights violations, and economic instability. Refugees, migrants, and stateless populations are among the most vulnerable groups in the world, facing numerous challenges in their pursuit of safety, security, and dignity. While the terms "refugees," "migrants," and "stateless" are often used interchangeably, each group has distinct legal definitions, rights, and needs that require tailored humanitarian responses.

11.6.1 Refugees: Forced displacement and protection

Refugees are individuals who are forced to flee their home country due to a well-founded fear of persecution on account of race, religion, nationality, membership in a particular social group, or political opinion (UNHCR,

1951). The 1951 Refugee Convention and its 1967 Protocol provide the legal framework for refugee protection, outlining the rights of refugees and the responsibilities of host countries. Refugees are entitled to non-refoulement, meaning they cannot be returned to a country where their life or freedom would be threatened (UNHCR, 1951). However, despite these protections, refugees often face immense challenges, including a lack of access to basic services such as healthcare, education, and legal support, as well as the threat of exploitation and violence in host countries.

The number of refugees worldwide has reached alarming levels, with the United Nations High Commissioner for Refugees (UNHCR) reporting that there were over 26 million refugees globally by the end of 2022 (UNHCR, 2023). Conflicts in Syria, Afghanistan, South Sudan, and Ukraine, as well as ongoing instability in parts of Africa and Asia, have contributed to this staggering number. Refugees often live in precarious conditions in refugee camps or urban areas, where they face overcrowding, limited access to employment, and a lack of legal recognition. The psychological toll of forced displacement is also significant, with many refugees experiencing trauma from violence, loss, and separation from their families (Fazel et al., 2005).

11.6.2 Migrants: Voluntary and forced movements

Migrants, in contrast to refugees, are individuals who choose to move to another country for reasons such as better economic opportunities, family reunification, or improved living conditions. Migration is a complex phenomenon that can be voluntary or forced. While voluntary migrants typically seek better opportunities, forced migrants are often driven by factors such as poverty, environmental degradation, or political instability, which make their migration a necessity rather than a choice (Castles, 2004). The International Organization for Migration (IOM) estimates that there are over 280 million international migrants worldwide, with many moving from low-income countries to higher-income countries in search of work or safety (IOM, 2022).

Migrants face a unique set of challenges, particularly in the context of immigration policies that vary significantly across countries. In many cases, migrants are subject to restrictive immigration laws, which can limit their access to legal protections, healthcare, and social services. Migrants may also face discrimination, exploitation, and marginalization in host countries, particularly if they are undocumented or live in precarious legal situations (Sullivan, 2018). Despite these challenges, migration can also offer significant benefits, including economic opportunities, social mobility, and access to better living conditions. However, the benefits of migration are often unevenly distributed, and many migrants face significant barriers to integration and upward mobility in their new communities (OECD, 2019).

11.6.3 Stateless populations: The invisible people

Statelessness refers to the condition in which an individual is not considered a national by any country, leaving them without the legal rights and protections afforded by nationality (UNHCR, 2023). Stateless individuals often face extreme vulnerability, as they are denied access to basic services such as healthcare, education, and employment, and are often excluded from political and social participation. Statelessness can arise due to a variety of factors, including discrimination based on ethnicity, religion, or gender, as well as gaps in national laws or policies that fail to recognize certain groups as citizens (Hathaway, 2008). The plight of stateless populations is often compounded by their lack of legal identity, making it difficult for them to access legal recourse or seek protection from the international community.

The UNHCR estimates that there are 4.2 million stateless individuals worldwide, although the actual number may be higher due to the challenges in identifying stateless populations (UNHCR, 2023). Statelessness is particularly prevalent in regions such as Southeast Asia, the Middle East, and parts of Africa, where ethnic and political conflicts have led to the marginalization of certain groups. For example, the Rohingya people in Myanmar have long been denied citizenship and have been subjected to violent persecution, forcing many to flee to neighboring countries as refugees (Zawacki, 2017). Stateless individuals are often caught in a cycle of exclusion, with limited prospects for integration or legal recognition, and their lack of nationality makes them highly vulnerable to exploitation, trafficking, and human rights abuses (UNHCR, 2023).

11.6.4 The intersections of refugees, migrants, and statelessness

While refugees, migrants, and stateless populations are distinct groups, their experiences often overlap, and many individuals may belong to more than one of these categories. For example, a migrant may become a refugee if they are forced to flee their host country due to persecution or conflict. Similarly, stateless individuals may be forced to migrate in search of a country that will recognize their legal status. The intersection of these categories highlights the complex nature of displacement and the need for comprehensive, flexible, and rights-based approaches to migration and refugee protection (Betts, 2013).

The international community has made considerable progress in addressing the needs of refugees and migrants through frameworks such as the Global Compact on Refugees (UNHCR, 2018) and the Global Compact for Safe, Orderly, and Regular Migration (UN, 2018). These agreements aim to enhance international cooperation and provide a framework for protecting the rights of displaced individuals. However, the implementation of these agreements remains inconsistent, and many countries continue to prioritize national security and economic interests over the protection of refugees and migrants. Additionally, the issue of statelessness remains under-addressed, with many countries failing to grant citizenship or legal recognition to marginalized groups.

The experiences of refugees, migrants, and stateless populations are shaped by a complex interplay of legal, social, and political factors. While refugees flee their countries due to persecution, migrants move in search of better opportunities, and stateless individuals are denied nationality altogether. All three groups face significant challenges in their pursuit of safety, security, and dignity, and their plight requires urgent attention from the international community. Addressing the needs of these populations requires a comprehensive approach that prioritizes human rights, legal protections, and international cooperation. As the number of displaced individuals continues to rise, it is essential that the global community work together to ensure that refugees, migrants, and stateless individuals are treated with dignity and respect and are provided with the support they need to rebuild their lives.

11.7 The Role of International Humanitarian Efforts

International humanitarian efforts are critical in addressing the needs of vulnerable populations affected by conflict, natural disasters, and other crises. These efforts are driven by a range of actors, including intergovernmental organizations, non-governmental organizations (NGOs), and various international agencies. The goal of these efforts is to provide life-saving assistance, protect human rights, and promote long-term recovery and resilience in the aftermath of crises. Humanitarian aid is often provided in the form of food, water, shelter, healthcare, and protection services, but it also involves advocating for the rights of displaced individuals, promoting peace and security, and addressing the root causes of conflict and suffering.

11.7.1 International humanitarian law and principles

At the core of international humanitarian efforts is the framework of International Humanitarian Law (IHL), which seeks to limit the effects of armed conflict and protect individuals who are not participating in hostilities, such as civilians, prisoners of war, and the wounded (International Committee of the Red Cross [ICRC], 2020). IHL is grounded in the Geneva Conventions of 1949 and their Additional Protocols, which provide rules on the conduct of warfare and the protection of non-combatants. The principles of humanity, impartiality, neutrality, and independence are fundamental to humanitarian action, ensuring that aid is provided based on need, without discrimination or political bias, and without interference in the internal affairs of states (Sphere Project, 2018).

International humanitarian organizations, such as the United Nations High Commissioner for Refugees (UNHCR), the World Food Programme (WFP), and the International Federation of Red Cross and Red Crescent Societies (IFRC), play a pivotal role in delivering humanitarian assistance in crises. These organizations operate in accordance with IHL, ensuring that aid is delivered in a manner that respects the dignity and rights of affected populations. Humanitarian workers are also bound by ethical standards that prioritize the well-being of affected communities, and efforts are made to ensure that aid is provided in a transparent, accountable, and effective manner (Barnett, 2011).

11.7.2 Coordination and multilateralism in humanitarian efforts

One of the key challenges in international humanitarian efforts is ensuring effective coordination among the many actors involved. Humanitarian crises often attract the involvement of a wide range of organizations, including UN agencies, international NGOs, local civil society groups, and donor governments. While this diversity of actors can bring valuable expertise and resources, it can also lead to fragmentation, duplication of efforts, and inefficiencies (Slim, 2015). To address these challenges, the international community has developed coordination mechanisms, such as the UN Office for the Coordination of Humanitarian Affairs (OCHA), which works to ensure that humanitarian aid is delivered in a coordinated and efficient manner. OCHA plays a significant role in organizing humanitarian responses, allocating resources, and facilitating communication between humanitarian actors (OCHA, 2021).

Multilateralism is also a key feature of international humanitarian efforts, as it allows for the pooling of resources and expertise from various countries and organizations. The United Nations system, for example, facilitates the coordination of humanitarian efforts through its various agencies and specialized bodies, such as the UNHCR, UNICEF, and the WFP.

Additionally, international partnerships between governments, NGOs, and the private sector are increasingly common, as they allow for a more holistic and sustainable approach to humanitarian assistance (Graham & Phillips, 2016). These partnerships can help to address the complex and interrelated needs of affected populations, such as providing food, water, shelter, healthcare, and education, while also supporting long-term recovery and development efforts.

11.7.3 Challenges and criticisms of international humanitarian efforts

Despite the significant contributions of international humanitarian efforts, these initiatives face numerous challenges and criticisms. One of the primary challenges is the political nature of humanitarian aid, as humanitarian organizations often operate in politically sensitive environments where access to affected populations is restricted by governments or armed groups. Humanitarian organizations may be forced to negotiate with parties to a conflict or may face restrictions on the movement of aid workers and supplies (Terry, 2002). In some cases, humanitarian assistance has been politicized or used as a tool for advancing the interests of powerful states or international organizations, undermining the neutrality and impartiality of aid (Chouliaraki, 2013).

Another criticism of international humanitarian efforts is the perceived ineffectiveness of aid in addressing the root causes of crises. While humanitarian aid can provide immediate relief, it often fails to address the underlying political, social, and economic factors that contribute to displacement and suffering. For example, aid efforts in conflict zones may focus on providing food and shelter without addressing the underlying causes of violence, such as political repression, ethnic tensions, or resource competition. In some cases, humanitarian aid may even exacerbate existing tensions or contribute to the prolongation of conflict by providing resources that are diverted or exploited by warring parties (Duffield, 2001).

Furthermore, there are concerns about the sustainability of humanitarian aid efforts. While humanitarian organizations are often effective in providing short-term relief, there is a growing recognition that longer-term solutions are needed to address the structural challenges faced by affected populations. This includes supporting efforts to rebuild communities, promote economic development, and strengthen local governance and institutions. The shift from emergency relief to development assistance, often referred to as the "humanitarian-development nexus," is an ongoing challenge that requires greater coordination and integration between humanitarian and development actors (Slim, 2015).

11.7.4 The future of international humanitarian efforts

As global crises continue to evolve, the future of international humanitarian efforts will depend on the ability of the international community to adapt and respond to new challenges. Climate change, for example, is expected to exacerbate existing humanitarian crises, as rising sea levels, extreme weather events, and resource scarcity contribute to displacement and conflict (Biermann et al., 2012). The growing number of refugees, migrants, and stateless individuals will also require a more comprehensive and integrated approach to humanitarian assistance, with a focus on addressing both immediate needs and long-term solutions.

Moreover, technological innovations are likely to play an increasingly important role in humanitarian efforts. Advances in data collection, communication, and logistics have the potential to improve the efficiency and effectiveness of humanitarian responses. For example, satellite imagery and drone technology can be used to assess damage and deliver aid in hard-to-reach areas, while digital platforms can facilitate the distribution of cash transfers and other forms of assistance (UNHCR, 2020). However, the use of technology also raises ethical concerns, such as the potential for surveillance and data privacy violations, which must be carefully managed to ensure that the rights of affected populations are respected.

International humanitarian efforts are essential in providing relief to populations affected by conflict, natural disasters, and other crises. While these efforts have made significant strides in alleviating human suffering, they face numerous challenges, including political obstacles, inefficiencies, and criticisms about their long-term impact. To improve the effectiveness of humanitarian assistance, it is crucial to strengthen coordination among humanitarian actors, ensure that aid is provided in a neutral and impartial manner, and focus on addressing the root causes of crises. As the world faces increasingly complex and interconnected challenges, the international community must work together to ensure that humanitarian efforts are effective, sustainable, and grounded in the principles of human dignity and rights.

References

Betts, A. (2013). *The refugee crisis and international politics*. Oxford University Press.

Collier, P. (2000). *Economic causes of civil conflict and their implications for policy*. World Bank.

Crisp, J. (2010). The impact of protracted refugee situations on host countries: A review of the evidence. *Journal of Refugee Studies, 23*(3), 304–323. https://doi.org/10.1093/jrs/feq026

Gleditsch, N. P. (2012). Climate change and conflict. *Journal of Peace Research, 49*(1), 3–13. https://doi.org/10.1177/0022343311431288

Machel, G. (1996). *The impact of armed conflict on children*. United Nations.

Silove, D. (2004). The psychological effects of torture, mass violence, and refugee trauma: A review. *Journal of Traumatic Stress, 17*(1), 5–21. https://doi.org/10.1023/B:JOTS.0000014671.00278.8e

Sommers, M. (2001). *Children and war: A review of the literature*. UNICEF.

UNHCR. (2023). *Global trends: Forced displacement in 2023*. United Nations High Commissioner for Refugees. https://www.unhcr.org/globaltrends

Zetter, R. (2007). More labels, fewer refugees: Remaking the refugee label in an era of globalization. *Journal of Refugee Studies, 20*(2), 172–192. https://doi.org/10.1093/jrs/fem011

Bohmelt, T., Twardzik, M., & Bormann, N. (2014). The economic consequences of civil wars: A systematic review. *International Interactions, 40*(4), 552–576. https://doi.org/10.1080/03050629.2014.904707

Collier, P. (2000). *Economic causes of civil conflict and their implications for policy*. World Bank.

Crisp, J. (2010). The impact of protracted refugee situations on host countries: A review of the evidence. *Journal of Refugee Studies, 23*(3), 304–323. https://doi.org/10.1093/jrs/feq026

Fazel, M., Von Hoebel, J., & Luntamo, M. (2005). Mental health consequences of war and displacement in conflict-affected populations. *Lancet, 365*(9465), 100–106. https://doi.org/10.1016/S0140-6736(04)16457-7

Kaldor, M. (2013). *New and old wars: Organized violence in a global era*. Stanford University Press.

Machel, G. (1996). *The impact of armed conflict on children*. United Nations.

Mollica, R. F., Wyshak, G., & de Marneffe, D. (2004). The psychosocial effects of war and violence on refugee children. *Pediatrics, 113*(6), 1606–1612. https://doi.org/10.1542/peds.113.6.1606

Murray, C. J. L., Richards, M. A., & King, G. (2012). *Global burden of disease and injury series*. Harvard University Press.

Sommers, M. (2001). *Children and war: A review of the literature*. UNICEF.

UNHCR. (2023). *Global trends: Forced displacement in 2023*. United Nations High Commissioner for Refugees. https://www.unhcr.org/globaltrends

WHO. (2017). *World health statistics 2017: Monitoring health for the SDGs*. World Health Organization. https://www.who.int/gho/publications/world_health_statistics/2017/en

Betts, A. (2013). *Survival migration: A new protection framework*. Global Governance,19(3),227–246.https://doi.org/10.1163/19426720-01903002

Castles, S. (2004). The factors that make and unmake migration policies. *International Migration Review, 38*(3), 852–884. https://doi.org/10.1111/j.1747-7379.2004.tb00204.x

Fazel, M., Von Hoebel, J., & Luntamo, M. (2005). Mental health consequences of war and displacement in conflict-affected populations. *Lancet, 365*(9465), 100–106. https://doi.org/10.1016/S0140-6736(04)16457-7

Hathaway, J. C. (2008). *The rights of refugees under international law*. Cambridge University Press.

International Organization for Migration (IOM). (2022). *World Migration Report 2022*. International Organization for Migration. https://www.iom.int/wmr

OECD. (2019). *International migration and its impact on economic growth and inequality*. OECD Publishing. https://doi.org/10.1787/9789264312477-en

Sullivan, M. (2018). Migration and vulnerability: The experiences of migrants in Europe. *European Journal of Migration and Law, 20*(1), 1–21. https://doi.org/10.1163/15718166-12340043

UNHCR. (1951). *Convention relating to the status of refugees*. United Nations. https://www.unhcr.org/1951-refugee-convention.html

UNHCR. (2023). *Global trends: Forced displacement in 2023*. United Nations High Commissioner for Refugees. https://www.unhcr.org/globaltrends

UNHCR. (2018). *Global compact on refugees*. United Nations High Commissioner for Refugees. https://www.unhcr.org/global-compact-on-refugees.html

United Nations. (2018). *Global compact for safe, orderly, and regular migration*. United Nations. https://www.un.org/pga/72/2018/07/13/final-draft-global-compact-for-migration/

Zawacki, M. (2017). The plight of the Rohingya refugees: Statelessness and the international community's failure. *International Journal of Refugee Law, 29*(4), 545–563. https://doi.org/10.1093/ijrl/eez023

Barnett, M. (2011). *Empire of humanity: A history of humanitarianism*. Cornell University Press.

Biermann, F., Boas, I., & Nabel, J. E. M. (2012). Climate change and human security: Impacts, vulnerability, and adaptation. *Global Environmental Change, 22*(1), 1–8. https://doi.org/10.1016/j.gloenvcha.2011.11.004

Chouliaraki, L. (2013). *The Ironic Spectator: Solidarity in the Age of Post-Humanitarianism*. Wiley-Blackwell.

Duffield, M. (2001). *Global governance and the new wars: The merging of development and security*. Zed Books.

Graham, H., & Phillips, D. (2016). *Humanitarian crises and international responses: The role of multilateralism*. Journal of International Affairs, 70(2), 45–63.

International Committee of the Red Cross (ICRC). (2020). *International humanitarian law: Overview*. https://www.icrc.org/en/international-humanitarian-law/overview

OCHA. (2021). *Humanitarian response overview*. United Nations Office for the Coordination of Humanitarian Affairs. https://www.unocha.org/

Sphere Project. (2018). *The Sphere Handbook: Humanitarian Charter and Minimum Standards in Humanitarian Response*. Sphere Association.

Slim, H. (2015). *Humanitarian ethics: A guide to the morality of aid in war and disaster*. Oxford University Press.

Terry, F. (2002). *Condemned to repeat? The paradox of humanitarian action*. Cornell University Press.

UNHCR. (2020). *Innovation and technology in humanitarian action*. United Nations High Commissioner for Refugees. https://www.unhcr.org/innovation/

12

Gender Inequality and Social Justice

Gender inequality remains one of the most persistent and pervasive issues in societies worldwide. It manifests in various forms, including disparities in education, employment, healthcare, political participation, and social and cultural rights. Gender inequality is not only a violation of human rights but also a significant barrier to achieving social justice, which advocates for equal access to resources, opportunities, and protections for all individuals, regardless of their gender. The intersection of gender inequality and social justice involves addressing these disparities while also recognizing the complex ways in which gender intersects with other forms of inequality, such as race, class, ethnicity, and disability.

12.1 The Roots of Gender Inequality

The roots of gender inequality can be traced to deeply entrenched social, cultural, and economic systems that have historically subordinated women and marginalized gender minorities. Patriarchy, a social system in which men hold primary power and authority, has been a foundational structure in many societies, perpetuating gender roles and norms that limit women's opportunities and freedom (Connell, 2005). These patriarchal structures are reinforced by cultural traditions, religious teachings, and legal systems that position men as the primary breadwinners, decision-makers, and heads of households. At the same time, women are often relegated to the domestic sphere and denied access to public life and economic resources (Walby, 1990). Gender inequality is further exacerbated by intersectional factors such as race, class, and sexuality, which compound the disadvantages faced by women and gender minorities, making their experiences of discrimination and exclusion more complex and multifaceted (Crenshaw, 1991).

DOI: 10.1201/9788743808787-15

At the institutional level, gender inequality is often perpetuated by discriminatory laws and policies that limit women's rights and freedoms. In many countries, women face legal barriers to property ownership, inheritance, and credit access, undermining their economic independence and security (Duflo, 2012). Additionally, women are disproportionately represented in low-wage and informal sectors, where they are often subjected to exploitative labor practices, lack of job security, and limited opportunities for career advancement (Kabeer, 2015). These structural inequalities in economic systems contribute to the gender pay gap, which remains a significant issue in both developed and developing countries (Blau & Kahn, 2017).

12.2 Gender Inequality and Social Justice

Social justice is a concept that emphasizes the fair distribution of resources, opportunities, and rights among all members of society. In the context of gender inequality, social justice calls for the dismantling of discriminatory systems and practices that perpetuate gender-based disparities. Achieving gender justice requires not only addressing the symptoms of inequality but also tackling the root causes, such as patriarchal norms, gender-based violence, and economic exploitation (Young, 2006). Social justice advocates for policies and practices that promote gender equality in all spheres of life, including education, healthcare, employment, and political participation. It recognizes the importance of women's agency and autonomy in shaping their own lives and futures.

One key aspect of gender justice is recognizing women and gender minorities as equal participants in political and public life. Historically, women have been excluded from decision-making processes at the national and local levels, leading to policies that do not adequately address their needs or concerns (Phillips, 1995). In many countries, women still face significant barriers to political participation, including cultural norms that discourage women's involvement in politics and structural obstacles such as lack of access to education, resources, and networks (Htun & Weldon, 2018). Social justice requires actively promoting women's political representation and leadership, ensuring that their voices are heard, and their perspectives are considered in formulating public policies.

12.3 Intersectionality and Gender Justice

The concept of intersectionality, coined by Kimberlé Crenshaw (1991), plays a crucial role in understanding gender inequality within the framework of social justice. Intersectionality recognizes that individuals experience multiple, intersecting forms of oppression and discrimination based on their gender, race, class, sexuality, and other social identities. For example, Black women and women of color often face compounded discrimination due to both their gender and race, leading to unique experiences of marginalization that cannot be fully understood by looking at gender or race in isolation (Crenshaw, 1991). Intersectional approaches to gender justice emphasize considering these overlapping identities and experiences when designing policies and interventions to address gender inequality.

For instance, in the context of labor rights, women of color, migrant women, and working-class women often face specific challenges that differ from those experienced by middle-class or white women. These challenges include lower wages, poorer working conditions, and limited access to labor protections. An intersectional approach to social justice would call for policies addressing these specific barriers while promoting broader gender equality (Choo & Ferree, 2010). Similarly, gender-based violence, a critical issue in the fight for gender justice, disproportionately affects marginalized groups, including LGBTQ+ individuals, women of color, and indigenous women. Addressing gender-based violence requires an intersectional approach that recognizes the different ways in which violence is experienced and the specific support needs of diverse groups (Berg & Millard, 2018).

12.4 The Role of Feminism in Achieving Social Justice

Feminism has been a driving force in fighting gender equality and social justice. Feminist movements have advocated for women's rights, gender equality, and the dismantling of patriarchal structures that perpetuate inequality. Feminism is not a monolithic movement but rather encompasses a diverse range of perspectives and approaches, including liberal feminism, radical feminism, socialist feminism, and intersectional feminism (Tong, 2009). While these different strands of feminism may have varying priorities and strategies, they share a common goal of achieving social justice for women and gender minorities.

Feminist movements have played a crucial role in advancing legal and policy reforms to promote gender equality. For example, the feminist movement has been instrumental in advocating for women's right to vote, access to education, reproductive rights, and protection from gender-based violence. Feminist activism has also highlighted the importance of addressing women's economic and social inequalities, advocating for equal pay, access to healthcare, and the redistribution of resources (Fraser, 2013). In recent years, feminist movements have increasingly focused on global issues such as climate change, migration, and human rights, recognizing the interconnectedness of gender justice and broader social justice struggles (Nash, 2019).

Gender inequality is a fundamental issue of social justice, requiring a comprehensive approach that addresses both the structural and cultural factors that perpetuate discrimination and marginalization. Achieving gender justice requires the dismantling of patriarchal systems, the promotion of women's political and economic empowerment, and the recognition of the intersecting forms of oppression that shape women's experiences. Feminist movements have advocated for gender equality and social justice, pushing for legal reforms, social change, and the recognition of women's agency and autonomy. By adopting an intersectional approach and addressing the root causes of gender inequality, societies can move closer to achieving true social justice for all individuals, regardless of their gender.

12.5 Gender-Based Violence and Discrimination

Gender-based violence (GBV) and discrimination are deeply interwoven societal issues that disproportionately affect women and gender minorities across the globe. These forms of violence and discrimination are rooted in power imbalances, gender norms, and structural inequalities, which perpetuate cycles of harm, marginalization, and inequality. Gender-based violence encompasses a wide range of harmful behaviors, including physical, sexual, emotional, and psychological abuse. At the same time, gender discrimination refers to unequal treatment based on an individual's gender, leading to systemic disadvantage and oppression. Both GBV and discrimination not only violate basic human rights but also hinder progress toward achieving gender equality and social justice.

12.5.1 The scope of gender-based violence

Gender-based violence is a global phenomenon that affects individuals of all genders, although women and girls are the most vulnerable. According to the World Health Organization (WHO), approximately one in three women worldwide have experienced physical or sexual violence in their lifetime, often at the hands of an intimate partner (WHO, 2021). GBV can take many forms, including intimate partner violence, sexual harassment, human trafficking, female genital mutilation (FGM), child marriage, and honor killings. These forms of violence are not isolated incidents but are symptomatic of broader societal structures that condone and perpetuate inequality, sexism, and patriarchal norms (UN Women, 2020).

Intimate partner violence (IPV) is one of the most common and devastating forms of GBV. It involves physical, emotional, and sexual abuse within intimate relationships and has severe consequences for victims, including physical injury, mental health issues, and economic instability (Garcia-Moreno et al., 2015). Sexual violence, including rape and sexual assault, is another widespread form of GBV that disproportionately affects women, particularly in conflict zones and in contexts where power dynamics are skewed (Berg & Millard, 2018). The pervasiveness of GBV highlights the urgent need for comprehensive legal frameworks, social support systems, and education to address the root causes of violence and provide victims with the necessary support for recovery and justice.

12.5.2 Gender discrimination and its impact

Gender discrimination refers to the unequal treatment of individuals based on their gender, often leading to exclusion, marginalization, and denial of rights. It manifests in various forms, including disparities in employment, education, healthcare, political representation, and legal rights. Women and gender minorities are particularly vulnerable to discrimination, with women facing systemic barriers to accessing education, securing employment, and achieving economic independence (Blau & Kahn, 2017). For example, women are often paid less than men for the same work, a phenomenon known as the gender pay gap, which persists across many countries and sectors (Blau & Kahn, 2017).

Discrimination also extends to the political realm, where women and gender minorities are underrepresented in decision-making positions. Despite progress in some areas, women continue to be excluded from leadership roles in politics, business, and other sectors, which perpetuates gender inequality and limits their influence on policies that affect their lives (Htun & Weldon, 2018). Furthermore, women of color, migrant women, and LGBTQ+ individuals often face compounded discrimination due to the intersectionality of gender with race, ethnicity, and sexual orientation (Crenshaw, 1991). This intersectional discrimination exacerbates the challenges faced by marginalized groups, making them more vulnerable to violence, poverty, and social exclusion.

12.5.3 The root causes of gender-based violence and discrimination

The root causes of gender-based violence and discrimination are deeply embedded in societal norms, power structures, and historical inequalities. Patriarchy, a system of male dominance, has historically relegated women and gender minorities to subordinate roles, positioning them as objects of control and subjugation (Connell, 2005). Patriarchal ideologies reinforce rigid gender roles, which dictate how individuals should behave based on their gender and punish those who do not conform to these expectations. For example, women are often expected to be caregivers and homemakers, while men are socialized to be providers and protectors. These gender norms limit opportunities for both men and women, restricting freedom of choice and reinforcing unequal power dynamics (Walby, 1990).

Cultural practices and traditions also play a significant role in perpetuating gender-based violence and discrimination. In some societies, harmful practices such as female genital mutilation (FGM), child marriage, and honor-based violence are justified by cultural or religious beliefs despite their harmful consequences (UNICEF, 2020). These practices not only violate the rights of women and girls but also perpetuate gender inequality by reinforcing the notion that women's bodies are the property of men or the state. Addressing these issues requires challenging harmful cultural norms and promoting alternative, rights-based frameworks that empower women and gender minorities.

12.5.4 The consequences of gender-based violence and discrimination

The consequences of gender-based violence and discrimination are far-reaching, affecting individuals, families, communities, and societies. For individuals, the physical, emotional, and psychological effects of GBV can be devastating. Survivors of intimate partner violence, sexual assault, and other forms of abuse often experience long-term trauma, including depression, anxiety, post-traumatic stress disorder (PTSD), and suicidal ideation (Campbell, 2002). The stigma surrounding GBV also makes it difficult for survivors to seek help, as they may fear social ostracism, further violence, or lack of support (Berg & Millard, 2018). Moreover, GBV often leads to economic instability, as victims may be forced to leave their homes, lose their jobs, or face discrimination in the workplace (Kabeer, 2015).

On a societal level, gender-based violence and discrimination hinder economic development and social progress. When women and gender minorities are excluded from education, employment, and leadership roles, societies miss out on the full potential of their human capital. Studies have shown that gender equality positively correlates with economic growth, as empowering women leads to higher productivity, innovation, and social cohesion (Duflo, 2012). Moreover, gender-based violence and discrimination contribute to social instability and conflict, as they perpetuate cycles of inequality and resentment. Addressing these issues is essential for achieving sustainable development, social justice, and peace.

12.5.5 Moving towards solutions

Addressing gender-based violence and discrimination requires a multifaceted approach that involves legal reforms, social change, and the empowerment of women and gender minorities. Legal frameworks must be strengthened to protect victims of GBV and ensure that perpetrators are held accountable. This includes enacting and enforcing laws that criminalize all forms of gender-based violence, including domestic violence, sexual harassment, and trafficking, as well as providing legal remedies for victims (UN Women, 2020). Additionally, social support systems, such as shelters, counseling services, and legal aid, must be expanded to provide survivors with the resources they need to rebuild their lives.

Education and awareness-raising campaigns are also crucial in challenging harmful gender norms and promoting gender equality. By educating both men and women about the negative consequences of GBV and discrimination, societies can foster a culture of respect, equality, and nonviolence (Heise et al., 2019). Furthermore, engaging men and boys in the fight for gender equality is essential, as they play a key role in challenging patriarchal norms and promoting positive models of masculinity (Flood, 2015).

Gender-based violence and discrimination are critical issues that require urgent attention and action. They are rooted in patriarchal power structures, cultural practices, and systemic inequalities that marginalize women and gender minorities. The consequences of GBV and discrimination are far-reaching, affecting individuals' well-being, economic stability, and social participation. Achieving gender equality and social justice requires a comprehensive approach that includes legal reforms, social support systems, education, and cultural change. By addressing the root causes of gender-based violence and discrimination and promoting gender equality, societies can create a more just and equitable world for all.

12.6 Empowering Women and Marginalized Groups

Empowering women and marginalized groups is essential for fostering an inclusive, equitable, and just society. Empowerment refers to the process of enabling individuals to gain control over their lives, make informed decisions, and participate fully in economic, political, and social spheres. Women and marginalized groups, including racial and ethnic minorities, LGBTQ+ individuals, and people with disabilities, often face systemic barriers to equality. Empowerment efforts must address these barriers and create an environment where individuals from these groups can thrive.

12.6.1 Women's empowerment and its impact

Women's empowerment is widely recognized as a key driver of social and economic development. Empowering women leads to improved outcomes not only for women themselves but also for their families, communities, and societies. According to Kabeer (2005), empowerment involves the expansion of people's ability to make choices and control their resources, thus enhancing their participation in decision-making processes. A critical component of women's empowerment is education, which provides women with the skills and knowledge necessary to improve their economic and social standing. Access to education is linked to better health outcomes, higher income levels, and increased political participation (Psacharopoulos & Patrinos, 2018). However, gender-based barriers to education, such as early marriage, cultural norms, and gender-based violence, still prevent many women from accessing educational opportunities, particularly in developing regions (UNESCO, 2020).

In addition to education, economic empowerment is a central aspect of women's empowerment. When women have access to financial resources, equal employment opportunities, and entrepreneurship opportunities, they can contribute significantly to their households and communities (Duflo, 2012). Economic independence allows women to make decisions regarding their health, family, and future, which in turn leads to greater gender equality and social stability. The International Labour Organization (2019) has emphasized the importance of gender-sensitive labor policies that promote equal pay for equal work, support women's participation in the workforce, and create opportunities for women to enter traditionally male-dominated sectors.

12.6.2 Empowering marginalized groups: Intersectionality and inclusion

Empowerment efforts must also extend to other marginalized groups, as their experiences of discrimination are often shaped by multiple, intersecting factors. For instance, racial and ethnic minorities, LGBTQ+ individuals, and people with disabilities face unique challenges that require tailored approaches to empowerment. Crenshaw (1991) introduced the concept of intersectionality to explain how various forms of discrimination – such as racism, sexism, and homophobia – interact and compound the experiences of marginalized individuals. Empowering these groups requires addressing the specific forms of exclusion they face and creating inclusive systems that promote their full participation in society.

For example, Black, Indigenous, and people of color (BIPOC) often face compounded discrimination that limits their access to education, healthcare, and employment (Williams & Mohammed, 2009). Addressing these inequities requires comprehensive policy reforms, including anti-racism education, affirmative action, and policies that ensure equal access to resources and opportunities. Similarly, LGBTQ+ individuals, particularly transgender and non-binary people, face higher rates of violence, discrimination, and mental health challenges (Meyer, 2003). Legal reforms that protect the rights of LGBTQ+ people, along with social programs that provide support and resources, are essential for promoting their empowerment.

People with disabilities also face significant barriers to empowerment, including physical, social, and attitudinal barriers that prevent them from fully participating in society. Empowerment for people with disabilities involves ensuring equal access to education, employment, and healthcare, as well as promoting societal attitudes that value their contributions and recognize their rights (Shakespeare, 2014). Creating accessible environments and providing accommodations in public spaces, schools, and workplaces are key steps in promoting the inclusion and empowerment of people with disabilities.

Empowering women and marginalized groups is crucial for creating equitable and just societies. By addressing the systemic barriers that hinder the full participation of these groups, we can foster an environment where all individuals have the opportunity to thrive. Empowerment is not only a human rights issue but also a critical factor in achieving sustainable development and social progress. As such, it is essential to continue advocating for policies and practices that promote gender equality, racial justice, and inclusion for all marginalized groups.

12.7 Advocacy for LGBTQ+ Rights

Advocacy for LGBTQ+ rights has become a crucial aspect of the global human rights movement, aiming to secure equality, dignity, and respect for individuals who identify as lesbian, gay, bisexual, transgender, queer, or any other non-heteronormative gender or sexual identity. Over the past few decades, significant strides have been made in advocating for the rights of LGBTQ+ individuals, but challenges remain in achieving full equality and combating discrimination. Advocacy for LGBTQ+ rights not only addresses issues of legal equality but also seeks to change societal attitudes, promote inclusion, and ensure access to essential services such as healthcare, education, and employment.

12.7.1 Legal protections and recognition

One of the primary focuses of LGBTQ+ advocacy has been securing legal protections against discrimination. Many countries have passed laws that protect LGBTQ+ individuals from discrimination in areas such as employment, housing, and public accommodations. However, the extent of legal protection varies widely around the world. According to the International Lesbian, Gay, Bisexual, Trans, and Intersex Association (ILGA, 2020), while some countries, such as Canada and many European nations, have made significant progress in extending legal protections to LGBTQ+ individuals, others, particularly in parts of Africa and the Middle East, continue to criminalize same-sex relationships and gender non-conformity. Advocacy groups such as the Human Rights Campaign (HRC) and ILGA work tirelessly to push for legal reforms that ensure equal rights for LGBTQ+ individuals globally.

In addition to anti-discrimination laws, LGBTQ+ advocacy has also focused on securing marriage equality and the recognition of same-sex relationships. Marriage equality is a significant milestone in the fight for LGBTQ+ rights, as it ensures that same-sex couples have access to the same legal rights, benefits, and protections as heterosexual couples. In countries like the United States, Australia, and South Africa, marriage equality has been achieved through legal reforms, but the fight for recognition continues in many other parts of the world (Badgett, 2014). Furthermore, the recognition of gender identity for transgender individuals, including the right to legally change one's gender marker on official documents, remains a key issue in many countries, with advocacy groups pushing for reforms that protect transgender rights (Grant et al., 2011).

12.7.2 Social and cultural advocacy

Beyond legal protections, LGBTQ+ advocacy also focuses on changing societal attitudes toward gender and sexual diversity. Social stigma and discrimination remain pervasive in many societies, leading to higher rates of mental health issues, violence, and exclusion for LGBTQ+ individuals. According to the World Health Organization (WHO, 2015), LGBTQ+ people, especially transgender individuals, are at an increased risk of experiencing mental health challenges, including depression, anxiety, and suicide, often due to societal rejection and discrimination. Advocacy efforts aim to raise awareness about these issues, challenge harmful stereotypes, and promote acceptance and inclusion.

Education plays a critical role in changing attitudes and reducing discrimination. Programs that promote LGBTQ+ inclusion in schools and workplaces are essential for fostering environments where LGBTQ+ individuals feel safe and supported. Research by Kosciw et al. (2018) highlights the importance of inclusive curricula that reflect the diversity of sexual orientations and gender identities, as well as the need for anti-bullying policies that protect LGBTQ+ students from harassment. These educational initiatives are crucial in reducing the social isolation and marginalization that many LGBTQ+ individuals experience, particularly during adolescence.

12.7.3 Health and well-being

Another significant area of LGBTQ+ advocacy is improving access to healthcare services. LGBTQ+ individuals, particularly transgender and non-binary people, often face barriers to accessing quality healthcare, including discrimination from healthcare providers, lack of culturally competent care, and legal and financial barriers. The American Medical Association (AMA, 2019) emphasizes the need for healthcare systems to be inclusive and affirming of LGBTQ+ individuals, with specific attention to the unique needs of transgender people, including access to gender-affirming care. Advocacy for LGBTQ+ rights in healthcare also includes addressing the higher rates of HIV/AIDS among gay and bisexual men, as well as promoting mental health services that are sensitive to the needs of LGBTQ+ individuals.

In addition, LGBTQ+ advocacy seeks to address the intersectional challenges faced by individuals who belong to multiple marginalized groups, such as LGBTQ+ people of color or LGBTQ+ individuals with disabilities. These individuals often face compounded discrimination and barriers to accessing resources, and advocacy efforts must address these intersectional issues to ensure that all LGBTQ+ individuals are included in the fight for equality (Bowleg, 2012).

12.7.4 Global perspectives and challenges

While progress has been made in many countries, LGBTQ+ individuals continue to face significant challenges in other parts of the world. In some regions, particularly in sub-Saharan Africa, the Middle East, and parts of Asia, LGBTQ+ people still face criminalization, violence, and social exclusion. Advocacy organizations, including the United Nations, Amnesty International, and ILGA, have been working to raise awareness about the human rights violations faced by LGBTQ+ individuals in these regions. However, cultural, religious, and political factors often create significant barriers to progress. In some countries, LGBTQ+ advocacy is met with hostility, and activists risk imprisonment, violence, and even death for their efforts (Global Justice Project, 2020). Despite these challenges, international pressure and solidarity from global LGBTQ+ communities continue to push for change.

Advocacy for LGBTQ+ rights is a multifaceted effort that encompasses legal reforms, social and cultural change, healthcare access, and global solidarity. While significant progress has been made in many countries, there is still much work to be done to ensure that LGBTQ+ individuals can live free from discrimination, violence, and exclusion. By continuing to advocate for legal protections, societal acceptance, and healthcare access, and by addressing the intersectional challenges faced by marginalized groups within the LGBTQ+ community, we can move closer to achieving full equality and human dignity for all individuals, regardless of their sexual orientation or gender identity.

References

American Medical Association (AMA). (2019). *LGBTQ+ health and the role of physicians*. https://www.ama-assn.org

Badgett, M. V. L. (2014). *The economic case for marriage equality*. The Williams Institute. https://williamsinstitute.law.ucla.edu

Berg, L., & Millard, A. (2018). Gender-based violence and intersectionality: A critical review of research and policy. *Social Politics: International Studies in Gender, State & Society, 25*(4), 459–482. https://doi.org/10.1093/sp/jxy029

Blau, F. D., & Kahn, L. M. (2017). The gender wage gap: Extent, trends, and explanations. *Journal of Economic Literature, 55*(3), 789–865. https://doi.org/10.1257/jel.20160995

Bowleg, L. (2012). *The problem with the phrase "women and minorities": Intersectionality—An important theoretical framework for public health*. American Journal of Public Health, 102(7), 1267–1273. https://doi.org/10.2105/AJPH.2012.300750

Campbell, J. C. (2002). Health consequences of intimate partner violence. *The Lancet, 359*(9314), 1331–1336. https://doi.org/10.1016/S0140-6736(02)08336-8

Choo, H. Y., & Ferree, M. M. (2010). Practicing intersectionality in sociological research: A critical analysis of inclusions, interactions, and institutions in the study of inequality. *Sociological Theory, 28*(2), 129–149. https://doi.org/10.1111/j.1467-9558.2010.01370.x

Connell, R. (2005). *Masculinities* (2nd ed.). University of California Press.

Crenshaw, K. (1991). Mapping the margins: Intersectionality, identity politics, and violence against women of color. *Stanford Law Review, 43*(6), 1241–1299. https://doi.org/10.2307/1229039

Duflo, E. (2012). Women empowerment and economic development. *Journal of Economic Literature, 50*(4), 1051–1079. https://doi.org/10.1257/jel.50.4.1051

Flood, M. (2015). Engaging men and boys in violence prevention. *The Lancet, 385*(9977), 2134–2135. https://doi.org/10.1016/S0140-6736(15)60223-4

Fraser, N. (2013). *Fortunes of feminism: From state-managed capitalism to neoliberal crisis*. Verso.

Garcia-Moreno, C., Hegarty, K., d'Oliveira, A. F. P., Koziol-McLain, J., Colombini

Global Justice Project. (2020). *Human rights violations against LGBTQ+ people in the Middle East and North Africa*. https://globaljusticeproject.org

Grant, J. M., Mottet, L. A., Tanis, J., Harrison, J., Herman, J. L., & Keisling, M. (2011). *Injustice at every turn: A report of the National Transgender Discrimination Survey*. National Center for Transgender Equality and the National Gay and Lesbian Task Force. https://transequality.org/sites/default/files/docs/resources/NTDS_Report.pdf

Htun, M., & Weldon, S. L. (2018). The civic origins of progressive policy change: Combating discrimination in the United States and the European Union. *American Political Science Review, 112*(3), 585–602. https://doi.org/10.1017/S0003055418000219

International Labour Organization (ILO). (2019). *Women and the future of work*. https://www.ilo.org

Kabeer, N. (2005). *Gender equality and women's empowerment: A critical analysis of the third millennium development goal*. Gender & Development, 13(1), 13–24.

Kabeer, N. (2015). *Gender, poverty, and inequality: A brief history of feminist approaches in development*. Routledge.

Meyer, I. H. (2003). *Prejudice, social stress, and mental health in lesbian, gay, and bisexual populations: Conceptual issues and research evidence*. Psychological Bulletin, 129(5), 674–697.

Phillips, A. (1995). *The politics of presence: The political representation of gender, ethnicity, and race*. Oxford University Press.

Psacharopoulos, G., & Patrinos, H. A. (2018). *Returns to investment in education: A decennial review of global literature*. Education Economics, 26(5), 445–458.

Shakespeare, T. (2014). *Disability rights and wrongs revisited*. Routledge.

Tong, R. (2009). *Feminist thought: A more comprehensive introduction* (2nd ed.). Westview Press.

UNESCO. (2020). *Global education monitoring report 2020: Inclusion and education: All means all*. United Nations Educational, Scientific and Cultural Organization.

Walby, S. (1990). *Theorizing patriarchy*. Blackwell.

Williams, D. R., & Mohammed, S. A. (2009). *Discrimination and racial disparities in health: Evidence and needed research*. Journal of Behavioral Medicine, 32(1), 20–47.

Young, I. M. (2006). *Justice and the politics of difference*. Princeton University Press.

13

Education in Crisis

Education systems worldwide are facing unprecedented challenges that threaten to undermine the progress made over decades. The COVID-19 pandemic exacerbated these challenges, disrupting learning and highlighting the deep inequities in educational access and quality. According to UNESCO (2020), more than 1.6 billion students in over 190 countries were affected by school closures, which led to significant learning losses, particularly among marginalized communities. While many countries have implemented remote learning strategies, the digital divide has left millions of students without access to the necessary technology or internet connectivity, further deepening educational inequalities (Becker et al., 2020).

In addition to the digital divide, the global education crisis is marked by other systemic issues such as overcrowded classrooms, insufficient teacher training, and inadequate educational resources. These challenges are particularly acute in low-income countries, where funding for education is often limited. According to the World Bank (2020), the lack of investment in education in many developing nations has resulted in a poor-quality education system that fails to equip students with the skills needed for the modern workforce. As a result, many young people are not prepared for the demands of the labor market, contributing to high rates of unemployment and underemployment.

Another critical issue in the education crisis is the growing impact of climate change on educational systems. Extreme weather events, such as floods, hurricanes, and droughts, are increasingly disrupting schools and educational infrastructure, particularly in vulnerable regions. A report by the Global Partnership for Education (2021) highlights that climate-related disasters have led to the destruction of school buildings and the displacement of students, further hindering access to education. Additionally, climate change disproportionately affects girls, who are often the first to be pulled out of school in times of crisis, as families prioritize boys' education or face financial constraints (UNICEF, 2020).

DOI: 10.1201/9788743808787-16

The education crisis also intersects with issues of social justice and inequality. Discrimination based on gender, ethnicity, disability, and socio-economic status continues to marginalize certain groups, limiting their access to quality education. Girls in conflict zones, for example, face significant barriers to education, including violence, early marriage, and the destruction of schools (UNICEF, 2019). In many parts of the world, marginalized communities, including indigenous populations and refugees, also face systemic barriers to education, further perpetuating cycles of poverty and inequality (Save the Children, 2020).

Addressing the education crisis requires a multifaceted approach that includes increased investment in education, equitable access to resources, and targeted interventions for marginalized groups. The United Nations Sustainable Development Goal 4 (SDG 4) emphasizes the importance of inclusive, equitable, and quality education for all, and calls for urgent action to address the challenges facing education systems globally (UN, 2015). Governments, international organizations, and civil society must collaborate to ensure that education is not only accessible but also of high quality, preparing future generations to thrive in an increasingly complex and interconnected world.

13.1 Barriers to Universal Education

Universal education is a fundamental human right and a cornerstone for sustainable development. However, despite global efforts to ensure that all children have access to quality education, significant barriers persist, particularly in low-income countries and marginalized communities. These barriers are multifaceted, involving social, economic, cultural, and political factors that prevent millions of children from receiving an education. According to the United Nations Educational, Scientific and Cultural Organization (UNESCO, 2020), approximately 258 million children and youth worldwide are out of school, with the majority living in conflict zones, rural areas, or poverty-stricken environments. Understanding these barriers is crucial for addressing educational inequities and achieving the goal of universal education.

One of the primary barriers to universal education is poverty. Families living in poverty often struggle to meet basic needs, making education a lower priority. In many cases, children are forced to work to support their families, leading to high dropout rates and limited educational attainment (UNICEF, 2019). This issue is particularly pronounced in sub-Saharan Africa and South Asia, where a significant proportion of the population lives below the poverty line. The World Bank (2020) emphasizes that education costs, including tuition fees, uniforms, and transportation, can be prohibitive for poor families, further exacerbating educational inequality. Additionally, poverty often correlates with poor infrastructure, such as inadequate school buildings, lack of teaching materials, and insufficient access to technology, which negatively impacts the quality of education (Save the Children, 2020).

Another critical barrier is gender inequality. In many parts of the world, girls face greater challenges than boys in accessing education. Cultural norms, early marriage, and gender-based violence are significant obstacles to girls' education, particularly in countries affected by conflict or traditional patriarchal systems. According to the Global Partnership for Education (2021), approximately 130 million girls worldwide are out of school due to gender-based discrimination. In some regions, families prioritize boys' education over girls', believing that girls are destined for domestic roles. Furthermore, in conflict zones, girls are often subjected to violence, including sexual assault and abduction, making school attendance dangerous and sometimes impossible (UNICEF, 2020). Gender inequality in education perpetuates cycles of poverty and limits economic opportunities for women, which in turn affects societal development.

In addition to gender inequality, children with disabilities face significant barriers to education. Despite international frameworks like the Convention on the Rights of Persons with Disabilities, many educational systems are not equipped to accommodate children with physical, sensory, or intellectual disabilities. According to the World Health Organization (WHO, 2018), around 15% of the world's population lives with some form of disability, and many of these individuals face discrimination and exclusion from educational systems. In low-income countries, there is often a lack of specialized teachers, accessible learning materials, and inclusive school infrastructure. The United Nations (2019) reports that children with disabilities are more likely to be out of school than their peers, and those who do attend often face significant challenges in accessing quality education.

Conflict and displacement also present significant barriers to universal education. Armed conflicts, political instability, and natural disasters disrupt education systems and displace millions of children, preventing them from attending school. The United Nations High Commissioner for Refugees (UNHCR, 2020) estimates that there are over 80 million displaced people globally, many of whom are children. These children often live in refugee camps or informal settlements where access to education is limited or non-existent. In conflict zones, schools are often destroyed, and teachers may be displaced or killed, further exacerbating the educational crisis. Even when education is available in these contexts, it is often of poor quality, with limited resources and inadequate facilities (Save the Children, 2020).

Lastly, systemic issues within educational policies and governance also hinder the achievement of universal education. In many countries, particularly those with weak governance structures, there is inadequate investment in education. According to the Global Education Monitoring Report (UNESCO, 2020), insufficient government funding for education leads to overcrowded classrooms, underpaid teachers, and a lack of educational materials. Corruption and mismanagement further exacerbate these issues, preventing resources from reaching the schools and communities that need them most. Moreover, political instability and lack of political will can result in the neglect of education as a priority, perpetuating inequalities and limiting the potential for social mobility.

Addressing the barriers to universal education requires a comprehensive and coordinated approach that tackles the underlying social, economic, and political factors. It is essential to increase investment in education, particularly in marginalized and conflict-affected regions, and to prioritize gender equality and inclusivity for children with disabilities. International organizations, governments, and civil society must work together to create an education system that is equitable, inclusive, and accessible to all children, regardless of their socio-economic status, gender, or physical ability. Only through such concerted efforts can the world move closer to achieving the goal of universal education and ensuring that every child has the opportunity to reach their full potential.

13.2 The Digital Divide and Knowledge Inequality

The digital divide refers to the gap between individuals, communities, and countries who have access to information and communication technologies (ICT) and those who do not. This divide has profound implications for knowledge inequality, as access to digital tools and the internet increasingly determines one's ability to acquire, share, and use information. In an era where the internet is central to education, economic participation, and social engagement, the digital divide perpetuates existing inequalities and hinders efforts toward inclusive development (Van Dijk, 2020). Knowledge inequality, in turn, is shaped by disparities in access to digital resources, which limit opportunities for learning, economic advancement, and participation in the global knowledge economy.

One of the primary factors contributing to the digital divide is socio-economic status. People in low-income communities or developing countries often lack the financial resources to afford internet access, computers, or smartphones. According to the International Telecommunication Union

(ITU, 2020), nearly half of the global population remains offline, with the majority of the unconnected living in rural areas or low-income nations. This lack of access means that individuals in these communities are often excluded from online educational resources, job opportunities, and vital information that could improve their quality of life. For example, during the COVID-19 pandemic, the shift to online learning highlighted the stark contrast between students who had access to digital devices and reliable internet and those who did not. This digital exclusion exacerbated educational inequalities, particularly in regions with high poverty rates (UNESCO, 2020).

In addition to economic barriers, geographical location plays a significant role in the digital divide. Rural and remote areas often face challenges in accessing reliable internet infrastructure. According to the World Bank (2020), broadband coverage in rural areas is significantly lower than in urban centers, leading to limited opportunities for digital engagement. This lack of infrastructure not only affects individuals' ability to access information but also impedes their participation in the digital economy, where skills in technology and online platforms are increasingly essential for employment. Moreover, in many low-income countries, the cost of internet access remains prohibitively high, further entrenching digital inequality (Bertot et al., 2016).

Gender also intersects with the digital divide, creating additional barriers for women and girls in many parts of the world. In many developing countries, women are less likely to own mobile phones or have access to the internet compared to men. The International Telecommunication Union (ITU, 2019) reports that globally, there are 250 million fewer women than men who use the internet. This gender gap is particularly pronounced in regions like South Asia and Sub-Saharan Africa, where cultural norms and gender-based violence restrict women's mobility and access to technology. As a result, women are disproportionately excluded from the digital world, limiting their access to educational resources, economic opportunities, and social networks. Bridging the digital gender divide is crucial for achieving gender equality and empowering women to participate fully in the digital economy (Gurumurthy & Chami, 2018).

Furthermore, the digital divide contributes to knowledge inequality by limiting access to quality education and learning materials. Digital technologies have the potential to democratize education by providing access to vast online resources, open educational platforms, and virtual learning environments. However, individuals without access to the internet or digital devices are unable to benefit from these opportunities. This is particularly evident in the context of higher education, where many universities and institutions now offer online courses, degrees, and research opportunities. Those without reliable internet access or digital literacy skills are effectively excluded from these opportunities, which can have long-term consequences for their career prospects and social mobility (Selwyn, 2016). As a result, the digital divide not only perpetuates educational inequalities but also reinforces broader socio-economic disparities.

The digital divide also exacerbates knowledge inequality in the workplace. As digital technologies become integral to many industries, individuals without digital skills or access to technology are at a significant disadvantage. In sectors like finance, healthcare, and technology, digital literacy is increasingly a prerequisite for employment. According to the World Economic Forum (2020), the rise of automation and artificial intelligence has led to a growing demand for workers with advanced digital skills, further marginalizing those who lack access to the necessary tools and training. This trend is particularly evident in low-income and rural areas, where educational and professional development opportunities are limited. The lack of access to digital tools and training perpetuates a cycle of inequality, where those without digital skills are unable to compete in the modern workforce, while those with digital access gain a competitive advantage.

The digital divide is a key driver of knowledge inequality, creating barriers to education, economic participation, and social engagement for millions of people around the world. Addressing this divide requires a multi-faceted approach that includes improving infrastructure, reducing the cost of internet access, and ensuring that marginalized groups, including women and rural populations, have equal access to digital technologies. By bridging the digital divide, we can create a more equitable knowledge economy, where everyone has the opportunity to learn, innovate, and participate in the digital world.

13.3 Reimagining Education for the 21st Century

The rapid pace of technological advancements, globalization, and the changing nature of work have necessitated a fundamental rethinking of education systems worldwide. Traditional models of education, which have largely been based on a one-size-fits-all approach, are increasingly seen as inadequate for preparing students for the complexities and challenges of the 21st century. In response to these evolving demands, there is a growing call for reimagining education to foster critical thinking, creativity, adaptability, and digital literacy. This vision of education focuses on equipping students not only with academic knowledge but also with the skills necessary to navigate an increasingly interconnected and rapidly changing world (Saavedra & Opfer, 2012).

One of the central tenets of reimagining education is the shift from a focus on rote memorization and standardized testing to a more holistic, learner-centered approach. The traditional educational model, which emphasizes a fixed curriculum and uniform testing, is increasingly seen as insufficient in preparing students for the complexities of modern life. Instead, there is a growing emphasis on personalized learning, which tailors educational experiences to individual needs, interests, and learning styles. Personalized learning aims to empower students to take ownership of their education, fostering self-direction and intrinsic motivation (Horn & Staker, 2015). By integrating technology into the learning process, personalized learning can provide students with access to a wealth of resources and tools that cater to their unique learning paths.

In addition to personalized learning, the integration of technology is a key component of reimagining education for the 21st century. Digital tools and online platforms have the potential to transform how students learn, collaborate, and access information. According to the Organisation for Economic Co-operation and Development (OECD, 2018), the effective use of technology in education can enhance learning outcomes, foster collaboration, and prepare students for the digital economy. Online learning platforms, virtual classrooms, and educational apps can provide students with opportunities for flexible, self-paced learning, while also enabling teachers to track progress and provide targeted support. Furthermore, technology can facilitate access to global knowledge networks, allowing students to engage with peers and experts from around the world, thereby fostering a more globalized perspective on education (Zhao, 2012).

However, the integration of technology in education must be approached thoughtfully to avoid exacerbating existing inequalities. The digital divide, which refers to the gap between those who have access to technology and those who do not, remains a significant barrier to educational equity. In many low-income communities and developing countries, students lack access to the devices and internet connectivity needed to fully participate in digital learning environments (Van Dijk, 2020). This divide not only limits access to educational resources but also perpetuates social and economic inequalities. As such, reimagining education for the 21st century requires a concerted effort to ensure that all students, regardless of their socio-economic background, have access to the tools and resources necessary for success in the digital age (Bertot et al., 2016).

In addition to technological integration, reimagining education for the 21st century involves a shift in the skills and competencies that are prioritized in the curriculum. Traditional education systems have long focused on knowledge acquisition in subjects such as mathematics, science, and literature. While these subjects remain important, there is an increasing recognition that students must also develop a broader set of skills to thrive in the modern world. These include critical thinking, problem-solving, creativity, collaboration, and communication (Saavedra & Opfer, 2012). The ability to think critically and adapt to changing circumstances is essential in a world characterized by rapid technological advancements and complex global challenges. As such, reimagining education involves a shift towards fostering these "21st-century skills" alongside traditional academic content.

Furthermore, reimagining education requires a focus on social-emotional learning (SEL) and character development. As the world becomes more interconnected, the ability to work collaboratively with people from diverse backgrounds and cultures is increasingly important. SEL programs aim to teach students skills such as empathy, emotional regulation, and resilience, which are essential for success in both personal and professional life (Durlak et al., 2011). By incorporating SEL into the curriculum, education systems can help students develop the emotional intelligence needed to navigate complex social and professional environments. This focus on the whole child – cognitive, emotional, and social – aligns with the growing recognition that education should prepare students not only for academic success but also for life beyond school (CASEL, 2020).

Finally, reimagining education for the 21st century requires a shift towards lifelong learning. In a rapidly changing world, the skills and knowledge required for success are constantly evolving. As a result, education must no longer be viewed as a one-time event that ends after graduation. Instead, lifelong learning must become a central component of educational systems. This involves creating opportunities for individuals to continue learning throughout their lives, whether through formal education, online courses, or informal learning networks. According to the World Economic Forum (2020), lifelong learning is essential for adapting to the changing demands of the workforce and for ensuring that individuals remain competitive in an increasingly automated and digital economy.

Reimagining education for the 21st century involves a comprehensive shift in how education is delivered, what is taught, and how learning is structured. By focusing on personalized learning, integrating technology, fostering 21st-century skills, promoting social-emotional learning, and supporting lifelong learning, education systems can better prepare students to navigate the complexities of the modern world. However, this vision of education can only be realized if efforts are made to address existing inequalities, ensuring that all students have access to the tools and resources they need to succeed.

As we move forward, it is crucial to continue rethinking and reshaping education to meet the demands of the 21st century.

References

Becker, S. A., Freeman, A., & Karp, K. (2020). *The digital divide in education: A pandemic-era challenge*. Journal of Educational Technology, 42(3), 65–74. https://doi.org/10.1016/j.jedu.2020.05.005

Bertot, J. C., Jaeger, P. T., & McClure, C. R. (2016). Public libraries and the internet: The role of public libraries in providing internet access and services to the underserved. *Information Technology and Libraries, 35*(4), 13–23. https://doi.org/10.6017/ital.v35i4.10232

CASEL. (2020). *What is social and emotional learning?*. Collaborative for Academic, Social, and Emotional Learning. https://casel.org/what-is-sel/

Global Partnership for Education. (2021). *Education and climate change: The impact of climate-related disasters on education*. https://www.globalpartnership.org

Global Partnership for Education. (2021). *Gender equality and education*. https://www.globalpartnership.org

Gurumurthy, A., & Chami, N. (2018). *Bridging the gender digital divide: A policy framework*. GenderIT. https://www.genderit.org

Horn, M. B., & Staker, H. (2015). *Blended: Using disruptive innovation to improve schools*. Jossey-Bass.

International Telecommunication Union (ITU). (2019). *Measuring the gender digital divide*. https://www.itu.int/en/ITU-D/Statistics/Pages/stat/default.aspx

International Telecommunication Union (ITU). (2020). *The state of broadband: Broadband as a foundation for sustainable development*. https://www.itu.int/en/ITU-D/Statistics/Pages/stat/default.aspx

Organisation for Economic Co-operation and Development (OECD). (2018). *The future of education and skills: Education 2030*. OECD. https://www.oecd.org/education/2030-project/

Saavedra, A. R., & Opfer, V. D. (2012). Teaching and learning 21st century skills: Lessons from the learning sciences. *The International Society for Research in Education and Development*, 3(2), 1–17. https://doi.org/10.1007/s11423-012-9261-2

Save the Children. (2020). *Education in emergencies: The impact of conflict on children's education*. https://www.savethechildren.org

Selwyn, N. (2016). *Education and technology: Key issues and debates*. Bloomsbury Publishing.
UNESCO. (2020). *COVID-19 educational disruption and response*. https://en.unesco.org/covid19/educationresponse
UNESCO. (2020). *Global education monitoring report 2020: Inclusion and education: All means all*. https://en.unesco.org/gem-report
UNHCR. (2020). *Global trends: Forced displacement in 2019*. https://www.unhcr.org
UNICEF. (2019). *The state of the world's children 2019: Children, education, and conflict*. https://www.unicef.org
UNICEF. (2020). *Climate change and education: Protecting children's futures*. https://www.unicef.org
United Nations Educational, Scientific and Cultural Organization (UNESCO). (2020). *Global education monitoring report 2020: Inclusion and education: All means all*. https://en.unesco.org/gem-report
United Nations. (2015). *Transforming our world: The 2030 agenda for sustainable development*. https://www.un.org/sustainabledevelopment/education/
United Nations. (2019). *Convention on the Rights of Persons with Disabilities*. https://www.un.org/disabilities
Van Dijk, J. A. (2020). *The digital divide*. Polity Press.
World Bank. (2020). *The digital divide and its impact on development*. https://www.worldbank.org
World Bank. (2020). *The education crisis in low-income countries: Challenges and solutions*. https://www.worldbank.org
World Economic Forum. (2020). *The future of jobs report 2020*. https://www.weforum.org/reports/the-future-of-jobs-report-2020
World Economic Forum. (2020). *The future of jobs report 2020*. World Economic Forum. https://www.weforum.org/reports/the-future-of-jobs-report-2020
World Health Organization. (2018). *Disability and health*. https://www.who.int/news-room/fact-sheets/detail/disability-and-health
Zhao, Y. (2012). *World class learners: Educating creative and entrepreneurial students*. Corwin Press.

PART IV

Political and Economic Crises

14

Political and Economic Crises

Political and economic crises are deeply interconnected phenomena that have profound implications for societies worldwide. These crises often emerge from complex, multifaceted issues such as governance failures, corruption, inequality, and global interdependencies. Political crises can be triggered by weak institutions, lack of accountability, and political instability, while economic crises frequently stem from mismanagement of resources, financial system vulnerabilities, and external shocks. The interrelationship between political and economic instability can exacerbate the effects of each, leading to a vicious cycle of decline, which is difficult to resolve without comprehensive reforms and international cooperation.

Political crises typically arise from systemic failures within the state apparatus, including corruption, authoritarianism, and lack of effective governance. When governments are unable to meet the needs of their populations, political instability ensues, often manifesting in protests, revolutions, or coups. For example, the Arab Spring of 2011, which saw widespread uprisings across the Middle East and North Africa, was driven by public dissatisfaction with authoritarian regimes, corruption, and economic stagnation (Bellin, 2012). Political instability often leads to economic downturns, as investor confidence declines, trade relationships are disrupted, and the ability to enact sound economic policies is hindered. Similarly, in countries like Venezuela, political instability has been a significant factor in the country's severe economic crisis, with hyperinflation, food shortages, and a collapse of public services (Corrales & Penfold, 2015).

Economic crises, on the other hand, typically emerge from systemic imbalances, such as unsustainable debt levels, market speculation, or external economic shocks. The 2008 global financial crisis, for instance, was precipitated by the collapse of the housing bubble in the United States and the subsequent failure of financial institutions that were deeply interconnected with global markets. The crisis led to widespread unemployment, poverty,

DOI: 10.1201/9788743808787-18

and a deep recession in many countries. Economic crises can trigger political crises by undermining the legitimacy of governments that are seen as responsible for the mismanagement of economic policies. For instance, austerity measures imposed in response to the Eurozone debt crisis led to widespread protests in countries like Greece and Spain, as citizens blamed their governments for the economic hardship (Stiglitz, 2016).

One of the most significant challenges in addressing political and economic crises is the role of inequality. Economic inequality often exacerbates political instability, as marginalized groups are less likely to have access to resources, education, or political power. In many cases, political elites may perpetuate economic inequality to maintain their power, creating a cycle of disenfranchisement and unrest. This dynamic was evident during the 2011 Occupy Wall Street movement, which highlighted the growing disparity between the wealthy elite and the rest of society, particularly in the aftermath of the 2008 financial crisis (Piketty, 2014). As inequality deepens, political systems can become more polarized, with citizens losing faith in democratic processes and turning to populist or extremist movements.

Globalization has also played a significant role in shaping both political and economic crises. The interconnectedness of global markets means that economic crises in one region can quickly spread to others, as evidenced by the 1997 Asian financial crisis and the 2008 global recession. Similarly, political crises in one country can have far-reaching effects, particularly in the context of international relations and security. The ongoing political crisis in Syria, for example, has led to regional instability, the displacement of millions of refugees, and the involvement of various international actors with conflicting interests (Friedman, 2014). As nations become more interconnected, the impact of political and economic crises can be amplified, requiring coordinated responses from international organizations and governments.

The role of international institutions in mitigating political and economic crises is critical. Organizations such as the International Monetary Fund (IMF), the World Bank, and the United Nations play a crucial role in providing financial assistance, facilitating diplomacy, and promoting stability in crisis-ridden regions. However, these institutions are often criticized for their approach to crisis management, particularly in developing countries. For instance, the IMF's structural adjustment programs, which have been implemented in countries facing economic crises, have been criticized for imposing austerity measures that disproportionately affect the poor and vulnerable (Stiglitz, 2002). In many cases, the solutions offered by international institutions may not adequately address the root causes of political and economic crises, leading to further instability.

Addressing political and economic crises requires a multifaceted approach that includes both short-term interventions and long-term structural reforms. In the short term, governments may need to implement emergency measures, such as providing financial assistance, stabilizing currency values, or negotiating ceasefires in conflict zones. However, these measures are often insufficient without addressing the underlying causes of instability, such as corruption, inequality, and lack of political accountability. Long-term reforms may include strengthening democratic institutions, promoting inclusive economic growth, and addressing global imbalances in trade and finance. For example, countries like Iceland, which faced a severe economic crisis in 2008, have implemented significant reforms to strengthen their financial systems and increase transparency in government (Gupta & Niskanen, 2014).

In conclusion, political and economic crises are deeply intertwined, with each exacerbating the effects of the other. Political instability often leads to economic downturns, while economic crises can undermine the legitimacy of governments and fuel political unrest. Inequality, globalization, and the role of international institutions are key factors in shaping these crises and their resolution. Addressing political and economic crises requires both immediate interventions and long-term structural reforms aimed at promoting stability, transparency, and inclusive growth. As the world becomes increasingly interconnected, the ability to address these crises in a coordinated and comprehensive manner will be crucial for ensuring global peace and prosperity.

14.1 The Global Economic Crisis

The global economic crisis, often referred to as the Great Recession, which began in 2007–2008, had profound and far-reaching consequences for the global economy, political structures, and social systems. It originated primarily in the financial sector, particularly in the United States, where the collapse of the housing market and the subsequent failure of major financial institutions led to a widespread economic downturn. The crisis quickly spread to other regions due to the interconnectedness of global markets, affecting both developed and developing economies. Understanding the causes, consequences, and responses to the global economic crisis is crucial for comprehending the vulnerabilities of the modern global economy and the challenges in creating sustainable economic systems.

The immediate cause of the global economic crisis was the bursting of the housing bubble in the United States. Leading up to the crisis, banks and financial institutions had engaged in risky lending practices, offering subprime mortgages to borrowers with poor credit histories. These mortgages were then bundled into securities and sold to investors worldwide. When housing prices began to fall, many borrowers defaulted on their loans, and the value of mortgage-backed securities plummeted, triggering a chain reaction that led to the collapse of major financial institutions like Lehman Brothers. This collapse sent shockwaves through the global financial system, as banks became reluctant to lend, leading to a credit freeze and a sharp contraction in economic activity (Gorton, 2010).

The global economic crisis was not confined to the financial sector. It quickly spread to other areas of the economy, leading to a severe recession in many countries. Unemployment rates surged, particularly in countries like the United States and Spain, where millions of people lost their jobs due to the contraction of industries like construction, manufacturing, and finance. In addition to unemployment, the crisis led to a significant decline in global trade, as demand for goods and services plummeted. The International Monetary Fund (IMF) estimated that global trade contracted by 12.2% in 2009, the largest decline since World War II (IMF, 2009). The crisis also led to a sharp decline in consumer confidence, further exacerbating the economic downturn.

The social consequences of the global economic crisis were significant. In many countries, the economic downturn led to increased poverty and inequality, as people lost their jobs, homes, and savings. The World Bank reported that the crisis pushed an additional 64 million people into extreme poverty, reversing years of progress in poverty reduction (World Bank, 2010). In addition, the crisis disproportionately affected marginalized groups, including women, racial minorities, and low-income populations, who were more vulnerable to economic shocks. For example, in the United States, African American and Hispanic communities experienced higher rates of unemployment and home foreclosures compared to white populations (Shapiro, Meschede, & Osoro, 2013).

The global economic crisis also had significant political implications. In many countries, the crisis led to a loss of faith in the ability of governments and financial institutions to manage the economy effectively. In the United States, the crisis contributed to the election of President Barack Obama, who campaigned on a platform of economic recovery and reform. Globally, the crisis sparked debates about the role of government in the economy, with some advocating for more regulation of financial markets, while others called for a reduction in government intervention. The crisis also led to the rise of populist movements in many countries, as people expressed frustration with the political establishment and the perceived failure of elites to protect their interests (Mudde, 2019).

The response to the global economic crisis was multifaceted, involving both national and international efforts to stabilize the economy and prevent further economic collapse. In the United States, the federal government implemented a series of measures, including the Troubled Asset Relief Program (TARP), which provided financial assistance to struggling banks, and the American Recovery and Reinvestment Act (ARRA), which aimed to stimulate economic growth through infrastructure investment and tax cuts. The Federal Reserve also took aggressive actions to lower interest rates and inject liquidity into the financial system. Similarly, the European Union implemented a range of measures, including bailouts for countries like Greece, Ireland, and Portugal, which faced severe financial crises as a result of the global downturn (Buiter, 2010).

One of the key lessons from the global economic crisis was the importance of financial regulation. In the years following the crisis, many countries introduced new regulations aimed at preventing a similar crisis in the future. The Dodd-Frank Wall Street Reform and Consumer Protection Act, passed in the United States in 2010, aimed to reduce the risk of financial instability by increasing oversight of financial institutions and introducing measures to protect consumers from predatory lending practices. Similarly, the Basel III framework, developed by the Basel Committee on Banking Supervision, introduced stricter capital requirements for banks to ensure that they could withstand financial shocks (Basel Committee on Banking Supervision, 2011). While these regulatory reforms have been credited with improving financial stability, they have also been criticized for not going far enough in addressing the underlying causes of the crisis, such as excessive risk-taking and the failure of financial institutions to account for systemic risk (Admati & Hellwig, 2013).

In conclusion, the global economic crisis of 2007–2008 was a watershed moment in the history of the global economy. It highlighted the vulnerabilities of interconnected financial systems and the dangers of unchecked risk-taking in financial markets. The crisis had far-reaching social, political, and economic consequences, including increased poverty and inequality, a loss of faith in political institutions, and a reassessment of the role of government in managing the economy. The response to the crisis involved a combination of government intervention, financial regulation, and international cooperation. While significant progress has been made in addressing some of the issues raised by the crisis, many challenges remain, particularly in terms of ensuring that the global economy is resilient to future shocks and that the benefits of economic growth are more equitably distributed.

14.2 Unemployment and Economic Disruptions

Unemployment and economic disruptions are central concerns in contemporary economies, with far-reaching consequences for individuals, communities, and entire nations. The global economic landscape has witnessed a series of disruptions, including recessions, technological advancements, and the COVID-19 pandemic, all of which have exacerbated unemployment and led to significant economic instability. Understanding the causes and consequences of unemployment, as well as the broader economic disruptions that fuel it, is essential for developing effective policy responses and promoting sustainable economic recovery.

Unemployment can be driven by various factors, including cyclical downturns, technological changes, and structural shifts in the economy. During periods of economic recession, businesses often reduce their workforce to cope with declining demand for goods and services, leading to cyclical unemployment. For example, during the global economic crisis of 2007-2008, millions of jobs were lost as businesses faced financial strain, with unemployment rates reaching historically high levels in many countries (Bureau of Labor Statistics, 2009). Technological advancements, particularly automation and artificial intelligence, have also contributed to structural unemployment, as machines replace human labor in certain industries. This phenomenon is particularly evident in manufacturing sectors, where automation has led to the displacement of workers in both developed and developing economies (Brynjolfsson & McAfee, 2014). Additionally, globalization has created economic disruptions by outsourcing jobs to countries with lower labor costs, further exacerbating unemployment in certain regions (Autor, Dorn, & Hanson, 2016).

The consequences of unemployment extend beyond financial hardship for individuals and families. Unemployment often leads to a cascade of social and psychological effects, including increased stress, mental health issues, and a decline in overall well-being. Research has shown that individuals who experience long-term unemployment are more likely to suffer from depression, anxiety, and other mental health disorders (Paul & Moser, 2009). Moreover, high unemployment rates can contribute to social instability, as individuals who are unable to find work may become disillusioned with the political system and may turn to radical ideologies or social unrest (Blanchard & Leigh, 2013). Economic disruptions caused by widespread unemployment can also have long-term effects on social mobility, as individuals who face prolonged joblessness may find it increasingly difficult to re-enter the workforce, leading to a cycle of poverty and inequality (Moffitt, 2015).

The COVID-19 pandemic has highlighted the vulnerability of labor markets to global disruptions. The pandemic led to unprecedented levels of unemployment as businesses shuttered and entire sectors, such as hospitality, travel, and retail, were severely impacted by lockdown measures and social distancing protocols. According to the International Labour Organization (2020), the pandemic resulted in an 8.8% decline in global working hours in 2020, equivalent to the loss of 255 million full-time jobs. This surge in unemployment was particularly devastating for vulnerable populations, including low-wage workers, women, and minorities, who were more likely to be employed in sectors that were hit hardest by the pandemic (Chung, 2020). The economic disruptions caused by the pandemic also exposed the inadequacies of social safety nets in many countries, as millions of workers faced job losses without sufficient access to unemployment benefits or healthcare.

In response to rising unemployment and economic disruptions, governments have implemented various policy measures aimed at mitigating the impact of job losses and stimulating economic recovery. During the 2007–2008 financial crisis, governments worldwide enacted stimulus packages, including direct financial assistance to individuals and businesses, to stabilize labor markets and promote economic growth. For example, the United States implemented the American Recovery and Reinvestment Act (ARRA) in 2009, which included provisions for unemployment benefits, job training, and public infrastructure projects (Blinder, 2013). Similarly, during the COVID-19 pandemic, governments introduced wage subsidies, unemployment benefits, and direct cash transfers to support workers and businesses affected by the crisis (OECD, 2020). These measures were essential in preventing a more severe economic collapse and mitigating the social impacts of widespread unemployment.

However, despite these efforts, long-term unemployment and economic disruptions remain persistent challenges. Structural changes in the economy, such as the rise of automation and the decline of traditional industries, require new approaches to workforce development and job creation. The future of work will likely involve a shift toward knowledge-based and service-oriented sectors, which may require workers to acquire new skills and adapt to changing technological landscapes (Frey & Osborne, 2017). Furthermore, addressing unemployment in the face of economic disruptions requires a comprehensive approach that includes not only economic policies but also social and psychological support for individuals facing job loss. Initiatives such as job retraining programs, mental health services, and the promotion of social entrepreneurship can play a critical role in supporting individuals and communities affected by unemployment.

In conclusion, unemployment and economic disruptions are interconnected phenomena that have profound implications for individuals, societies, and economies. While cyclical downturns, technological advancements, and global crises such as the COVID-19 pandemic can lead to job losses and economic instability, effective policy responses and social support systems are essential for mitigating the impact of unemployment. As economies continue to evolve, it is crucial to develop strategies that promote job creation, workforce adaptability, and social inclusion to ensure a sustainable and resilient future.

14.3 The Challenges of Globalization and Trade

Globalization and trade have reshaped the global economic landscape, offering unprecedented opportunities for economic growth, innovation, and interconnectedness. However, these phenomena also present significant challenges that can exacerbate inequalities, undermine local industries, and create environmental and social risks. Understanding the complexities of globalization and trade is essential for developing policies that promote equitable development, sustainable practices, and social cohesion in an increasingly interconnected world.

One of the most significant challenges of globalization is the widening gap between rich and poor nations. While globalization has spurred economic growth in many developing countries, it has also led to increasing wealth disparities both within and between countries (Piketty, 2014). Multinational corporations and global financial institutions often benefit disproportionately from globalization, while local industries in developing nations struggle to compete with cheaper, mass-produced goods from more industrialized economies. For example, the liberalization of trade policies in the 1990s, such as the North American Free Trade Agreement (NAFTA), led to significant job losses in sectors like agriculture and manufacturing in countries like Mexico, as local producers were unable to compete with the influx of subsidized goods from the United States (Harrison, 2013). This phenomenon, often referred to as "race to the bottom," occurs when countries lower labor standards, environmental regulations, and wages to attract foreign investment, further deepening inequality and exploitation (Rodrik, 2011).

Another critical challenge of globalization is the erosion of local cultures and traditions. As global trade encourages the spread of Western consumer goods, media, and values, local cultures and industries often face pressures to conform to global standards. This cultural homogenization can result in the loss of indigenous knowledge, languages, and traditions, which are vital components of cultural identity (Tomlinson, 1999). Additionally, the dominance of multinational corporations, particularly in the food, fashion, and entertainment industries, can overshadow local businesses, leading to a decline in cultural diversity and the marginalization of small-scale producers (Scholte, 2000). While globalization has undoubtedly facilitated the exchange of ideas and cultural practices, it has also contributed to the dominance of a Western-centric worldview, which can undermine the richness and diversity of global cultures.

Environmental degradation is another significant challenge associated with globalization and trade. The expansion of global trade has led to increased production and consumption, which, in turn, has driven higher levels of resource extraction, waste generation, and pollution. The global supply chains that characterize modern trade often rely on cheap, unsustainable practices, including deforestation, overfishing, and the exploitation of natural resources in developing countries (Sachs, 2015). For example, the global demand for palm oil, a key ingredient in many consumer products, has contributed to widespread deforestation in Southeast Asia, threatening biodiversity and exacerbating climate change (Vijay et al., 2016). Moreover, the environmental costs of globalization are often externalized, meaning that the negative impacts of trade are borne by the poorest and most vulnerable populations, who are least equipped to address these challenges (Stiglitz, 2002).

Globalization and trade also have significant implications for labor markets. While trade liberalization can create new job opportunities in some sectors, it can also lead to job displacement in others. The offshoring of manufacturing jobs to countries with lower labor costs is one of the most visible consequences of globalization. This trend has led to the decline of traditional industries in developed nations, particularly in sectors like textiles, electronics, and automotive manufacturing (Autor, Dorn, & Hanson, 2013). As companies seek to minimize labor costs by outsourcing production to countries with weaker labor protections, workers in high-income countries face job insecurity, wage stagnation, and declining working conditions. In contrast, workers in low-income countries often face exploitation, poor working conditions, and limited labor rights, as multinational corporations prioritize profit over worker welfare (Klein, 2000). This global "race to the bottom" has raised concerns about the future of work, labor standards, and the equitable distribution of the benefits of trade.

Finally, the governance of globalization and trade remains a significant challenge. International trade agreements, such as those brokered by the World Trade Organization (WTO), often prioritize economic growth and market liberalization over social and environmental concerns. Critics argue that these agreements disproportionately benefit wealthy nations and multinational corporations, while neglecting the needs of developing countries and marginalized communities (Chang, 2002). For example, the WTO's intellectual property rules have been criticized for limiting access to affordable medicines in developing countries, as patents prevent the production of generic drugs (Kapczynski, 2012). Additionally, trade agreements often lack sufficient provisions for environmental protection, labor rights, and social justice, leading to a situation where economic growth comes at the expense of social and environmental well-being (Rodrik, 2011). The challenge, therefore, is to reform global trade governance to ensure that trade policies are inclusive, equitable, and sustainable.

In conclusion, while globalization and trade have facilitated economic growth, technological innovation, and cultural exchange, they also present significant challenges that must be addressed to ensure a more equitable and sustainable global system. The widening wealth gap, environmental degradation, cultural homogenization, labor exploitation, and governance issues all require comprehensive solutions that prioritize social justice, environmental sustainability, and the well-being of marginalized communities. By reimagining global trade and investing in policies that promote inclusive growth, fair labor standards, and environmental protection, it is possible to mitigate the negative impacts of globalization and create a more equitable and sustainable global economy.

14.4 Building Inclusive Economies

Building inclusive economies is essential for fostering long-term social stability, reducing inequality, and ensuring that all individuals have access to the opportunities necessary to achieve economic well-being. An inclusive economy promotes equitable growth, where the benefits of economic development are shared broadly across different social, ethnic, and economic groups. It is characterized by policies that ensure equal access to resources, opportunities, and outcomes for all members of society, regardless of their gender, race, income level, or geographic location (Stiglitz, 2012). Achieving inclusive economic growth requires addressing systemic inequalities, improving access to education and healthcare, and creating robust social safety nets that provide support for the most vulnerable populations.

A fundamental aspect of building inclusive economies is addressing income inequality. Income inequality has been rising globally, with the wealthiest individuals and corporations capturing a disproportionate share of economic gains, while the poorest groups continue to experience stagnating or declining incomes (Piketty, 2014). This growing disparity undermines social cohesion and impedes economic development, as it limits access to opportunities for large segments of the population. To build an inclusive economy, governments must implement progressive tax policies, promote fair wages, and invest in social programs that provide a safety net for those who are most in need (Atkinson, 2015). Additionally, ensuring that economic growth is inclusive requires creating opportunities for wealth generation among marginalized groups, such as women, minorities, and people with disabilities. This can be achieved through targeted programs that improve access to capital, training, and markets, thus enabling these groups to participate more fully in the economy (World Bank, 2020).

Education and skills development are central to fostering inclusive economies. Access to quality education is a key determinant of economic mobility, yet millions of people, particularly in low-income countries, remain excluded from educational opportunities due to poverty, gender discrimination, or geographical barriers (UNESCO, 2014). By investing in education systems that are inclusive, equitable, and responsive to the needs of diverse populations, countries can equip individuals with the skills necessary to participate in the modern economy. This includes not only formal education but also vocational training and lifelong learning opportunities that allow workers to adapt to changing job markets and technological advancements (OECD, 2019). Additionally, inclusive education policies should focus on reducing gender disparities in education and promoting the inclusion of marginalized groups, such as ethnic minorities and people with disabilities, to ensure that all individuals can access the skills and knowledge needed to thrive in an increasingly globalized economy.

Healthcare is another critical pillar of inclusive economies. Access to affordable and quality healthcare is a fundamental human right and essential for enabling individuals to contribute fully to the economy. Poor health can lead to lost productivity, higher healthcare costs, and reduced economic opportunities, particularly for vulnerable groups such as women, children, and the elderly (World Health Organization [WHO], 2013). Building inclusive economies requires expanding access to healthcare services, improving the quality of care, and addressing the social determinants of health, such as nutrition, sanitation, and housing. Universal health coverage (UHC) is an important goal for achieving inclusive economic growth, as it ensures that all individuals, regardless of their income or social status, can access the healthcare services they need without facing financial hardship (Kutzin, 2013).

Social protection systems are also integral to building inclusive economies. These systems provide a safety net for individuals who are unemployed, disabled, elderly, or otherwise unable to participate in the workforce. Social protection programs, such as unemployment benefits, pensions, and disability allowances, play a crucial role in reducing poverty and inequality by ensuring that individuals and families have a basic level of income security (Barrientos, 2013). In addition to income support, social protection systems can include access to public services such as healthcare, education, and housing, which are essential for promoting social inclusion and economic mobility. Strengthening social protection systems is particularly important in the context of globalization, as it helps mitigate the risks associated with economic shocks, technological change, and market volatility (Devereux & Sabates-Wheeler, 2004).

Environmental sustainability is another key component of building inclusive economies. Economic growth that depletes natural resources or causes environmental degradation ultimately undermines the long-term prospects for inclusive development. Inclusive economies must prioritize sustainable development practices that balance economic growth with environmental stewardship, ensuring that future generations inherit a planet capable of supporting human life and prosperity (Sachs, 2015). This includes promoting green technologies, renewable energy, and sustainable agriculture, as well as addressing the environmental impacts of industrialization and urbanization. Furthermore, policies that integrate environmental sustainability into economic planning can create new opportunities for job creation and innovation, particularly in sectors such as clean energy, waste management, and conservation (Stiglitz, 2017).

Finally, inclusive economies require strong institutions that can promote good governance, ensure the rule of law, and protect human rights. Effective institutions are essential for creating an environment in which economic opportunities are accessible to all and where corruption, discrimination, and injustice are minimized (Acemoglu & Robinson, 2012). Strengthening democratic institutions, promoting transparency, and ensuring that public policies are inclusive and responsive to the needs of all citizens are critical for achieving inclusive economic growth. Moreover, fostering social trust and cooperation between governments, businesses, and civil society organizations is essential for building an economy that is both inclusive and sustainable.

In conclusion, building inclusive economies is a multifaceted challenge that requires addressing inequality, investing in education and healthcare, expanding social protection systems, promoting environmental sustainability, and strengthening institutions. By creating an economic system that provides opportunities for all individuals to participate and prosper, societies can foster long-term social stability, reduce poverty, and promote sustainable development. Achieving inclusive economic growth requires comprehensive policies that prioritize equity, social justice, and the well-being of marginalized populations, ensuring that the benefits of economic development are shared broadly across society.

References

Acemoglu, D., & Robinson, J. A. (2012). *Why nations fail: The origins of power, prosperity, and poverty*. Crown Business.

Admati, A. R., & Hellwig, M. F. (2013). *The bankers' new clothes: What's wrong with banking and what to do about it*. Princeton University Press.

Atkinson, A. B. (2015). *Inequality: What can be done?* Harvard University Press.

Autor, D. H., Dorn, D., & Hanson, G. H. (2013). The China shock: Learning from labor-market adjustment to large changes in trade. *Annual Review of Economics, 8*, 205–240. https://doi.org/10.1146/annurev-economics-080315-015041

Autor, D. H., Dorn, D., & Hanson, G. H. (2016). The China shock: Learning from labor-market adjustment to large changes in trade. *Annual Review of Economics, 8*, 205–240. https://doi.org/10.1146/annurev-economics-080315-015041

Barrientos, A. (2013). Social protection and poverty. *International Journal of Social Welfare, 22*(4), 342–354. https://doi.org/10.1111/ijsw.12016

Basel Committee on Banking Supervision. (2011). *Basel III: A global regulatory framework for more resilient banks and banking systems*. Bank for International Settlements.

Bellin, E. (2012). *Reconsidering the robustness of authoritarianism in the Middle East: Lessons from the Arab Spring*. Comparative Politics, 44(2), 127–149.

Blanchard, O., & Leigh, D. (2013). Growth forecast errors and fiscal multipliers. *American Economic Review, 103*(3), 117–120. https://doi.org/10.1257/aer.103.3.117

Blinder, A. S. (2013). *After the music stopped: The financial crisis, the response, and the work ahead*. Penguin Press.

Brynjolfsson, E., & McAfee, A. (2014). *The second machine age: Work, progress, and prosperity in a time of brilliant technologies*. W. W. Norton & Company.

Buiter, W. H. (2010). *The global economic crisis: Causes, cures, and systemic implications*. Journal of Globalization and Development, 1(2), 1–42.

Bureau of Labor Statistics. (2009). *The employment situation: November 2009*. U.S. Department of Labor. https://www.bls.gov/news.release/archives/empsit_12042009.pdf

Chang, H. J. (2002). *Kicking away the ladder: Development strategy in historical perspective*. Anthem Press.

Chung, H. (2020). The COVID-19 pandemic and gender equality. *International Labour Review, 159*(3), 213–236. https://doi.org/10.1111/ilr.12174

Corrales, J., & Penfold, M. (2015). *Venezuela: The rise and fall of an oil state*. Brookings Institution Press.

Devereux, S., & Sabates-Wheeler, R. (2004). Transformative social protection. *Institute of Development Studies*.

Frey, C. B., & Osborne, M. A. (2017). The future of employment: How susceptible are jobs to computerization? *Technological Forecasting and Social Change, 114*, 254–280. https://doi.org/10.1016/j.techfore.2016.08.019

Friedman, T. L. (2014). *The world is flat 3.0: A brief history of the twenty-first century*. Picador.

Gorton, G. (2010). *Slapped by the invisible hand: The panic of 2007*. Oxford University Press.

Gupta, A. K., & Niskanen, W. A. (2014). *The Icelandic financial crisis: A study in the resilience of democratic institutions*. Cato Journal, 34(3), 391–409.

Harrison, A. (2013). *Globalization and poverty*. University of Chicago Press.

IMF. (2009). *World economic outlook: Crisis and recovery*. International Monetary Fund.

International Labor Organization. (2020). *ILO monitor: COVID-19 and the world of work. Sixth edition*. International Labor Organization. https://www.ilo.org/global/topics/coronavirus/impacts-and-responses/lang--en/index.htm

Kapczynski, A. (2012). The access to medicines controversy: A history of the global debate over intellectual property, access to medicines, and public health. *Health and Human Rights, 14*(2), 1–22.

Klein, N. (2000). *No logo: Taking aim at the brand bullies*. Picador.

Kutzin, J. (2013). Health financing for universal coverage and health system performance: Concepts and implications for policy. *Bulletin of the World Health Organization, 91*(8), 602–611. https://doi.org/10.2471/BLT.12.113985

Moffitt, R. A. (2015). *The US social safety net and the Great Recession*. Brookings Institution Press.

Mudde, C. (2019). *The far right today*. Polity.

OECD. (2019). *Skills for the digital economy: A global view*. OECD Publishing.

OECD. (2020). *The impact of COVID-19 on jobs and income*. Organisation for Economic Co-operation and Development. https://www.oecd.org/coronavirus/en/

Paul, K. I., & Moser, K. (2009). Unemployment impairs mental health: Meta-analyses. *Journal of Vocational Behavior, 74*(3), 264–282. https://doi.org/10.1016/j.jvb.2009.01.003

Piketty, T. (2014). *Capital in the twenty-first century*. Harvard University Press.

Rodrik, D. (2011). *The globalization paradox: Democracy and the future of the world economy*. W. W. Norton & Company.

Sachs, J. D. (2015). *The age of sustainable development*. Columbia University Press.

Scholte, J. A. (2000). *Globalization: A critical introduction*. Palgrave Macmillan.

Shapiro, T. M., Meschede, T., & Osoro, S. (2013). *The roots of the widening racial wealth gap: Explaining the black-white economic divide*. Institute on Assets and Social Policy.

Stiglitz, J. E. (2002). *Globalization and its discontents*. W.W. Norton & Company.

Stiglitz, J. E. (2012). *The price of inequality: How today's divided society endangers our future*. W.W. Norton & Company.

Stiglitz, J. E. (2016). *The Euro: How a common currency threatens the future of Europe*. W.W. Norton & Company.

Stiglitz, J. E. (2017). *Globalization and its discontents revisited: Anti-globalization in the era of Trump*. W.W. Norton & Company.

Tomlinson, J. (1999). *Globalization and culture*. University of Chicago Press.

UNESCO. (2014). *Education for all global monitoring report 2013/4: Teaching and learning: Achieving quality for all*. UNESCO.

Vijay, V., Pimm, S. L., Jenkins, C. N., & Smith, S. E. (2016). The impacts of oil palm on recent deforestation and biodiversity loss. *PLoS ONE, 11*(7), e0159668. https://doi.org/10.1371/journal.pone.0159668

WHO. (2013). *Health systems strengthening and universal health coverage*. World Health Organization.

World Bank. (2010). *The impact of the global financial crisis on poverty in developing countries*. World Bank.

15

Political Instability and Governance

Political instability is a critical factor that can undermine a country's economic development, social cohesion, and democratic processes. It refers to situations where governments are unable to maintain control over the political system due to internal conflicts, corruption, weak institutions, or external pressures. Political instability often manifests in the form of coups, civil wars, widespread protests, or other forms of unrest that disrupt the normal functioning of governance. These disruptions hinder the ability of governments to implement effective policies, maintain law and order, and provide public goods and services to their citizens (Acemoglu & Robinson, 2012). Furthermore, political instability is often associated with corruption, human rights violations, and the breakdown of legal and social institutions, all of which can exacerbate poverty and inequality.

One of the main consequences of political instability is the erosion of governance structures. Governance refers to the processes, structures, and institutions through which decisions are made and implemented in a society (Grindle, 2004). Effective governance requires strong institutions that are capable of enforcing the rule of law, ensuring accountability, and promoting transparency. In politically unstable environments, these institutions are often weak or non-functional, leading to a lack of trust in government and a diminished capacity to address the needs of the population. For example, corruption may flourish in unstable political systems, as politicians and bureaucrats exploit their positions for personal gain, thereby undermining the effectiveness of public administration (Kaufmann, Kraay, & Mastruzzi, 2010). As a result, citizens may become disillusioned with the political process, which can lead to a cycle of instability and governance failure.

Another key aspect of political instability is its impact on economic development. Political instability can create an uncertain environment that discourages both domestic and foreign investment. Investors are typically reluctant to invest in countries where the political environment is volatile, as

DOI: 10.1201/9788743808787-19

the risks of losing capital or facing expropriation are high. This can lead to a decline in economic growth, as businesses are less likely to expand, and job creation is limited (Alesina, Ozler, Roubini, & Swagel, 1996). Furthermore, political instability often leads to the diversion of resources away from development projects and social services toward security and defense expenditures. In conflict-prone regions, for example, governments may allocate a significant portion of their budgets to military spending, further exacerbating poverty and underdevelopment (Collier, 2007).

Political instability also contributes to social fragmentation and conflict. In many cases, instability is driven by ethnic, religious, or ideological divisions within society. When these divisions are not effectively managed, they can escalate into violent conflicts, leading to widespread displacement, loss of life, and the destruction of infrastructure. For example, the Rwandan genocide in 1994 was a result of long-standing ethnic tensions that were exacerbated by political instability and weak governance (Prunier, 1995). Similarly, the ongoing conflict in Syria has been fueled by a combination of political repression, sectarian divisions, and external interventions, leading to one of the most devastating humanitarian crises of the 21st century (Fisk, 2008). In such environments, governance becomes even more challenging, as governments struggle to maintain control over diverse and often fragmented societies.

In addition to its economic and social consequences, political instability can also undermine democratic processes. In countries where political instability is prevalent, there is often a tendency toward authoritarianism or the erosion of democratic norms. Leaders may consolidate power by undermining democratic institutions, suppressing opposition, and curbing civil liberties. For instance, in many countries experiencing political instability, elections are often manipulated or rigged, leading to the marginalization of opposition parties and the stifling of political dissent (Diamond, 2008). In some cases, political instability can lead to the collapse of democratic institutions altogether, as evidenced by the rise of military juntas or the establishment of autocratic regimes in countries such as Myanmar (Hlaing, 2017). This deterioration of democratic governance not only undermines political stability but also weakens the rule of law and the protection of human rights.

To address political instability and improve governance, it is essential to strengthen the institutions that underpin the political system. This includes promoting transparency, accountability, and the rule of law, as well as ensuring that political processes are inclusive and representative of the diverse interests within society. Effective governance requires the development of strong legal frameworks that can address corruption, protect human rights, and ensure that the government is accountable to its citizens. Furthermore, international organizations and development agencies can play a critical role in supporting political stability by providing technical assistance, promoting democratic reforms, and facilitating dialogue between conflicting groups (UNDP, 2016). While political instability may be difficult to eliminate entirely, efforts to improve governance and strengthen institutions can help mitigate its negative effects and create a more stable and prosperous society.

In conclusion, political instability poses significant challenges to governance, economic development, and social cohesion. It erodes trust in institutions, hinders economic growth, and exacerbates social divisions. Addressing political instability requires comprehensive efforts to strengthen governance structures, promote transparency and accountability, and foster inclusive political processes. By addressing the root causes of instability and investing in the development of strong, democratic institutions, countries can work toward achieving long-term political stability and sustainable development.

15.1 The Rise of Populism and Authoritarianism

The rise of populism and authoritarianism has been a defining feature of global politics in recent years. Populism refers to political movements that claim to represent the will of the "common people" against the perceived elite, often framing political issues in terms of a dichotomy between the people and the establishment (Mudde & Kaltwasser, 2017). Authoritarianism, on the other hand, refers to a system of governance characterized by centralized power, limited political freedoms, and the suppression of opposition (Linz, 2000). The two phenomena often intersect, with populist leaders adopting authoritarian tactics to consolidate power and undermine democratic institutions. The rise of populism and authoritarianism has been driven by a combination of economic, social, and political factors, and has had profound implications for the functioning of democracies around the world.

One of the primary drivers of the rise of populism is economic inequality and dissatisfaction with the existing political order. Many populist movements have gained traction in response to perceived economic marginalization, particularly in the wake of the 2008 global financial crisis. Economic dislocation, rising unemployment, and stagnating wages have fueled resentment toward elites and established political parties, which are often blamed for failing to address the needs of the working class (Rooduijn, 2018). Populist leaders have capitalized on this discontent by promising to restore power to the people and challenge the influence of elites, often by framing themselves as outsiders who are untainted by the corruption of the political establishment

(Mudde, 2004). This rhetoric resonates with voters who feel left behind by globalization and the neoliberal economic policies that have dominated much of the Western world in recent decades.

The rise of populism is also closely linked to cultural and identity politics. In many cases, populist leaders have exploited fears about immigration, national identity, and cultural change to galvanize support. Populist movements often frame issues such as immigration and multiculturalism as threats to the cultural and economic well-being of the "native" population, positioning themselves as defenders of national identity against external forces (Inglehart & Norris, 2016). This has been particularly evident in Europe and the United States, where populist parties have capitalized on growing fears about the impact of immigration on social cohesion and job security. By appealing to cultural anxieties, populist leaders have been able to mobilize large segments of the population who feel that their way of life is under threat.

Authoritarianism, meanwhile, has been fueled by the erosion of democratic norms and institutions. In many countries, populist leaders have used their electoral victories as a mandate to weaken democratic institutions and concentrate power in the hands of the executive. This has included efforts to undermine the independence of the judiciary, restrict media freedom, and limit political opposition (Foa & Mounk, 2016). In some cases, populist leaders have used authoritarian tactics to silence dissent, including the use of violence, intimidation, and legal persecution of political opponents. The rise of populism has thus been accompanied by a gradual erosion of democratic norms, with populist leaders often positioning themselves as the defenders of democracy while simultaneously undermining the very institutions that sustain it.

The global rise of populism and authoritarianism has been facilitated by the increasing polarization of political discourse. In many countries, political debates have become more divisive, with populist leaders framing issues in terms of an "us versus them" mentality. This has contributed to a breakdown in social cohesion, as citizens become more polarized and distrustful of one another (Pew Research Center, 2017). Populist leaders often exploit this polarization by using inflammatory rhetoric to stoke divisions and rally their base, while demonizing political opponents and the media. This polarization is further exacerbated by the rise of social media, which has created echo chambers where individuals are exposed primarily to information that reinforces their existing beliefs, further entrenching ideological divisions (Tufekci, 2017).

The rise of populism and authoritarianism has profound implications for the future of democracy. While populist leaders often frame their rise as a response to the failure of traditional political elites, their policies and actions can undermine the very foundations of democratic governance. By concentrating power in the hands of the executive, undermining checks and balances, and eroding the independence of institutions, populist and authoritarian leaders can weaken democratic institutions and make it more difficult for citizens to hold their leaders accountable (Mounk, 2018). This trend is particularly concerning in countries where democratic norms are still fragile, as the erosion of democratic institutions can set the stage for the rise of more authoritarian regimes.

In conclusion, the rise of populism and authoritarianism is a complex and multifaceted phenomenon that has been driven by economic, social, and political factors. Populist leaders have capitalized on economic inequality, cultural anxieties, and dissatisfaction with the political establishment to gain support, while authoritarianism has been fueled by the erosion of democratic norms and institutions. The rise of populism and authoritarianism poses significant challenges to democracy, as it can lead to the concentration of power, the weakening of democratic institutions, and the erosion of political freedoms. To address this challenge, it is essential to strengthen democratic institutions, promote political inclusivity, and foster a more informed and engaged citizenry.

15.2 Corruption and the Erosion of Trust

Corruption is a pervasive issue that undermines the integrity of institutions, erodes public trust, and impedes societal development. It can be defined as the abuse of power for personal gain, and it exists in various forms, including bribery, embezzlement, nepotism, and favoritism (Rose-Ackerman, 1999). Corruption affects both the public and private sectors, influencing the behavior of political leaders, bureaucrats, business executives, and ordinary citizens. The consequences of corruption are far-reaching, as it not only distorts economic systems but also weakens democratic institutions, exacerbates inequality, and fosters political instability (Transparency International, 2020). One of the most significant consequences of corruption is the erosion of trust between citizens and their governments, which has profound implications for the functioning of democratic societies.

The relationship between corruption and the erosion of trust is well-documented in academic literature. Corruption undermines the legitimacy of political institutions by creating a perception that government actions are driven by personal interests rather than the public good. When citizens perceive that their leaders are corrupt, they are less likely to trust them or the institutions they represent (Rothstein & Stolle, 2008). Trust in government is a cornerstone of democratic governance, as it fosters political stability, encourages civic participation, and enhances the effectiveness of public policies (Putnam, 2000). When corruption is widespread, citizens may become disillusioned with the political process, leading to lower levels of political engagement, voter turnout, and support for democratic institutions (Miller, 2017). This erosion of trust can create a vicious cycle, where citizens' lack of trust in government leads to weaker institutions, which in turn allows corruption to thrive.

The economic impact of corruption further exacerbates the erosion of trust. Corruption distorts market mechanisms by diverting public resources away from essential services and infrastructure projects, such as healthcare, education, and transportation. This misallocation of resources leads to inefficiencies, increased costs, and a lack of public goods, which disproportionately affects marginalized groups (Gupta, Davoodi, & Tiongson, 2002). In addition, corruption increases the cost of doing business, as companies may need to pay bribes to secure contracts or navigate bureaucratic hurdles (Kaufmann, 2005). This creates an uneven playing field, where those with political connections or financial resources have an advantage, further deepening inequality and social divisions. As a result, citizens may feel that the system is rigged against them, reinforcing their distrust in both the government and market.

Corruption also weakens the rule of law and undermines the effectiveness of public institutions. In countries where corruption is widespread, the legal system may be compromised, with judges and law enforcement officers accepting bribes or turning a blind eye to illegal activities (La Porta et al., 1999). This lack of accountability erodes the public's confidence in the ability of institutions to deliver justice and protect their rights. The absence of a functioning legal system can also lead to an increase in organized crime and violence, further destabilizing society and reducing trust in both the government and the justice system (Klitgaard, 1988). In such environments, citizens may resort to informal or illegal means to resolve disputes, further undermining the legitimacy of formal institutions.

The media plays a crucial role in exposing corruption and holding public officials accountable. However, in many countries, the media itself is subject to corruption, either through direct bribery or through government control (Carothers, 2006). This creates a situation where the public is unable to access accurate information about government activities, further eroding trust in political institutions. In some cases, the media may even contribute to the spread of corruption by sensationalizing scandals or focusing on individual wrongdoers rather than addressing the systemic nature of corruption (Harrison, 2010). This lack of transparency can make it difficult for citizens to hold their leaders accountable, reinforcing the perception that corruption is an inevitable part of political life.

The erosion of trust due to corruption can have long-term consequences for democratic governance and social cohesion. When trust in government is low, citizens may become more susceptible to populist or authoritarian leaders who promise to restore order and eliminate corruption. These leaders often use rhetoric that appeals to public frustration with the political establishment, but their solutions may undermine democratic norms and institutions. For example, populist leaders may concentrate power in the executive branch, weaken checks and balances, and curtail press freedom in the name of fighting corruption (Mudde & Kaltwasser, 2017). While these leaders may gain short-term popularity, their actions can further erode trust in government and deepen the political and social divisions within society.

Addressing corruption and restoring trust in government requires a multifaceted approach. Strengthening democratic institutions, improving transparency, and ensuring accountability are key steps in combating corruption and rebuilding public trust (Treisman, 2000). Civil society organizations, the media, and international institutions can play a critical role in exposing corruption and advocating for reforms. Moreover, promoting civic education and encouraging political participation can help citizens better understand the importance of trust in government and the need for integrity in public life (Rose-Ackerman, 2008). Ultimately, rebuilding trust in government requires a commitment to good governance, the rule of law, and the protection of human rights.

15.3 Strengthening Democratic Institutions

Strengthening democratic institutions is a fundamental aspect of fostering stable, inclusive, and effective governance. Democratic institutions, which include the executive, legislature, judiciary, and civil society, form the backbone of any democratic system. These institutions ensure that power is exercised in a transparent, accountable, and equitable manner, safeguarding the rights of citizens and promoting the public good. However, the effectiveness of democratic institutions is often challenged by issues such as corruption, political polarization, and weak rule of law. As such, strengthening these institutions requires comprehensive reforms aimed at enhancing transparency, accountability, inclusivity, and public trust.

One of the key strategies for strengthening democratic institutions is enhancing transparency. Transparency in government operations, decision-making processes, and public spending is essential for ensuring that citizens have access to the information needed to hold their leaders accountable. Transparent institutions are less likely to engage in corrupt practices and are more likely to gain the trust of the public (Treisman, 2000). For example, the establishment of independent audit institutions, open data platforms, and freedom of information laws can increase the transparency of government actions, making it easier for citizens and civil society organizations to monitor and scrutinize the activities of public officials (Bertot, Jaeger, & Grimes, 2010). In addition, transparent electoral processes are critical for ensuring that elections are free, fair, and credible, which is a cornerstone of democratic legitimacy.

Accountability is another critical component of strong democratic institutions. In a healthy democracy, public officials are held accountable for their actions through mechanisms such as regular elections, checks and balances, and independent oversight bodies. Accountability helps to prevent the abuse of power and ensures that government decisions align with the needs and interests of the public. Strengthening accountability involves not only improving the functioning of institutions such as parliaments, courts, and regulatory bodies but also empowering citizens to actively engage in the democratic process (Schedler, 1999). This can be achieved through measures such as strengthening civil society organizations, protecting press freedom, and promoting political participation, which enable citizens to hold their leaders accountable for their actions.

The rule of law is another cornerstone of democratic governance. A strong and independent judiciary is essential for ensuring that laws are applied fairly and consistently, protecting citizens' rights and freedoms. The rule of law also ensures that no one is above the law, including political leaders and public officials. Strengthening the rule of law involves reforming legal systems to ensure that they are accessible, impartial, and free from political interference (Miller, 2017). Judicial independence, access to justice, and the protection of human rights are all critical elements of the rule of law that must be safeguarded to ensure that democratic institutions function effectively. Furthermore, the rule of law supports economic development by providing a stable legal environment in which businesses can operate and individuals can resolve disputes peacefully.

In addition to transparency, accountability, and the rule of law, inclusivity is a key factor in strengthening democratic institutions. Inclusive democratic institutions ensure that all segments of society, including marginalized and vulnerable groups, have a voice in the political process. This involves removing barriers to political participation, such as discrimination, voter suppression, and unequal access to education and resources (Sen, 1999). Inclusive institutions also promote social justice by ensuring that policies and decisions reflect the diverse needs and interests of the population. For example, gender quotas, affirmative action programs, and other forms of positive discrimination can help to increase the representation of women, ethnic minorities, and other underrepresented groups in political decision-making (Krook & O'Brien, 2012). By fostering inclusivity, democratic institutions can better address the needs of all citizens and promote social cohesion.

The role of civil society in strengthening democratic institutions cannot be overstated. Civil society organizations, including non-governmental organizations (NGOs), advocacy groups, and grassroots movements, play a vital role in holding governments accountable, promoting transparency, and ensuring that the voices of marginalized groups are heard. A vibrant civil society fosters political engagement and encourages citizens to participate in democratic processes. Moreover, civil society organizations can serve as watchdogs, monitoring government actions and advocating for policy reforms that promote democratic values (Diamond, 1999). Strengthening civil society involves creating an enabling environment for NGOs and activists to operate freely, without fear of repression or censorship.

Finally, the international community has an important role to play in strengthening democratic institutions. International organizations, such as the United Nations, World Bank, and regional bodies like the European Union, can provide technical assistance, financial support, and diplomatic pressure to help countries build and strengthen democratic institutions (Fukuyama, 2004). Additionally, international human rights standards and conventions can serve as benchmarks for assessing the effectiveness of democratic institutions and ensuring that governments uphold their commitments to democracy, human rights, and the rule of law.

In conclusion, strengthening democratic institutions is essential for ensuring that democracy functions effectively and that citizens' rights and freedoms are protected. This requires comprehensive reforms aimed at enhancing transparency, accountability, inclusivity, and the rule of law. By fostering strong democratic institutions, societies can promote good governance, reduce corruption, and create a more equitable and just political system. Moreover, strengthening democratic institutions is a continuous process that requires the active participation of citizens, civil society, and the international community.

References

Acemoglu, D., & Robinson, J. A. (2012). *Why nations fail: The origins of power, prosperity, and poverty*. Crown Business.

Alesina, A., Ozler, S., Roubini, N., & Swagel, P. (1996). Political instability and economic growth. *Journal of Economic Growth, 1*(2), 189–211. https://doi.org/10.1007/BF00138863

Bertot, J. C., Jaeger, P. T., & Grimes, J. M. (2010). *Using ICTs to create a culture of transparency: E-government and social media as openness and anti-corruption tools for societies. Government Information Quarterly, 27*(3), 264–271. https://doi.org/10.1016/j.giq.2010.03.001

Carothers, T. (2006). *The media and democracy: A global perspective*. Carnegie Endowment for International Peace.

Collier, P. (2007). *The bottom billion: Why the poorest countries are failing and what can be done about it*. Oxford University Press.

Diamond, L. (1999). *Developing democracy: Toward consolidation*. Johns Hopkins University Press.

Diamond, L. (2008). The spirit of democracy: The struggle to build free societies throughout the world. *Henry Holt and Company*.

Fisk, R. (2008). *The great war for civilization: The conquest of the Middle East*. Alfred A. Knopf.

Foa, R. S., & Mounk, Y. (2016). The democratic disconnect. *Journal of Democracy, 27*(3), 5–17. https://doi.org/10.1353/jod.2016.0043

Fukuyama, F. (2004). *State-building: Governance and world order in the 21st century*. Cornell University Press.

Grindle, M. S. (2004). *Good enough governance: Poverty reduction and reform in developing countries*. Governance, 17(4), 525–548. https://doi.org/10.1111/j.1468-0491.2004.00263.x

Gupta, S., Davoodi, H., & Tiongson, E. R. (2002). *Corruption and the provision of health care and education services*. In H. H. Tanzi & L. Schaltegger (Eds.), *Public sector economics: Theory and practice* (pp. 121–134). Springer.

Harrison, M. (2010). *Corruption and the media: A comparative perspective*. International Journal of Press/Politics, 15(3), 1–25. https://doi.org/10.1177/1940161210374430

Hlaing, K. (2017). Myanmar: The crisis of governance. *Asian Journal of Comparative Politics, 3*(3), 157–168. https://doi.org/10.1177/2057891117737682

Inglehart, R., & Norris, P. (2016). *Trump, Brexit, and the rise of populism: Economic have-nots and cultural backlash*. Harvard Kennedy School.

Kaufmann, D. (2005). *Myths and realities of governance and corruption*. World Bank Institute.

Kaufmann, D., Kraay, A., & Mastruzzi, M. (2010). The worldwide governance indicators: Methodology and analytical issues. *World Bank Policy Research Working Paper No. 5430*. https://doi.org/10.1596/1813-9450-5430

Kaufmann, D., Kraay, A., & Mastruzzi, M. (2010). The worldwide governance indicators: Methodology and analytical issues. *World Bank Policy Research Working Paper No. 5430*. https://doi.org/10.1596/1813-9450-5430

Klitgaard, R. (1988). *Controlling corruption*. University of California Press.

Krook, M. L., & O'Brien, D. Z. (2012). *All the president's men? The appointment of female cabinet ministers worldwide. The Journal of Politics, 74*(3), 840–854. https://doi.org/10.1017/S0022381612000340

La Porta, R., Lopez-de-Silanes, F., Shleifer, A., & Vishny, R. W. (1999). *The quality of government. Journal of Law, Economics, and Organization, 15*(1), 222–279. https://doi.org/10.1093/jleo/15.1.222

Linz, J. J. (2000). Totalitarian and authoritarian regimes. *Lynne Rienner Publishers*.

Miller, W. L. (2017). *Corruption and the erosion of trust in government. Political Studies Review, 15*(1), 1–16. https://doi.org/10.1177/1478929916671044

Mounk, Y. (2018). *The people vs. democracy: Why our freedom is in danger and how to save it*. Harvard University Press.

Mudde, C. (2004). The populist zeitgeist. *Government and Opposition, 39*(4), 541–563. https://doi.org/10.1111/j.1477-7053.2004.00135.x

Mudde, C., & Kaltwasser, C. R. (2017). *Populism: A very short introduction*. Oxford University Press.

Pew Research Center. (2017). *The politics of populism: The rise of populist parties in Europe*. Pew Research Center. https://www.pewresearch.org

Prunier, G. (1995). *The Rwanda crisis: History of a genocide*. Columbia University Press.

Putnam, R. D. (2000). *Bowling alone: The collapse and revival of American community*. Simon & Schuster.

Rooduijn, M. (2018). A populist Zeitgeist? *The Journal of Political Ideologies, 23*(4), 317–334. https://doi.org/10.1080/13569317.2018.1537157

Rose-Ackerman, S. (1999). *Corruption and government: Causes, consequences, and reform*. Cambridge University Press.

Rose-Ackerman, S. (2008). *Corruption and democracy. International Political Science Review, 29*(4), 413–431. https://doi.org/10.1177/0192512108093811

Rothstein, B., & Stolle, D. (2008). *The state and social capital: An institutional approach. Comparative Politics, 40*(4), 447–464. https://doi.org/10.2307/20434199

Schedler, A. (1999). *Conceptualizing accountability. Democratization, 6*(1), 1–28. https://doi.org/10.1080/13510349908403544

Sen, A. (1999). *Development as freedom*. Oxford University Press.

Treisman, D. (2000). *The causes of corruption: A cross-national study. Journal of Public Economics, 76*(3), 399–457. https://doi.org/10.1016/S0047-2727(00)00092-4

Tufekci, Z. (2017). *Twitter and tear gas: The power and fragility of networked protest*. Yale University Press.

UNDP. (2016). *Human development report 2016: Human development for everyone*. United Nations Development Programme.

16

The Crisis of Knowledge and Misinformation

The crisis of knowledge and misinformation has become one of the most pressing challenges of the 21st century. As the digital age accelerates, the accessibility of information has dramatically increased, but so has the spread of misinformation and disinformation. Misinformation refers to false or inaccurate information that is spread without malicious intent, while disinformation is deliberately fabricated to deceive. The rapid dissemination of both types of information, particularly through social media platforms, has led to a crisis in how knowledge is produced, shared, and consumed. This crisis not only undermines public trust in institutions but also exacerbates societal divisions and hampers effective decision-making on critical issues, such as public health, climate change, and political stability.

One of the primary causes of the crisis of knowledge is the proliferation of digital platforms that enable the rapid spread of information. While the internet has democratized access to knowledge, it has also created an environment where anyone can publish information without rigorous fact-checking or oversight. Social media platforms, in particular, have become breeding grounds for misinformation, as they prioritize sensational content that generates engagement over accuracy (Friggeri, Galstyan, & Gummadi, 2014). The algorithms used by these platforms amplify emotionally charged and polarizing content, which often leads to the spread of false or misleading information. This has been particularly evident in recent years with the spread of misinformation related to the COVID-19 pandemic, vaccine hesitancy, and political polarization (Pennycook & Rand, 2018).

The crisis of knowledge is further compounded by the erosion of trust in traditional sources of information, such as the media, academic institutions, and government bodies. In an era of "fake news" and alternative facts, many people have become skeptical of expert opinions and institutional

DOI: 10.1201/9788743808787-20

authority. This mistrust has been fueled by the increasing fragmentation of the media landscape, where individuals can curate their information to align with their preexisting beliefs, creating echo chambers that reinforce misinformation (Sunstein, 2018). For example, the rise of partisan news outlets and the decline of traditional journalism have contributed to a situation where the public is often unable to distinguish between credible sources and those with biased or misleading agendas. This undermines the role of knowledge as a tool for informed decision-making and social cohesion.

Another significant factor contributing to the crisis of knowledge is the growing influence of populist leaders who exploit misinformation to further their political agendas. Populist rhetoric often thrives on the dissemination of false or misleading information that plays on people's fears and prejudices. Leaders who embrace populism frequently challenge the legitimacy of established knowledge systems, presenting themselves as champions of the "common people" against so-called corrupt elite. This has been seen in various global contexts, from the rise of "Brexit" in the United Kingdom to the spread of anti-science sentiments in the United States (Larsen & Olofsson, 2018). By casting doubt on expert knowledge and promoting alternative facts, populist leaders contribute to the erosion of public trust in science, media, and democratic institutions.

The consequences of the crisis of knowledge and misinformation are far-reaching and profound. One of the most immediate consequences is the public's inability to make informed decisions on critical issues. For instance, misinformation about climate change has led to confusion and inaction on one of the most pressing global challenges of our time. Despite overwhelming scientific consensus on the human causes of climate change, misinformation campaigns funded by vested interests have sought to undermine public support for climate action (Oreskes & Conway, 2010). Similarly, misinformation about vaccines has contributed to declining vaccination rates, putting public health at risk and hindering efforts to control preventable diseases (Larson et al., 2014).

Misinformation also contributes to the deepening of social divisions and political polarization. As people become more entrenched in their beliefs, they are less likely to engage in constructive dialogue or consider alternative perspectives. This "filter bubble" effect, where individuals are exposed only to information that confirms their existing views, exacerbates polarization and undermines social cohesion (Pariser, 2011). In democratic societies, this polarization can erode the ability to build consensus on important issues, leading to gridlock and a breakdown in governance. Moreover, the spread of misinformation can lead to the scapegoating of marginalized groups, as false narratives are used to justify discrimination and exclusion (Farkas & Neumayer, 2017).

Addressing the crisis of knowledge and misinformation requires a multifaceted approach. One critical step is improving media literacy and critical thinking skills, so that individuals can better assess the credibility of information they encounter online. Educational programs that teach people how to identify reliable sources, fact-check claims, and understand the mechanics of misinformation can help to build resilience against false narratives (Lewandowsky, Ecker, & Cook, 2017). Additionally, social media platforms must take greater responsibility for the content shared on their sites. This includes implementing more effective content moderation, promoting accurate information, and reducing the spread of harmful misinformation (Pennycook & Rand, 2018). Governments and international organizations can also play a role by supporting fact-checking initiatives and encouraging transparency in information sharing.

In conclusion, the crisis of knowledge and misinformation poses a significant threat to societal well-being, democratic governance, and global cooperation. The rapid spread of false information, coupled with the erosion of trust in traditional knowledge sources, has created a volatile information environment that undermines informed decision-making and exacerbates social divisions. To address this crisis, it is essential to promote media literacy, strengthen the credibility of information sources, and hold digital platforms accountable for the content they disseminate. Only through a concerted, collaborative effort can we hope to restore the integrity of knowledge and ensure that it serves as a foundation for informed, responsible decision-making in the 21st century.

16.1 The Spread of Fake News and Propaganda

The spread of fake news and propaganda has become a significant global issue, amplified by the rapid advancement of digital technologies and the proliferation of social media platforms. Fake news refers to false or misleading information presented as legitimate news, often intended to deceive or manipulate public opinion (Lazer et al., 2018). Propaganda, on the other hand, is a more deliberate and systematic attempt to influence attitudes and behaviors through biased or misleading information, often serving political, ideological, or commercial purposes (Jowett & O'Donnell, 2018). Together, these phenomena have profound implications for democracy, social cohesion, and the integrity of information ecosystems.

The digital revolution has dramatically increased the speed and reach of information dissemination. Social media platforms such as Facebook, Twitter, and YouTube allow information, whether accurate or false, to be shared widely and instantaneously (Tandoc, Lim, & Ling, 2018). Algorithms that prioritize user engagement inadvertently amplify sensationalized or polarizing content, creating fertile ground for the spread of fake news and propaganda. Research shows that fake news spreads more rapidly than factual information due to its emotionally charged and surprising nature, which encourages users to share it further (Vosoughi, Roy, & Aral, 2018). This virality, combined with the difficulty of distinguishing between credible and false sources online, poses a significant challenge to information literacy.

The consequences of fake news and propaganda are particularly troubling in political and social contexts. During elections, for example, fake news campaigns have been used to manipulate voter perceptions, polarize electorates, and undermine trust in democratic institutions (Allcott & Gentzkow, 2017). Similarly, propaganda has been weaponized to influence public opinion on contentious issues such as immigration, climate change, and public health crises. For instance, misinformation surrounding COVID-19 led to vaccine hesitancy and the spread of conspiracy theories, undermining public health efforts and exacerbating the pandemic (Cinelli et al., 2020). Such cases highlight the broader societal risks posed by the unchecked dissemination of false information.

Addressing the spread of fake news and propaganda requires a multifaceted approach. Media literacy education is essential to equip individuals with the critical thinking skills necessary to evaluate the credibility of information sources (McGrew et al., 2018). Moreover, technology companies have a responsibility to implement measures such as fact-checking systems, content moderation, and algorithmic transparency to limit the spread of false information. Governments and policymakers must also strike a balance between combating misinformation and preserving freedom of expression. Collaborative efforts involving civil society, academia, and the private sector are crucial to mitigating the impacts of fake news and safeguarding the integrity of information systems.

In conclusion, the spread of fake news and propaganda represents a complex and pressing challenge in the digital age. It undermines trust in institutions, exacerbates social divisions, and disrupts democratic processes. Addressing this issue requires coordinated efforts across multiple stakeholders, emphasizing media literacy, technological solutions, and ethical information governance.

16.2 The Declining Trust in Science and Expertise

The declining trust in science and expertise has become a prominent issue in contemporary society, with significant implications for addressing global challenges such as climate change, public health crises, and technological advancements. Historically, science and expertise have been seen as pillars of progress, grounded in empirical evidence, peer review, and methodological rigor. However, recent trends indicate growing skepticism and mistrust among segments of the public, which undermines the role of science in informing policies and decision-making (Gauchat, 2012). This erosion of trust stems from multiple interconnected factors, including misinformation, political polarization, economic interests, and the democratization of information through digital media.

One key driver of declining trust in science is the politicization of scientific knowledge. Issues such as climate change, vaccination, and public health responses have become increasingly polarized, with scientific findings often interpreted through ideological or partisan lenses (Bolsen, Druckman, & Cook, 2014). For instance, climate change denial has been fueled by vested economic interests, particularly in fossil fuel industries, which have actively promoted doubt and misinformation to challenge the scientific consensus (Oreskes & Conway, 2010). Similarly, during the COVID-19 pandemic, public health measures, including mask mandates and vaccination campaigns, became points of contention, leading to skepticism toward scientific authorities. This politicization erodes trust by framing science as subjective or biased, rather than objective and evidence-based.

Another contributing factor is the proliferation of misinformation and disinformation through digital media. The rise of social media platforms has democratized access to information but has also created an environment where false or misleading content spreads rapidly (Vosoughi, Roy, & Aral, 2018). Unlike traditional media, which often relies on editorial oversight and fact-checking, social media algorithms prioritize content based on engagement rather than accuracy, amplifying sensationalized or emotionally charged misinformation. As a result, scientific findings are often distorted or misrepresented, leading to confusion and mistrust among the public. For example, misinformation regarding vaccines has contributed to vaccine hesitancy, even in the face of overwhelming scientific evidence demonstrating their safety and efficacy (Larson et al., 2018).

The changing nature of expertise also plays a role in declining trust. In the age of the internet, individuals increasingly rely on personal research and anecdotal evidence, challenging traditional hierarchies of knowledge and expertise (Collins & Evans, 2007). While this shift empowers individuals to seek information, it also blurs the line between credible experts and unqualified sources, creating a false equivalence between scientific consensus and fringe opinions. This phenomenon, sometimes referred to as the "death of expertise," undermines public confidence in scientific authorities and institutions (Nichols, 2017).

Moreover, historical and systemic factors have contributed to skepticism toward science, particularly among marginalized communities. Past scientific abuses, such as the Tuskegee Syphilis Study, have left lasting scars, leading to distrust in medical and scientific institutions (Washington, 2006). Additionally, the underrepresentation of diverse voices in science exacerbates perceptions of exclusion and bias, further eroding trust. Addressing these systemic issues requires inclusive approaches to science communication and engagement, ensuring that diverse perspectives are recognized and valued.

To rebuild trust in science and expertise, a multifaceted approach is necessary. Improving science communication, fostering transparency, and addressing misinformation are critical steps. Scientists and experts must engage with the public in accessible and relatable ways, emphasizing the process of scientific inquiry and its inherent self-correcting nature (Scheufele & Krause, 2019). Media literacy programs can also equip individuals with the skills to critically evaluate information sources and distinguish credible science from misinformation. Furthermore, policies that promote diversity and inclusion in scientific institutions can help rebuild trust among historically marginalized communities.

In conclusion, the declining trust in science and expertise is a complex issue shaped by political, social, and technological factors. While challenges such as misinformation, politicization, and changing perceptions of expertise persist, rebuilding trust requires collaborative efforts across scientific, educational, and policy domains. By fostering transparency, inclusivity, and effective communication, science can continue to serve as a cornerstone for addressing the world's most pressing challenges.

16.3 Media's Role in Shaping Perceptions

The media plays a critical role in shaping public perceptions, influencing how individuals interpret events, issues, and people. Through the dissemination of information, framing of narratives, and agenda-setting, media outlets can significantly impact societal beliefs, attitudes, and behaviors. Traditional media, such as newspapers, television, and radio, or digital platforms like social media and online news serves as a powerful intermediary between events and public understanding (McCombs & Shaw, 1972). However, this influence is a double-edged sword: while media can inform and educate, it can also distort reality, amplify biases, and contribute to misinformation, particularly in today's fragmented and fast-paced information environment.

One of the most significant ways media shapes perceptions is through framing, the process by which certain aspects of a story are emphasized while others are downplayed. Frames influence how audiences interpret and respond to information by highlighting specific themes or viewpoints (Entman, 1993). For example, media coverage of climate change can frame the issue as a scientific problem, an economic challenge, or a political debate, leading to varying public perceptions and levels of concern (Boykoff, 2011). Similarly, the framing of immigration issues often influences whether immigrants are viewed as contributors to society or as threats, depending on the tone and focus of the coverage (Esses, Medianu, & Lawson, 2013). Such framing decisions have profound implications for public opinion, policy preferences, and social attitudes.

Another key function of media is agenda-setting, where media outlets prioritize certain topics over others, thereby influencing what the public perceives as important (McCombs & Shaw, 1972). By determining which issues receive coverage and how prominently they are featured, media can shape societal priorities and political discourse. For instance, extensive media attention to terrorism or crime can create heightened perceptions of risk and insecurity, even if statistical evidence suggests that these threats are relatively low (Altheide, 2006). Similarly, the media's focus on celebrity culture or sensational stories can divert attention away from critical but less "newsworthy" issues, such as poverty, inequality, or environmental degradation.

The proliferation of digital media has further amplified the media's role in shaping perceptions, particularly through social media platforms. Unlike traditional media, which is subject to editorial oversight and fact-checking, digital media allows for the rapid dissemination of unfiltered content, including misinformation and disinformation (Vosoughi, Roy, & Aral, 2018). Social media algorithms, which prioritize content based on engagement metrics, often amplify polarizing or emotionally charged narratives, creating echo chambers where individuals are exposed primarily to information that aligns with their existing beliefs (Pariser, 2011). This phenomenon not only reinforces biases but also fosters divisions within society, as differing media narratives create divergent perceptions of reality.

Furthermore, the media's representation of marginalized groups significantly shapes societal attitudes toward these communities. Research has shown that stereotypes perpetuated through media portrayals contribute to bias, discrimination, and exclusion. For example, negative portrayals of racial minorities in news coverage can reinforce prejudices and perpetuate systemic inequities (Dixon & Williams, 2015). Similarly, the underrepresentation of women, LGBTQ+ individuals, and other marginalized groups in media narratives limits their visibility and perpetuates cultural biases (Gill, 2007). Addressing these disparities requires more inclusive and equitable approaches to media representation.

Despite its challenges, the media also holds immense potential to drive positive change by raising awareness, fostering empathy, and promoting social justice. For example, investigative journalism has historically played a critical role in uncovering corruption, human rights abuses, and environmental degradation, holding powerful actors accountable (Schudson, 2008). Additionally, media campaigns on issues such as public health, climate change, and social equity have mobilized public support for meaningful action.

In conclusion, the media's role in shaping perceptions is profound and multifaceted, influencing public understanding, attitudes, and behaviors through framing, agenda-setting, and representation. While media can educate and inform, it also has the potential to distort, polarize, and mislead. In the digital age, where information spreads rapidly and unfiltered, promoting media literacy, ethical journalism, and inclusive representation is critical to ensuring that media serves as a force for positive societal change.

References

Allcott, H., & Gentzkow, M. (2017). Social media and fake news in the 2016 election. *Journal of Economic Perspectives, 31*(2), 211–236. https://doi.org/10.1257/jep.31.2.211

Altheide, D. L. (2006). Terrorism and the politics of fear. *Cultural Studies ↔ Critical Methodologies, 6*(4), 415–439. https://doi.org/10.1177/1532708605285733

Bolsen, T., Druckman, J. N., & Cook, F. L. (2014). The influence of partisan motivated reasoning on public opinion. *Political Behavior, 36*(2), 235–262. https://doi.org/10.1007/s11109-013-9238-0

Boykoff, M. T. (2011). *Who speaks for the climate? Making sense of media reporting on climate change.* Cambridge University Press.

Cinelli, M., Quattrociocchi, W., Galeazzi, A., Valensise, C. M., Brugnoli, E., Schmidt, A. L., ... & Scala, A. (2020). The COVID-19 social media infodemic. *Scientific Reports, 10*(1), 1–10. https://doi.org/10.1038/s41598-020-73510-5

Collins, H., & Evans, R. (2007). *Rethinking expertise.* University of Chicago Press.

Dixon, T. L., & Williams, C. L. (2015). The changing misrepresentation of race and crime on network and cable news. *Journal of Communication, 65*(1), 24–39. https://doi.org/10.1111/jcom.12133

Entman, R. M. (1993). Framing: Toward clarification of a fractured paradigm. *Journal of Communication, 43*(4), 51–58. https://doi.org/10.1111/j.1460-2466.1993.tb01304.x

Esses, V. M., Medianu, S., & Lawson, A. S. (2013). Uncertainty, threat, and the role of the media in promoting the dehumanization of immigrants and refugees. *Journal of Social Issues, 69*(3), 518–536. https://doi.org/10.1111/josi.12027

Farkas, J., & Neumayer, C. (2017). *The role of social media in spreading misinformation and disinformation.* In M. T. M. R. T. (Ed.), *The Cambridge Handbook of Social Media and Politics* (pp. 167–185). Cambridge University Press. https://doi.org/10.1017/9781316338008.013

Friggeri, A., Galstyan, A., & Gummadi, K. P. (2014). *Rumor cascades. Proceedings of the 23rd International Conference on World Wide Web,* 1013–1024. https://doi.org/10.1145/2566486.2567981

Gauchat, G. (2012). Politicization of science in the public sphere: A study of public trust in the United States, 1974 to 2010. *American Sociological Review, 77*(2), 167–187. https://doi.org/10.1177/0003122412438225

Gill, R. (2007). *Gender and the media.* Polity Press.

Jowett, G. S., & O'Donnell, V. (2018). *Propaganda & persuasion* (7th ed.). SAGE Publications.

Larsen, H. L., & Olofsson, A. (2018). *Populism and anti-expert sentiments in Western democracies: The role of knowledge and trust. Journal of Political Ideologies, 23*(3), 238–258. https://doi.org/10.1080/13569317.2018.1510369

Larson, H. J., Jarrett, C., Eckersberger, E., Smith, D. M. D., & Paterson, P. (2018). Understanding vaccine hesitancy around vaccines and vaccination from a global perspective: A systematic review of published literature, 2007–2012. *Vaccine, 32*(19), 2150–2159. https://doi.org/10.1016/j.vaccine.2014.01.081

Larson, H. J., Jarrett, C., Eckersberger, E., Smith, D. M., & Paterson, P. (2014). *Understanding vaccine hesitancy around vaccines and vaccination from a global perspective: A systematic review of the literature. Vaccine, 32*(19), 2150–2159. https://doi.org/10.1016/j.vaccine.2014.01.081

Lazer, D. M. J., Baum, M. A., Benkler, Y., Berinsky, A. J., Greenhill, K. M., Menczer, F., ... & Zittrain, J. L. (2018). The science of fake news. *Science, 359*(6380), 1094–1096. https://doi.org/10.1126/science.aao2998

Lewandowsky, S., Ecker, U. K. H., & Cook, J. (2017). *Beyond Misinformation: Understanding and Coping with the "Post-Truth" Era. Journal of Applied Research in Memory and Cognition, 6*(4), 353–369. https://doi.org/10.1016/j.jarmac.2017.07.008

McCombs, M. E., & Shaw, D. L. (1972). The agenda-setting function of mass media. *Public Opinion Quarterly, 36*(2), 176–187. https://doi.org/10.1086/267990

McGrew, S., Breakstone, J., Ortega, T., Smith, M., & Wineburg, S. (2018). Can students evaluate online sources? Learning from assessments of civic online reasoning. *Theory & Research in Social Education, 46*(2), 165–193. https://doi.org/10.1080/00933104.2017.1416320

Nichols, T. M. (2017). *The death of expertise: The campaign against established knowledge and why it matters.* Oxford University Press.

Oreskes, N., & Conway, E. M. (2010). *Merchants of doubt: How a handful of scientists obscured the truth on issues from tobacco smoke to global warming.* Bloomsbury Press.

Pariser, E. (2011). *The filter bubble: What the Internet is hiding from you.* Penguin Press.

Pennycook, G., & Rand, D. G. (2018). *Fighting misinformation on social media using crowdsourced judgments of news source quality. Proceedings of the National Academy of Sciences, 115*(7), 1249–1254. https://doi.org/10.1073/pnas.1718832115

Scheufele, D. A., & Krause, N. M. (2019). Science audiences, misinformation, and fake news. *Proceedings of the National Academy of Sciences, 116*(16), 7662–7669. https://doi.org/10.1073/pnas.1805871115

Schudson, M. (2008). *Why democracies need an unlovable press.* Polity Press.

Sunstein, C. R. (2018). *#Republic: Divided democracy in the age of social media.* Princeton University Press.

Tandoc, E. C., Lim, Z. W., & Ling, R. (2018). Defining "fake news": A typology of scholarly definitions. *Digital Journalism, 6*(2), 137–153. https://doi.org/10.1080/21670811.2017.1360143

Vosoughi, S., Roy, D., & Aral, S. (2018). The spread of true and false news online. *Science, 359*(6380), 1146–1151. https://doi.org/10.1126/science.aap9559

Washington, H. A. (2006). *Medical apartheid: The dark history of medical experimentation on Black Americans from colonial times to the present.* Doubleday.

PART V

Cultural and Ethical Crises

17

Identity, Belonging, and Social Fragmentation

Cultural and ethical crises are becoming increasingly prevalent in contemporary societies due to the rapid pace of globalization, technological advancements, and shifting social norms. These crises often emerge when long-standing cultural values and ethical principles clash with modern developments, creating uncertainty, tension, and moral ambiguity. As societies become more interconnected, cultural identities are challenged, and ethical dilemmas arise, particularly in areas such as technology, economics, politics, and environmental sustainability. Addressing these crises is not a task for individual entities but a global challenge that requires a nuanced understanding of cultural diversity, ethical frameworks, and, most importantly, the need for global cooperation. It is only through collective efforts that we can effectively tackle these challenges.

One significant aspect of cultural crises is the erosion of cultural identity in the face of globalization. While globalization fosters economic integration and cultural exchange, it can also marginalize local traditions, languages, and practices, leading to cultural homogenization (Appadurai, 1996). The dominance of Western media, values, and consumer culture often overshadows indigenous knowledge systems and cultural expressions, resulting in the loss of cultural heritage and identity (Tomlinson, 1999). For example, Indigenous communities around the world face cultural crises as their land, traditions, and ways of life are threatened by economic development, urbanization, and climate change. This cultural erosion raises ethical concerns about cultural preservation, respect for diversity, and the rights of marginalized groups to maintain their unique identities.

Ethical crises, on the other hand, often arise from technological advancements and their unintended consequences. The rapid development of artificial intelligence, biotechnology, and digital surveillance technologies has

DOI: 10.1201/9788743808787-22

created ethical dilemmas regarding privacy, autonomy, and fairness (Floridi et al., 2018). For instance, the widespread use of surveillance systems raises questions about individual privacy and the potential for authoritarian control. In contrast, advances in genetic engineering prompt concerns about the ethical implications of altering human life. These developments challenge traditional ethical frameworks and demand new principles for responsible innovation and technology use. Ethical crises are further exacerbated by the unequal distribution of technological benefits, which often deepen existing social and economic inequalities.

Moreover, cultural and ethical crises intersect in the realm of environmental sustainability. The exploitation of natural resources and environmental degradation are often driven by short-term economic interests, disregarding ethical considerations for future generations and the planet's health. This cultural and ethical crisis reflects a disconnection between human societies and nature, particularly in industrialized and consumer-driven cultures (Latour, 2017). Indigenous communities, who often hold deep cultural and spiritual connections to nature, face ethical injustices as their lands are exploited without consent, leading to displacement and cultural loss (Whyte, 2018). The ethical principles of intergenerational justice, which emphasizes the responsibility of the current generation to leave a sustainable world for future generations, and environmental stewardship are essential to addressing these crises and fostering sustainable development.

In addition, the rise of cultural and ethical crises is closely tied to political polarization and social fragmentation. Political ideologies and cultural values often clash, leading to ethical dilemmas regarding freedom of expression, human rights, and social justice. For example, debates over immigration, gender equality, and LGBTQ+ rights highlight deep cultural divides and ethical conflicts within societies. These tensions are further amplified by media polarization and misinformation, which distort perceptions of cultural and ethical issues. The media, particularly in the digital age, plays a significant role in shaping public opinion and can exacerbate or alleviate cultural and ethical crises (Fukuyama, 2018). Such crises challenge societies to reconcile conflicting values while upholding principles of inclusivity, equity, and human dignity.

Societies must prioritize dialogue, ethical reflection, and cultural pluralism to address cultural and ethical crises. Inclusive dialogue bridges cultural divides and promotes mutual understanding and respect. Ethical frameworks such as Amartya Sen's Capabilities Approach emphasize the importance of human dignity, agency, and well-being in addressing ethical dilemmas (Sen, 1999). Fostering intercultural dialogue can help bridge cultural divides and promote mutual understanding and respect (UNESCO, 2009). Ethical education and media literacy are critical tools for navigating these crises, as they empower individuals to critically engage with cultural and ethical challenges in a globalized world.

In conclusion, cultural and ethical crises are complex and multifaceted challenges that arise from the interplay of globalization, technological advancements, and shifting societal norms. These crises threaten cultural diversity, raise profound ethical dilemmas, and exacerbate social and environmental injustices. Addressing them requires a commitment to ethical principles, cultural preservation, and inclusive dialogue to foster a more just, sustainable, and harmonious world. Potential solutions include promoting cultural diversity, enhancing ethical education, and fostering intercultural dialogue. These measures can help bridge cultural divides, promote mutual understanding and respect, and empower individuals to critically engage with cultural and ethical challenges in a globalized world.

Identity, belonging, and social fragmentation are deeply interconnected concepts that shape individual experiences and broader societal dynamics. Identity refers to how individuals and groups define themselves based on cultural, social, and personal factors, including ethnicity, gender, religion, and nationality (Hall, 1996). Belonging is the sense of connection and acceptance within a community or group, providing emotional security and purpose (Yuval-Davis, 2011). However, contemporary forces such as globalization, technological advancements, migration, and economic inequality have disrupted traditional forms of identity and belonging, contributing to social fragmentation. Social fragmentation refers to weakening social bonds and increasing social divisions, leading to alienation, polarization, and disconnection.

Globalization has significantly influenced identity and belonging by fostering cultural exchange while simultaneously challenging traditional cultural norms. As societies become more interconnected, individuals are exposed to diverse cultural influences, which can enrich their identities but may also create tensions and conflicts. For example, immigrants and diaspora communities often navigate multiple cultural identities, balancing their heritage with the dominant culture of their host societies (Berry, 2005). This process, while enriching, can lead to identity crises and struggles for belonging, particularly when individuals face discrimination or exclusion. Moreover, globalization has accelerated cultural homogenization, where dominant cultural values, often from Western societies, overshadow local traditions, leading to cultural erosion and a sense of loss among marginalized groups (Tomlinson, 1999).

Technological advancements, particularly the rise of digital communication and social media, have further complicated identity and belonging. On one hand, digital platforms enable individuals to form online communities, fostering connections based on shared interests, values, and identities. These virtual spaces can empower marginalized voices and provide a sense of belonging to individuals who may feel isolated offline (Turkle, 2011). However, the same technologies have contributed to social fragmentation by promoting ideological polarization and creating "filter bubbles," where individuals are exposed only to information that reinforces their existing beliefs (Pariser, 2011). This phenomenon exacerbates societal divisions, as people retreat into echo chambers, reducing opportunities for meaningful dialogue and mutual understanding.

Economic inequality is another significant driver of social fragmentation, exacerbating disparities in access to resources, opportunities, and social mobility. When individuals or groups feel excluded from economic and social systems, their sense of belonging is undermined, creating feelings of alienation and resentment (Piketty, 2014). For instance, marginalized communities may struggle to integrate into mainstream society due to systemic barriers, further deepening social divisions. This exclusion can manifest in social unrest, political extremism, and a decline in trust in institutions, all weakening societal cohesion (Putnam, 2000).

Additionally, political polarization has intensified social fragmentation, as ideological divides create "us versus them" dynamics that fracture communities. In many societies, debates over identity-related issues such as immigration, gender rights, and cultural representation have become highly politicized, fueling tensions and reducing social solidarity (Fukuyama, 2018). Political leaders and media outlets often exploit these divisions for their agendas, amplifying cultural and ideological conflicts. This process undermines collective identities and erodes the shared values necessary for societal cohesion.

Addressing the challenges of identity, belonging, and social fragmentation requires fostering inclusive societies that embrace diversity while promoting shared values and social justice. Policies that reduce economic inequality encourage intercultural dialogue, and strengthen community ties can help rebuild trust and social cohesion. Amartya Sen's *Capabilities Approach* emphasizes empowering individuals to achieve their full potential while respecting their cultural identities (Sen, 1999). Furthermore, promoting critical media literacy can mitigate the divisive effects of digital technologies by encouraging individuals to engage with diverse perspectives and bridge ideological divides (Pariser, 2011).

In conclusion, identity, belonging, and social fragmentation are critical issues in contemporary societies shaped by globalization, technological change, economic inequality, and political polarization. While these forces have created opportunities for cultural exchange and digital connectivity, they have disrupted traditional forms of identity and community, leading to increased social fragmentation. Addressing these challenges requires inclusive approaches that promote cultural diversity, reduce inequalities, and foster shared values to rebuild a sense of belonging and social cohesion.

17.1 The Impact of Globalization on Cultural Heritage

Globalization has had a profound and multifaceted impact on cultural heritage, encompassing both tangible and intangible dimensions of culture. Cultural heritage, which includes monuments, artifacts, traditions, languages, and rituals, reflects collective identity, historical continuity, and societal values (UNESCO, 2003). While globalization has facilitated cultural exchange, raised awareness, and promoted heritage preservation through international cooperation, it has posed significant threats, including cultural homogenization, commercialization, and the erosion of local traditions. The dual nature of globalization highlights both opportunities and challenges in safeguarding cultural heritage.

One of the positive impacts of globalization is the increased visibility and recognition of cultural heritage on a global scale. International organizations such as UNESCO have played a crucial role in preserving and promoting cultural heritage through programs like the *World Heritage Convention* (UNESCO, 2003). These initiatives provide financial support, technical expertise, and global platforms to protect heritage sites and intangible cultural practices. Moreover, globalization has enabled greater cultural exchange and dialogue, fostering mutual understanding and appreciation of diverse traditions. For instance, advancements in digital technologies have made it possible to document, archive, and share cultural practices, art, and languages with a global audience, enhancing their preservation and dissemination (Harrison, 2013).

However, globalization has also led to the modification and commercialization of cultural heritage. As cultural heritage becomes a marketable commodity in global tourism and entertainment industries, its authenticity and significance can be undermined. Heritage sites and cultural practices are often adapted to cater to tourist expectations, leading to the exploitation of local traditions and communities (Kirshenblatt-Gimblett, 1995). For example, traditional festivals and rituals may be altered or oversimplified to appeal to international audiences, stripping them of their cultural depth and meaning. This process, often called "heritage tourism," risks reducing cultural heritage to a spectacle or consumer product, prioritizing economic gain over cultural preservation (Steger, 2020).

Cultural homogenization is another significant consequence of globalization, driven by the dominance of Western cultural values, global media, and consumerism. Local languages, traditions, and knowledge systems are increasingly at risk as globalized lifestyles and mass culture become pervasive (Tomlinson, 1999). For example, indigenous languages are disappearing at an alarming rate due to the spread of dominant global languages, such as English, which are often considered essential for economic and social mobility (Crystal, 2000). This loss of linguistic diversity represents the erosion of cultural heritage and the disappearance of unique ways of understanding and interpreting the world.

Furthermore, globalization has intensified the challenges posed by urbanization and development, often threatening cultural heritage's physical preservation. Rapid urban expansion and infrastructure projects frequently destroy or neglect historical sites and monuments (Harrison, 2013). Economic pressures and a lack of resources exacerbate these challenges in developing regions, as governments prioritize economic growth over heritage conservation. This trend underscores the need for sustainable development approaches that balance economic progress with the protection of cultural heritage.

Global efforts to preserve cultural heritage must emphasize inclusive and community-centered approaches in response to these challenges. Local communities play a vital role in safeguarding their cultural heritage, as they are the custodians of traditional knowledge, practices, and values. Empowering these communities through education, participation, and equitable benefits from heritage tourism can help ensure that cultural heritage remains meaningful and sustainable (Smith, 2006). Additionally, policies that promote linguistic and cultural diversity, combined with technological innovations for heritage documentation, can help mitigate the adverse effects of globalization while preserving cultural identity for future generations.

Globalization has both enriched and endangered cultural heritage. While it has facilitated cultural exchange, technological advancements, and international collaboration, it has also contributed to cultural homogenization, commercialization, and the loss of local traditions. Addressing these challenges requires a balanced approach to prioritizing cultural diversity, community empowerment, and sustainable heritage management. By recognizing the value of cultural heritage as a cornerstone of identity and societal well-being, globalization can be harnessed to preserve and celebrate the richness of human culture.

17.2 Nationalism, Xenophobia, and Social Divisions

Nationalism, xenophobia, and social divisions have increasingly shaped political, social, and cultural landscapes across the globe. In contrast, nationalism can foster a sense of pride, identity, and unity within a nation; its exclusionary and extreme forms often contribute to social fragmentation, conflict, and marginalization of "outsiders" (Anderson, 2006). Xenophobia, the fear or hatred of foreigners and those perceived as different exacerbates these divides, leading to discrimination, violence, and erosion of social cohesion (Gingrich & Banks, 2006). These phenomena create a complex interplay that reinforces societal divisions and undermines inclusion, equity, and peace efforts.

Nationalism, particularly in its contemporary resurgence, often prioritizes the interests of a specific nation or cultural group over others. Benedict Anderson (2006) describes nations as "imagined communities" constructed through shared language, culture, and history. While this sense of collective identity can unify citizens, nationalism can also manifest in exclusionary forms, particularly when tied to ethnic or cultural superiority. In such cases, it can marginalize minority groups, immigrants, and other "outsiders," fostering social divisions. For instance, right-wing populist movements in Europe and North America have capitalized on nationalist rhetoric to promote policies restricting immigration and assert cultural homogeneity (Mudde, 2019). These movements often frame globalization, immigration, and multiculturalism as threats to national identity, reinforcing fear and resentment among dominant cultural groups.

Xenophobia, closely tied to nationalism, intensifies social divisions by creating a divide between "us" and "them." It often emerges during economic hardship, political instability, or cultural change when groups seek scapegoats for societal problems (Pettigrew et al., 2008). Xenophobic attitudes are evident in anti-immigrant sentiments and discriminatory policies that target refugees and marginalized communities. For example, the 2015 European refugee crisis saw a rise in xenophobic rhetoric and policies across several nations, fueled by fears of cultural displacement and economic competition (Wodak, 2015). Media narratives and political discourse often amplify these fears, portraying immigrants and refugees as threats to national security, cultural identity, and economic stability. Such depictions not only legitimize xenophobic attitudes but also exacerbate social polarization and undermine intercultural understanding.

The intersection of nationalism and xenophobia contributes to deepening social divisions, particularly in multicultural societies. Social divisions manifest as inequalities in access to resources, opportunities, and representation for marginalized groups. These divisions are reinforced through systemic discrimination, exclusionary policies, and cultural stigmatization. For example, racialized nationalism in colonial and post-colonial contexts has historically marginalized indigenous peoples and ethnic minorities, denying them equal rights and cultural recognition (Gingrich & Banks, 2006). In modern contexts, nationalist policies that prioritize the majority culture often lead to the erasure or suppression of minority languages, traditions, and identities, further exacerbating social fragmentation.

Furthermore, nationalism and xenophobia are often manipulated for political gain, as leaders use divisive rhetoric to consolidate power and mobilize support. By appealing to cultural and economic displacement fears, populist leaders create a sense of urgency and loyalty among their base while alienating and scapegoating others (Mudde, 2019). This strategy deepens social divisions and erodes democratic values such as pluralism, tolerance, and inclusivity. In extreme cases, it can lead to violence, ethnic conflict, and human rights abuses, as seen in various historical and contemporary examples of ethnic nationalism.

Addressing nationalism, xenophobia, and social divisions requires fostering inclusive national identities and promoting intercultural understanding. Educational initiatives, media literacy, and policies celebrating diversity can help challenge exclusionary narratives and reduce xenophobia (Pettigrew et al., 2008). Moreover, international cooperation and dialogue are essential in addressing the root causes of these issues, such as economic inequality, political instability, and cultural alienation. By prioritizing inclusion, equity, and shared humanity, societies can move beyond divisive ideologies and work toward greater social cohesion and justice.

17.3 Building Inclusive and Cohesive Societies

Building inclusive and cohesive societies is essential for promoting social justice, equality, and sustainable development. In an increasingly globalized world, societies are becoming more diverse, with individuals from various cultural, ethnic, religious, and socioeconomic backgrounds coexisting. The challenge lies in ensuring that all members of society feel valued, respected, and empowered, regardless of their differences. Inclusivity and social cohesion are about tolerance and creating environments where diverse groups can thrive, contribute, and live harmoniously. Achieving this requires a multidimensional approach that includes policy reforms, education, community engagement, and promoting shared values and social justice.

One of the fundamental elements of building inclusive societies is recognizing and protecting human rights. Inclusivity is grounded in the principles of equality and non-discrimination, ensuring that all individuals have equal access to opportunities, resources, and services (Sen, 1999). This involves addressing systemic inequalities, such as those based on race, gender, disability, and socioeconomic status, and creating policies that foster social inclusion. For example, affirmative action policies, anti-discrimination laws, and social welfare programs can help level the playing field for marginalized and disadvantaged groups. Additionally, promoting gender equality, racial justice, and the rights of indigenous peoples are critical components of creating a society where everyone can participate fully in social, economic, and political life (Nussbaum, 2011).

Education plays a crucial role in fostering inclusivity and social cohesion. Schools and educational institutions are places for academic learning and where individuals develop their understanding of social norms, values, and relationships. Multicultural education, which emphasizes the value of diversity and teaches students about different cultures, histories, and perspectives, is essential for promoting inclusivity (Banks, 2008). Moreover, education can challenge stereotypes, reduce prejudice, and encourage empathy, helping to build a more cohesive society. Programs that promote intercultural dialogue, conflict resolution, and community-building can also play a significant role in fostering mutual respect and understanding among diverse groups (Tajfel & Turner, 1986).

Community engagement is another critical factor in building inclusive societies. Individuals actively involved in decision-making are more likely to feel a sense of ownership and responsibility for their communities. Participatory governance models, which encourage the involvement of all members of society in shaping policies and practices, can help create more inclusive and cohesive communities. This approach ensures that marginalized and underrepresented groups' voices are heard and their needs are addressed in policy development (Putnam, 2000). Community-based organizations, grassroots movements, and local initiatives can also play a key role in fostering social cohesion by promoting dialogue, collaboration, and collective action among diverse groups.

In addition to policy reforms and education, fostering a shared sense of belonging is crucial for social cohesion. When individuals feel that they are part of a larger community, they are more likely to engage in pro-social behaviors and contribute to the well-being of others. Building a sense of belonging requires recognizing and celebrating individuals' diverse identities and experiences while emphasizing shared values and goals. For example, national identity can be inclusive based on shared values such as democracy, justice, and human rights rather than ethnic or cultural homogeneity (Kymlicka, 2007). Inclusive national narratives and symbols that reflect the diversity of the population can help foster a sense of belonging and unity while respecting the differences within society.

However, building inclusive and cohesive societies is not without its challenges. Resistance to inclusion can hinder progress, particularly from groups that feel threatened by social change or cultural diversity. Xenophobia, racism, and discrimination can create divisions and undermine efforts to build a more inclusive society (Mudde, 2019). Overcoming these challenges requires policy interventions and shifts in attitudes and behaviors. Promoting empathy, solidarity, and collective responsibility is essential for countering divisive ideologies and fostering a culture of inclusivity.

Building inclusive and cohesive societies is a complex and ongoing process that requires concerted efforts across multiple sectors, including policy, education, community engagement, and cultural change. By addressing systemic inequalities, promoting diversity, and fostering a shared sense of belonging, societies can create environments where all individuals, regardless of their background, can participate fully and contribute to the common good. While challenges remain, pursuing inclusivity and social cohesion is essential for building peaceful, just, and sustainable societies in the 21st century.

References

Anderson, B. (2006). *Imagined communities: Reflections on the origin and spread of nationalism* (Rev. ed.). Verso.

Appadurai, A. (1996). *Modernity at large: Cultural dimensions of globalization.* University of Minnesota Press.

Banks, J. A. (2008). *An introduction to multicultural education* (4th ed.). Pearson.

Berry, J. W. (2005). Acculturation: Living successfully in two cultures. *International Journal of Intercultural Relations, 29*(6), 697–712. https://doi.org/10.1016/j.ijintrel.2005.07.013

Crystal, D. (2000). *Language death.* Cambridge University Press.

Floridi, L., Cowls, J., Beltrametti, M., Chatila, R., Chazerand, P., Dignum, V., ... & Vayena, E. (2018). AI4People – An ethical framework for a good AI society: Opportunities, risks, principles, and recommendations. *Minds and Machines, 28*(4), 689–707. https://doi.org/10.1007/s11023-018-9482-5

Fukuyama, F. (2018). *Identity: The demand for dignity and the politics of resentment.* Farrar, Straus and Giroux.

Gingrich, A., & Banks, M. (2006). *Neo-nationalism in Europe and beyond: Perspectives from social anthropology.* Berghahn Books.

Hall, S. (1996). Introduction: Who needs 'identity'? In S. Hall & P. du Gay (Eds.), *Questions of cultural identity* (pp. 1–17). SAGE Publications.

Harrison, R. (2013). *Heritage: Critical approaches.* Routledge.

Kirshenblatt-Gimblett, B. (1995). Theorizing heritage. *Ethnomusicology, 39*(3), 367–380. https://doi.org/10.2307/924627

Kymlicka, W. (2007). *Multicultural odyssey: Navigating the new international politics of diversity.* Oxford University Press.

Latour, B. (2017). *Facing Gaia: Eight lectures on the new climatic regime.* Polity Press.

Mudde, C. (2019). *The far-right today.* Polity Press.

Nussbaum, M. (2011). *Creating capabilities: The human development approach.* Belknap Press.

Pariser, E. (2011). *The filter bubble: How the new personalized web changes what we read and think.* Penguin Press.

Pettigrew, T. F., Wagner, U., & Christ, O. (2008). Recent advances in intergroup contact theory. *International Journal of Intercultural Relations, 32*(3), 187–195. https://doi.org/10.1016/j.ijintrel.2007.12.001

Piketty, T. (2014). *Capital in the twenty-first century.* Harvard University Press.

Putnam, R. D. (2000). *Bowling alone: The collapse and revival of American community.* Simon & Schuster.

Sen, A. (1999). *Development as freedom.* Oxford University Press.

Smith, L. (2006). *Uses of heritage.* Routledge.

Steger, M. B. (2020). *Globalization: A concise introduction* (5th ed.). Oxford University Press.

Tajfel, H., & Turner, J. C. (1986). The social identity theory of intergroup behavior. In S. Worchel & W. G. Austin (Eds.), *Psychology of intergroup relations* (pp. 7–24). Nelson-Hall.

Tomlinson, J. (1999). *Globalization and culture.* University of Chicago Press.

Turkle, S. (2011). *Alone together: Why we expect more from technology and less from each other.* Basic Books.

UNESCO. (2003). *Convention for the safeguarding of intangible cultural heritage.* United Nations Educational, Scientific and Cultural Organization.

UNESCO. (2009). *Investing in cultural diversity and intercultural dialogue.* UNESCO Publishing.

Whyte, K. P. (2018). Indigenous science (fiction) for the Anthropocene: Ancestral dystopias and fantasies of climate change crises. *Environment and Planning E: Nature and Space, 1*(1–2), 224–242. https://doi.org/10.1177/2514848618777621

Wodak, R. (2015). *The politics of fear: What right-wing populist discourses mean.* SAGE Publications.

Yuval-Davis, N. (2011). *The politics of belonging: Intersectional contestations.* SAGE Publications.

18

Technology and Ethical Dilemmas

The rapid advancement of technology has transformed nearly every aspect of human life, from healthcare and communication to education and governance. While technological innovations have brought about significant benefits, they have also introduced complex ethical dilemmas that challenge traditional moral frameworks and societal norms. These ethical issues arise from the unintended consequences of new technologies, the potential for misuse, and the broader societal impacts of their implementation. As technology continues to evolve, it is essential to address these dilemmas to ensure that technological progress aligns with ethical principles such as justice, equity, and human dignity.

One of the most pressing ethical dilemmas in technology concerns privacy and data security. With the proliferation of digital technologies, vast amounts of personal data are collected, stored, and analyzed by governments, corporations, and other entities. While this data can be used to improve services, enhance user experiences, and drive innovation, it also raises significant concerns about surveillance, consent, and the potential for exploitation (Zuboff, 2019). The ethical question arises as to how much personal information should be collected, who has access to it, and how it should be protected. For instance, social media platforms and tech companies often collect data without explicit informed consent from users, leading to concerns about the violation of privacy rights and the manipulation of personal information for profit (O'Flaherty, 2020). Balancing the benefits of data-driven technologies with the need for privacy protection requires careful ethical considerations and regulatory frameworks that safeguard individual rights.

Another ethical dilemma related to technology is the growing reliance on artificial intelligence (AI) and automation. AI systems are increasingly used in decision-making processes across various sectors, including healthcare, criminal justice, and finance. While AI has the potential to improve efficiency, reduce human error, and enhance decision-making, it also raises

DOI: 10.1201/9788743808787-23

concerns about bias, accountability, and the displacement of human workers (Binns, 2018). AI algorithms are often trained on large datasets, which may reflect existing societal biases, leading to discriminatory outcomes. For example, AI systems used in predictive policing or hiring practices may inadvertently perpetuate racial or gender biases, reinforcing inequality and injustice (O'Neil, 2016). Furthermore, as AI systems become more autonomous, questions arise about who is responsible for the actions of these systems, especially in cases of harm or wrongdoing. The ethical challenge lies in ensuring that AI technologies are developed and deployed in ways that promote fairness, transparency, and accountability.

The ethical implications of biotechnology and genetic engineering also present significant dilemmas. Advances in gene editing technologies, such as CRISPR, have opened up new possibilities for treating genetic diseases, enhancing agricultural productivity, and even potentially altering human genetics. However, these technologies raise profound ethical questions about the boundaries of human intervention in nature. The ability to edit genes carries the risk of unintended consequences, such as the creation of genetic inequalities or the alteration of traits that could have long-term social and environmental impacts (Sandel, 2007). Additionally, the potential for "designer babies" and genetic enhancement has sparked debates about the moral implications of selecting traits such as intelligence, physical appearance, or abilities. The ethical dilemma centers on whether it is morally acceptable to manipulate the human genome and the potential societal consequences of such actions, including the exacerbation of social inequalities and the loss of genetic diversity.

Furthermore, the environmental impact of technology poses significant ethical challenges. The development and widespread use of technologies such as fossil fuel-based energy systems, electronic waste, and resource extraction have contributed to environmental degradation, climate change, and the depletion of natural resources (Borenstein, Herkert, & Herkert, 2017). As society becomes increasingly reliant on technology, it is essential to consider the ethical implications of technological development in relation to environmental sustainability. This includes evaluating the long-term environmental costs of technological innovations and ensuring that new technologies are developed with sustainability in mind. Ethical frameworks such as environmental justice call for a more equitable distribution of the benefits and burdens of technology, particularly for marginalized communities that are disproportionately affected by environmental harm (Schlosberg, 2007).

In conclusion, while technology has the potential to improve lives and solve global challenges, it also raises complex ethical dilemmas that require careful consideration. Issues such as privacy, AI bias, genetic engineering, and environmental sustainability must be addressed through thoughtful ethical frameworks, regulatory oversight, and public discourse. As technological advancements continue to shape the future, it is crucial to ensure that they are developed and deployed in ways that prioritize human well-being, social justice, and environmental sustainability. By grappling with these ethical dilemmas, society can harness the power of technology while minimizing its potential harm.

18.1 Artificial Intelligence and Automation

The rise of artificial intelligence (AI) and automation represents one of the most significant technological transformations of the 21st century. These technologies are reshaping industries, labor markets, and society, with the potential to bring about both profound benefits and complex challenges. AI, which refers to machines designed to simulate human intelligence, and automation, which involves the use of technology to perform tasks that would otherwise require human labor, are rapidly advancing across various sectors, including healthcare, manufacturing, transportation, and finance. However, the widespread implementation of AI and automation raises important ethical, economic, and social questions, particularly regarding employment, inequality, privacy, and the future of human work.

One of the primary concerns surrounding AI and automation is their impact on employment. Automation has the potential to replace human workers in many industries, especially those involving repetitive or manual tasks. According to Brynjolfsson and McAfee (2014), the automation of jobs in sectors such as manufacturing, retail, and logistics could lead to significant job displacement, particularly for low-skilled workers. This phenomenon, often referred to as "technological unemployment," raises concerns about the future of work and the economic implications of widespread job loss. While AI and automation can increase productivity and efficiency, they may also exacerbate inequality by disproportionately affecting workers in vulnerable sectors, leading to greater income disparity and social unrest (Frey & Osborne, 2017). Moreover, the displacement of workers could lead to a shift in labor markets, where high-skill jobs in AI and robotics may be in demand, but those without the necessary skills may struggle to find new employment opportunities.

However, proponents of AI and automation argue that these technologies can also create new job opportunities and enhance economic growth. As automation takes over routine tasks, workers may be freed up to engage in more creative, strategic, and complex roles that require human ingenuity and emotional intelligence. Furthermore, AI can help drive innovation in fields such as healthcare, education, and environmental sustainability, creating new industries and avenues for employment (Chui, Manyika, & Miremadi, 2016). For example, AI-powered tools can assist in diagnosing diseases, analyzing medical data, and developing personalized treatment plans, thus improving healthcare outcomes and creating demand for new healthcare roles. Similarly, automation can increase the efficiency of renewable energy systems, contributing to the transition to a more sustainable economy and creating green jobs. The challenge, however, is ensuring that workers are equipped with the skills needed to adapt to these new roles through education, training, and reskilling programs.

Another significant concern associated with AI and automation is the ethical implications of decision-making. As AI systems become more integrated into decision-making processes, particularly in areas such as criminal justice, hiring, and lending, questions arise about fairness, accountability, and transparency. AI algorithms are often trained on large datasets, which may contain biases that reflect existing societal inequalities (O'Neil, 2016). For example, predictive policing algorithms may disproportionately target minority communities, while hiring algorithms may favor certain demographic groups over others. These biases can perpetuate discrimination and reinforce existing social inequalities, raising ethical concerns about the fairness and justice of AI-driven decisions. Furthermore, the lack of transparency in AI algorithms – often referred to as the "black box" problem – makes it difficult for individuals to understand how decisions are being made and to challenge potentially harmful outcomes (Burrell, 2016). Addressing these ethical issues requires the development of AI systems that are transparent, accountable, and designed to minimize bias, as well as the establishment of regulations that ensure the responsible use of AI in decision-making.

AI and automation also raise significant privacy concerns, particularly regarding the collection and use of personal data. AI systems often rely on vast amounts of data to function effectively, which can include sensitive information such as health records, financial data, and online behaviors. The widespread use of AI in sectors like healthcare, finance, and social media raises the risk of data breaches, unauthorized surveillance, and misuse of personal information (Zuboff, 2019). For example, AI-powered surveillance systems used by governments and corporations could infringe on individuals' privacy rights and lead to the erosion of civil liberties. Ensuring the ethical use of AI requires robust data protection laws, informed consent practices, and mechanisms for individuals to control their personal information (O'Flaherty, 2020). Moreover, as AI systems become more autonomous, questions about accountability and liability in the event of harm or error arise. Who is responsible when an AI system causes harm, and how can individuals seek redress for such harm? These questions are central to the ethical deployment of AI technologies.

In conclusion, AI and automation are transforming the global landscape, offering both opportunities and challenges. While these technologies have the potential to enhance productivity, innovation, and efficiency, they also raise important ethical and societal concerns related to employment, fairness, privacy, and accountability. To fully realize the benefits of AI and automation while mitigating their risks, it is essential to develop ethical frameworks, regulatory policies, and education systems that ensure these technologies are deployed responsibly. As AI and automation continue to evolve, society must prioritize human well-being, equity, and social justice to ensure that technological progress benefits all members of society.

18.2 Privacy, Surveillance, and Digital Rights

In the digital age, the intersection of privacy, surveillance, and digital rights has become one of the most pressing issues in contemporary society. With the proliferation of digital technologies, personal data has become a commodity that is increasingly collected, stored, and analyzed by both private companies and governments. While this data can enhance user experiences, improve services, and drive innovation, it also raises significant concerns about privacy, autonomy, and the potential for surveillance. As digital technologies continue to evolve, the ethical, legal, and social implications of privacy and surveillance are critical areas of debate, particularly in relation to individual rights, data protection, and the balance between security and freedom.

One of the primary concerns surrounding privacy in the digital era is the widespread collection of personal data by tech companies, governments, and other entities. Digital platforms, such as social media, e-commerce sites, and search engines, collect vast amounts of data about individuals, including personal preferences, location, and behavior. This data is often used to target advertisements, optimize user experiences, and provide personalized services. However, the extensive collection of personal data raises significant concerns about consent, control, and the potential for misuse (Zuboff, 2019). Many users are unaware of the extent to which their data is collected or how it is used, leading to a lack of informed consent. Furthermore, data breaches and unauthorized access to personal information have become increasingly common, exposing individuals to risks such as identity theft and financial fraud (O'Flaherty, 2020). The ethical question arises as to how much control individuals should have over their own data and how privacy can be protected in an increasingly data-driven world.

Surveillance is another critical issue in the context of privacy and digital rights. Governments and corporations have increasingly turned to surveillance technologies, such as facial recognition, geolocation tracking, and online monitoring, to gather information about individuals' activities. While surveillance can be justified in the name of national security or crime prevention, it also raises concerns about civil liberties and the erosion of privacy rights (Lyon, 2007). For example, the use of mass surveillance by governments to monitor citizens' communications and movements can lead to a "chilling effect" on free speech and dissent, as individuals may alter their behavior if they believe they are being watched. Similarly, the widespread use of surveillance by corporations to track consumer behavior and preferences raises questions about the extent to which individuals' personal lives should be subject to scrutiny for commercial purposes (Andrejevic, 2014). The ethical dilemma lies in balancing the need for security and convenience with the protection of individual privacy and autonomy.

Digital rights, which encompass the rights of individuals to privacy, freedom of expression, and access to information in the digital realm, are also a central concern in discussions about privacy and surveillance. As digital technologies have become an integral part of daily life, the question of how to protect individuals' rights in the online space has gained prominence. The European Union's General Data Protection Regulation (GDPR), for example, represents a significant step toward safeguarding digital rights by giving individuals greater control over their personal data and ensuring that companies are held accountable for data protection (European Commission, 2018). Similarly, the right to be forgotten, which allows individuals to request the deletion of personal data from online platforms, has been enshrined in various legal frameworks as a way to protect individuals' privacy and digital identities (Custers et al., 2018). However, challenges remain in ensuring that digital rights are upheld globally, particularly in countries with less robust data protection laws and where government surveillance practices may infringe on individuals' freedoms.

Moreover, the increasing use of artificial intelligence (AI) and machine learning in surveillance and data collection has raised new concerns about the potential for discriminatory practices and bias. AI algorithms are often used to analyze vast amounts of data to predict behavior, identify threats, and personalize content. However, these algorithms can perpetuate existing biases in the data, leading to discriminatory outcomes in areas such as hiring, law enforcement, and credit scoring (O'Neil, 2016). For example, predictive policing algorithms that rely on historical crime data may disproportionately target marginalized communities, reinforcing racial and socioeconomic inequalities. Additionally, the lack of transparency in AI decision-making processes, often referred to as the "black box" problem, makes it difficult for individuals to understand how their data is being used or to challenge potentially harmful decisions (Burrell, 2016). These issues highlight the need for greater accountability and transparency in the use of AI in surveillance and data processing, as well as the importance of ensuring that digital rights are protected in an increasingly automated world.

In conclusion, privacy, surveillance, and digital rights are critical issues that require ongoing attention as digital technologies continue to evolve. The widespread collection of personal data, the increasing use of surveillance technologies, and the growing importance of digital rights raise significant ethical, legal, and social challenges. Protecting privacy in the digital age requires a careful balance between the benefits of data-driven technologies and the need to safeguard individual freedoms. Legal frameworks such as the GDPR represent important steps toward protecting digital rights, but global cooperation and ongoing efforts are necessary to ensure that privacy and autonomy are upheld in an increasingly interconnected world. As technology continues to advance, it is essential to prioritize the protection of digital rights and ensure that individuals have control over their personal information and the ability to navigate the digital space with confidence and security.

18.3 The Ethics of Genetic Engineering

Genetic engineering, the manipulation of an organism's genetic material to alter its characteristics, has raised significant ethical concerns since its inception. This technology encompasses a variety of practices, including gene editing, cloning, and the creation of genetically modified organisms (GMOs). While these advancements offer promising applications in medicine, agriculture, and environmental sustainability, they also provoke complex ethical dilemmas. The ethics of genetic engineering touch on a range of issues, including the potential risks to individuals and the environment, questions of consent and autonomy, social justice, and the broader implications for human identity and society.

One of the most prominent ethical concerns surrounding genetic engineering is the potential for unintended harm, particularly in the context of human genetic modifications. Techniques like CRISPR-Cas9, which allow for precise editing of genes, have revolutionized the field of medicine by offering the potential to cure genetic disorders such as cystic fibrosis and sickle cell anemia (Doudna & Charpentier, 2014). However, these technologies are not without risks. The possibility of off-target effects, where unintended genes are altered, raises concerns about long-term safety (Buchanan et al., 2000). Furthermore, the modification of germline cells – those that are passed on to future generations – has the potential to create lasting genetic changes that could have unforeseen consequences. Critics argue that genetic engineering should be approached with caution, as the full extent of its long-term effects is not yet understood (Sandel, 2004). In this context, the ethical responsibility of scientists and policymakers is to ensure that genetic interventions are safe and do not lead to harmful consequences for individuals or society.

Another key ethical issue is the question of consent, particularly in relation to genetic modifications made to embryos or germline cells. When genetic alterations are made to embryos, these changes are inherited by future generations, raising questions about whether it is ethical to make such decisions on behalf of individuals who cannot consent (Savulescu & Bortolotti, 2009). This concern is especially relevant in the case of non-therapeutic genetic modifications, such as selecting for physical or cognitive traits. The idea of creating "designer babies" through genetic engineering has sparked fears about the commodification of human life and the potential for genetic enhancement to exacerbate social inequalities (Kass, 2002). Critics argue that such practices could lead to a society where genetic differences are stigmatized, creating a new form of inequality based on genetic "perfection" (Habermas, 2003). On the other hand, proponents of genetic engineering argue that it could offer significant benefits, particularly in preventing debilitating genetic diseases, and that it could be seen as an extension of human progress and medical intervention (Parker et al., 2016).

The ethical debate surrounding genetic engineering also involves issues of social justice and equity. As gene-editing technologies become more advanced and accessible, there is concern that they may be available only to the wealthy, exacerbating existing social inequalities. For example, the high cost of gene therapies could make them inaccessible to low-income individuals or countries, leading to a genetic divide where only the privileged have access to genetic enhancements (Mackenzie, 2016). This raises important questions about fairness and justice in the distribution of genetic technologies. If genetic engineering becomes a tool for enhancing human abilities, there is a risk that it could reinforce existing power imbalances, creating a society where genetic "advantage" becomes another marker of social stratification. Addressing these concerns requires ensuring that genetic technologies are developed and distributed in a way that promotes equity and benefits all members of society, rather than deepening existing divisions.

Furthermore, genetic engineering has profound implications for human identity and the concept of what it means to be human. The ability to modify the genetic makeup of an individual or species challenges traditional notions of human nature and raises philosophical questions about the essence of life. Some ethicists argue that genetic engineering could lead to the loss of the diversity and unpredictability that define human existence, reducing individuals to a set of predetermined traits (Habermas, 2003). Others, however, contend that genetic engineering represents a natural extension of human progress, offering the potential to eliminate suffering and improve human well-being (Sandel, 2004). The ethical dilemma lies in determining the limits of human intervention in nature and balancing the potential benefits of genetic engineering with the preservation of human dignity and diversity.

In conclusion, the ethics of genetic engineering involves a delicate balance between the potential benefits of the technology and the risks it poses to individuals, society, and the environment. While genetic engineering holds great promise in fields such as medicine and agriculture, it also raises significant ethical concerns regarding safety, consent, equity, and human identity. As these technologies continue to evolve, it is essential that they are developed within a robust ethical framework that ensures their responsible use. This requires ongoing dialogue and careful consideration of the broader implications of genetic engineering, as well as the creation of policies and regulations that promote justice, safety, and the well-being of future generations.

References

Andrejevic, M. (2014). *Surveillance in the digital enclosure*. The Communication Review, 17(4), 335–350. https://doi.org/10.1080/10714421.2014.961302

Binns, R. (2018). *The ethics of artificial intelligence*. Oxford University Press.

Borenstein, J., Herkert, J. R., & Herkert, M. (2017). *The ethics of technology: A framework for ethical decision making*. MIT Press.

Brynjolfsson, E., & McAfee, A. (2014). *The second machine age: Work, progress, and prosperity in a time of brilliant technologies*. W. W. Norton & Company.

Buchanan, A., Brock, D. W., Daniels, N., & Wikler, D. (2000). *From chance to choice: Genetics and justice*. Cambridge University Press.

Burrell, J. (2016). *How the machine 'thinks': Understanding opacity in machine learning algorithms*. Big Data & Society, 3(1), 1–12. https://doi.org/10.1177/2053951715622512

Chui, M., Manyika, J., & Miremadi, M. (2016). *Where machines could replace humans—and where they can't (yet)*. McKinsey Quarterly.

Custers, B., Dechesne, F., Van der Hof, S., & Zwitter, A. (2018). *The right to be forgotten: A comparative analysis of the EU and US legal frameworks*. Computer Law & Security Review, 34(5), 1057–1070. https://doi.org/10.1016/j.clsr.2018.06.001

Doudna, J. A., & Charpentier, E. (2014). The new frontier of genome engineering with CRISPR-Cas9. *Science, 346*(6213), 1258096. https://doi.org/10.1126/science.1258096

European Commission. (2018). *General Data Protection Regulation (GDPR) compliance guidelines*. Retrieved from https://ec.europa.eu/info/law/law-topic/data-protection_en

Frey, C. B., & Osborne, M. A. (2017). *The future of employment: How susceptible are jobs to computerization?* Technological Forecasting and Social Change, 114, 254–280. https://doi.org/10.1016/j.techfore.2016.08.019

Habermas, J. (2003). *The future of human nature*. Polity Press.

Kass, L. R. (2002). The ethics of human cloning. *The New Atlantis, 1*(1), 9–22. https://www.thenewatlantis.com/publications/the-ethics-of-human-cloning

Lyon, D. (2007). *Surveillance studies: An overview*. Polity Press.

Mackenzie, C. (2016). *Genetic engineering and the social justice challenge*. Cambridge University Press.

O'Flaherty, K. (2020). *The ethical implications of data privacy in the digital age*. Journal of Information Ethics, 29(2), 102–115.

O'Neil, C. (2016). *Weapons of math destruction: How big data increases inequality and threatens democracy*. Crown Publishing Group.

Parker, M., Chisholm, L., & Latham, A. (2016). *Ethics of genetic engineering and the human genome*. Springer.

Sandel, M. J. (2007). *The case against perfection: Ethics in the age of genetic engineering*. Harvard University Press.

Savulescu, J., & Bortolotti, L. (2009). *Genetic enhancement: A new frontier in bioethics*. Oxford University Press.

Schlosberg, D. (2007). *Defining environmental justice: Theories, movements, and nature*. Oxford University Press.

Zuboff, S. (2019). *The age of surveillance capitalism: The fight for a human future at the new frontier of power*. PublicAffairs.

19

Global Wars and Conflict Crisis

Global wars and conflicts have profoundly shaped the course of human history, leaving indelible marks on societies, economies, and geopolitics. These crises, which span from world wars to regional conflicts, arise from a complex interplay of political, economic, cultural, and ideological factors. Their devastating impacts underscore the urgent need for cooperative solutions to address the root causes of violence and promote sustainable peace.

The two World Wars of the 20th century are among the most catastrophic conflicts in human history. World War I (1914–1918) resulted in the deaths of over 16 million people and set the stage for profound geopolitical changes, including the collapse of empires and the redrawing of national boundaries (Keegan, 1998). The Treaty of Versailles in 1919, while aimed at ensuring peace, created resentment that contributed to the outbreak of World War II (1939–1945), a conflict that claimed more than 70 million lives, including millions of civilians (Weinberg, 2005). The aftermath of these wars saw the establishment of the United Nations (UN) to prevent future conflicts and promote international cooperation. However, despite such efforts, the Cold War period (1947–1991) witnessed a proliferation of proxy wars, such as the Korean War (1950–1953) and the Vietnam War (1955–1975), as ideological rivalries between the United States and the Soviet Union played out on a global scale (Gaddis, 2005).

In the post-Cold War era, conflicts have increasingly been characterized by intrastate violence, ethnic tensions, and resource competition. Civil wars, such as those in Rwanda (1994) and Syria (2011–present), have led to mass atrocities, including genocide and the displacement of millions (Des Forges, 1999; Yassin-Kassab & Al-Shami, 2016). These conflicts highlight the failure of governance, economic disparities, and deep-seated social divisions as key drivers of modern warfare. Moreover, the global war on terror, initiated after the September 11 attacks in 2001, has reshaped international relations and led to protracted military engagements in Afghanistan,

DOI: 10.1201/9788743808787-24

Iraq, and other regions, often with unintended humanitarian consequences (Bacevich, 2016).

Global wars and conflicts exacerbate human suffering, disrupt economic development, and perpetuate cycles of poverty and inequality. According to the United Nations, conflicts today are the leading cause of displacement, with over 100 million people forcibly displaced worldwide as of 2022 (UNHCR, 2022). War-torn regions face the destruction of infrastructure, education systems, and healthcare services, which undermines long-term recovery and resilience. Additionally, conflicts over natural resources, such as oil, water, and arable land, are expected to intensify due to climate change, further complicating global peace efforts (Klare, 2012).

Addressing the global conflict crisis requires a multi-faceted approach that includes diplomacy, peacebuilding, and conflict resolution mechanisms. International organizations, such as the United Nations and regional bodies like the African Union, play a pivotal role in mediating disputes, deploying peacekeeping missions, and supporting post-conflict reconstruction. Promoting inclusive governance, economic development, and social justice is essential for addressing the root causes of conflict and fostering lasting peace (Boutros-Ghali, 1992). Furthermore, strengthening international cooperation and multilateral agreements can help mitigate the risks of future conflicts in an increasingly interconnected world.

19.1 Historical Overview of Global Wars and Conflicts

The 20th century marked an era of unprecedented global wars and conflicts that reshaped the political, social, and economic landscape of the world. Beginning with World War I (1914–1918), often referred to as "The Great War," this conflict stemmed from a complex interplay of factors, including militarism, alliances, imperialism, and nationalism. Triggered by the assassination of Archduke Franz Ferdinand of Austria-Hungary, World War I saw the Central powers (Germany, Austria-Hungary, and the Ottoman Empire) clash with the Allied powers (France, Britain, Russia, and later the United States). The war's devastating trench warfare and use of modern weaponry resulted in over 16 million deaths and widespread destruction. The Treaty of Versailles (1919), which placed heavy reparations on Germany, sowed the seeds for future conflict by fostering resentment and economic instability (Keegan, 2000).

World War II (1939–1945) emerged as a direct consequence of the unresolved tensions from World War I, compounded by the rise of fascist regimes in Germany, Italy, and Japan. Adolf Hitler's aggressive expansionism and the policy of appeasement by Western powers escalated into a global conflict that engulfed Europe, Asia, and the Pacific. The war was marked by large-scale atrocities, including the Holocaust, which resulted in the genocide of six million Jews. The conflict ended with the defeat of the Axis powers, the atomic bombings of Hiroshima and Nagasaki, and the establishment of the United Nations to promote international peace. World War II caused an estimated 70–85 million deaths, profoundly altering global geopolitics by paving the way for the Cold War and decolonization movements (Weinberg, 2005).

The Cold War era (1947–1991) introduced a new dimension to global conflict, characterized by ideological rivalry between the United States and the Soviet Union. Although the two superpowers avoided direct military confrontation, they engaged in a series of proxy wars in regions such as Asia, Africa, and Latin America. The Korean War (1950–1953) and the Vietnam War (1955–1975) exemplified the global struggle between capitalism and communism. In Korea, the division between the North (supported by the Soviet Union and China) and the South (backed by the United States) solidified into a lasting geopolitical rift. Similarly, the Vietnam War underscored the devastating human cost of Cold War interventions, with millions of casualties and widespread environmental destruction (Hobsbawm, 1994).

In addition to these conflicts, the Cold War fueled regional instabilities, including civil wars in Africa and Latin America. The Soviet-Afghan War (1979–1989) further highlighted the global nature of ideological battles, as the U.S. supported Afghan insurgents to counter Soviet influence. The Cold War's conclusion in 1991, marked by the dissolution of the Soviet Union, ushered in a new era of conflict dynamics.

The post-Cold War period witnessed a significant shift from inter-state wars to intra-state conflicts, driven by ethnic, religious, and resource-based tensions. Examples include the Rwandan genocide (1994), the wars in the former Yugoslavia, and ongoing conflicts in the Middle East, such as the Syrian Civil War. Globalization and the rise of terrorism, particularly after the 9/11 attacks in 2001, introduced asymmetric warfare and non-state actors as significant players in global conflicts (Kaldor, 2013).

This historical overview demonstrates how wars and conflicts have evolved, influenced by political ideologies, economic pressures, and technological advancements. While the World Wars were global in scope and scale, the Cold War and its aftermath emphasized regional and localized conflicts with far-reaching consequences. Understanding these historical patterns is crucial to addressing contemporary challenges and fostering sustainable peace.

19.2 Root Causes of Modern Conflicts

Modern conflicts are often driven by a complex interplay of political, economic, social, and environmental factors. Unlike the large-scale inter-state wars of the past, contemporary conflicts tend to be more localized, intra-state, and prolonged, often involving non-state actors. At the heart of these conflicts lie issues such as political instability, economic inequality, ethnic and religious divisions, and environmental challenges, which together create a volatile environment conducive to violence and war.

Political instability and failed governance are among the most significant contributors to modern conflicts. Weak institutions, corruption, and authoritarian regimes often lead to a breakdown of the social contract between the state and its citizens. This creates power vacuums that are exploited by insurgent groups, militias, or extremist organizations. For instance, the Syrian Civil War began in 2011 as a result of political repression and the government's violent crackdown on pro-democracy protests. The conflict escalated into a multi-faceted war involving regional and global powers, exacerbated by sectarian divisions and the rise of extremist groups like ISIS (Lynch, 2016). Failed governance in Syria not only led to internal displacement but also triggered a refugee crisis with significant international repercussions.

Economic inequality and poverty further exacerbate tensions, particularly in regions where wealth disparities are pronounced. Economic marginalization often fuels grievances among disadvantaged groups, leading to social unrest and violence. Sub-Saharan Africa provides numerous examples of conflicts rooted in economic disparities. For instance, the Niger Delta conflict in Nigeria stems from the exploitation of oil resources, where local communities suffer from environmental degradation and lack access to the wealth generated by oil extraction (Obi, 2010). Similarly, unequal distribution of resources and opportunities often intensifies ethnic and regional rivalries, creating fertile ground for insurgencies.

Ethnic and religious divisions play a critical role in modern conflicts, particularly when manipulated by political elites for their own gain. Ethnic tensions have fueled violence in places like Rwanda, where the 1994 genocide resulted in the deaths of approximately 800,000 Tutsis and moderate Hutus (Mamdani, 2001). In other cases, religious extremism has been a driving force, as seen in the rise of groups like Boko Haram in Nigeria and the Taliban in Afghanistan. These groups exploit socio-political grievances and use religious ideologies to recruit followers and justify violence.

Another emerging driver of conflict is resource scarcity and competition, particularly in the context of climate change. Natural resources such as oil, water, and arable land have become flashpoints for violence, especially in regions already experiencing political instability and economic hardship. Climate change exacerbates these challenges by increasing droughts, floods, and desertification, leading to food insecurity and forced migration. For example, the Syrian Civil War was partially preceded by a severe drought (2006–2010) that devastated agricultural livelihoods, displaced rural populations, and increased tensions in urban areas (Kelley et al., 2015). Similarly, in Darfur, Sudan, competition over scarce water and grazing resources contributed to conflict between pastoralist and agricultural communities, highlighting the intersection of environmental and social factors (Brown, 2013).

Geopolitical rivalries also continue to fuel modern conflicts, as major powers compete for influence in strategically important regions. The proxy wars in the Middle East, such as the conflicts in Yemen and Libya, reflect the involvement of regional and global actors pursuing their political and economic interests. These rivalries often prolong conflicts and complicate peace efforts, as external powers provide military and financial support to competing factions (Gause, 2014). Table 19.1. provides the major wars and conflicts throughout history.

The table provides an overview of significant wars and conflicts across history, detailing their time periods, major participants, and key descriptions. It begins with World War I (1914–1918), a global conflict centered in Europe, sparked by the assassination of Archduke Franz Ferdinand, involving the Allied powers against the Central powers. This was followed by World War II (1939–1945), a devastating global conflict marked by the Holocaust and the use of atomic weapons, where the Allied powers defeated the Axis powers. The Vietnam War (1955–1975) highlights the Cold War-era struggle between communist North Vietnam and anti-communist South Vietnam, supported by opposing global superpowers.

The table also covers the Korean War (1950–1953), a conflict between North and South Korea with significant involvement from the USA, China, and the UN. The prolonged Cold War (1947–1991) is characterized as a geopolitical standoff between the USA and USSR, marked by proxy wars and a nuclear arms race. Historical conflicts like the American Civil War (1861–1865), fought over slavery between the Union and Confederacy, and the Napoleonic Wars (1803–1815), led by Napoleon Bonaparte, are also included.

Additionally, the table mentions the ongoing Afghan War (2001–present), initiated by the US after the 9/11 attacks to combat terrorism, and the Iraq War (2003–2011), aimed at dismantling Saddam Hussein's regime. It also touches on the Israeli-Palestinian conflict (1948–present), a deeply rooted struggle over territory and sovereignty, and even speculates on a hypothetical World

Table 19.1 The major wars and conflicts throughout history.

War/Conflict	Time period	Major participants	Description
World War I	1914–1918	Allied powers (e.g., UK, France, Russia, USA) vs. Central powers (e.g., Germany, Austria-Hungary)	A global war centered in Europe, triggered by the assassination of Archduke Franz Ferdinand.
World War II	1939–1945	Allied pwers (e.g., USA, UK, USSR) vs. Axis powers (e.g., Germany, Japan, Italy)	A global conflict marked by the Holocaust, atomic bombings, and the defeat of Nazi Germany.
Vietnam War	1955–1975	North Vietnam (supported by USSR, China) vs. South Vietnam (supported by USA)	A Cold War-era conflict between communist North Vietnam and anti-communist South Vietnam.
Korean War	1950–1953	North Korea (supported by China, USSR) vs. South Korea (supported by USA, UN)	A war between communist North Korea and capitalist South Korea, involving significant US intervention.
Cold War	1947–1991	USA vs. USSR and their respective allies	A geopolitical tension between the USA and the USSR, marked by proxy wars, nuclear arms race, and ideological conflict.
American Civil War	1861–1865	Union vs. Confederacy	A war fought between the northern states (Union) and the southern states (Confederacy) primarily over slavery.
Napoleonic Wars	1803–1815	France (under Napoleon) vs. various European coalitions	A series of conflicts led by Napoleon Bonaparte, which reshaped Europe.
World War III (Theoretical)	N/A	N/A	A hypothetical future global conflict, often discussed in terms of nuclear warfare and global politics.
Afghan War	2001–Present	USA, NATO vs. Taliban, Al-Qaeda	A conflict initiated by the US in response to the 9/11 terrorist attacks, aimed at dismantling Al-Qaeda and the Taliban.
Iraq War	2003–2011	USA, UK vs. Iraq (Saddam Hussein's regime)	A controversial war was initiated by the US to remove Saddam Hussein from power, citing weapons of mass destruction.
Israeli Palestinian Conflict	1948–Present	Israel vs. Palestine, Arab States	A long-standing conflict over territory, political sovereignty, and religious significance.

Source: (Gaddis, 2005; Karnow, 1997; Overy, 2014; Stueck, 1995; Keegan, 1998).

War III, often discussed in the context of nuclear warfare. This compilation underscores the diverse causes, scales, and impacts of conflicts throughout history, reflecting their profound influence on global geopolitics and human society.

Modern conflicts are rarely driven by a single cause but rather by the convergence of political instability, economic inequality, ethnic and religious divisions, and environmental stressors. Climate change and resource competition increasingly serve as catalysts, intensifying pre-existing vulnerabilities in fragile states. Case studies such as the Syrian Civil War, conflicts in sub-Saharan Africa, and the Darfur crisis illustrate how multiple factors intersect to create complex and enduring crises. Addressing these root causes requires a multi-dimensional approach that promotes governance reforms, economic equity, environmental sustainability, and international cooperation.

19.3 Humanitarian Consequences of Wars and Conflicts

Wars and conflicts have devastating humanitarian consequences, causing immense suffering to individuals, communities, and nations. The impacts extend far beyond the battlefield, leaving lasting scars on societies. These consequences include displacement, loss of life, destruction of infrastructure, and long-term psychological and economic harm. The scale of the humanitarian toll is particularly evident in conflict zones where the provision of essential services such as healthcare, education, and humanitarian aid becomes nearly impossible.

One of the most immediate consequences of war is displacement. Conflicts force millions of people to flee their homes, creating refugees and internally displaced persons (IDPs). According to the United Nations High Commissioner for Refugees (UNHCR), by the end of 2022, over 108 million people were forcibly displaced worldwide due to conflict, persecution, and violence (UNHCR, 2023). Refugees face precarious living conditions in overcrowded camps or host countries, often lacking access to clean water, food, shelter, and healthcare. For example, the Syrian Civil War has displaced more than 13 million people since 2011, with over 6.8 million becoming refugees in neighboring countries like Turkey, Lebanon, and Jordan (Ferris & Kirisci, 2016). This displacement not only disrupts lives but also places immense strain on host nations' resources and infrastructure.

In addition to displacement, wars result in massive loss of life and physical destruction. Civilian casualties often outnumber military deaths, particularly in modern conflicts where urban warfare and aerial bombardments destroy residential areas, hospitals, and schools. The Rwandan genocide of 1994, for instance, resulted in the deaths of approximately 800,000 Tutsis and moderate Hutus within 100 days (Mamdani, 2001). Such events leave families shattered, with countless children orphaned and women widowed, creating intergenerational trauma and loss. Furthermore, the destruction of infrastructure, including roads, power grids, and healthcare systems, cripples a nation's ability to recover and rebuild after the conflict ends.

The humanitarian aid response in conflict zones is fraught with challenges. Providing food, medical care, and education to affected populations is often hindered by ongoing violence, restricted access, and targeted attacks on aid workers. Humanitarian organizations such as the International Committee of the Red Cross (ICRC) and Médecins Sans Frontières (Doctors Without Borders) play a critical role in delivering life-saving assistance, but their efforts are often constrained by security risks and limited resources (Slim, 2015). For instance, in Yemen, a conflict-driven humanitarian crisis has left 80% of the population in need of aid, yet delivering food and medicine is hampered by blockades and ongoing hostilities (Mundy, 2018).

The psychological impact of wars on individuals and societies is profound and long-lasting. Survivors of conflict, particularly children, often suffer from post-traumatic stress disorder (PTSD), anxiety, and depression. Exposure to violence, displacement, and the loss of loved ones can have lifelong effects on mental health and emotional well-being. In war-torn regions like Afghanistan, generations of children have grown up witnessing conflict, which disrupts their education and social development (Betancourt et al., 2013). The inability to access education further perpetuates cycles of poverty and violence, as young people are denied opportunities for a better future.

Wars also have significant economic consequences, leaving nations trapped in poverty and underdevelopment. The destruction of infrastructure, loss of productive labor, and disruption of trade result in economic stagnation. Post-conflict societies often face challenges in rebuilding their economies, as resources are diverted toward reconstruction and security. For instance, in Iraq and Syria, decades of conflict have devastated agricultural and industrial sectors, leading to widespread unemployment and poverty (World Bank, 2017). Additionally, the economic burden of hosting refugees falls heavily on neighboring countries, creating tensions and resource competition.

In conclusion, the humanitarian consequences of wars and conflicts are multifaceted, affecting millions of people worldwide. Displacement, loss of life, destruction of infrastructure, and psychological trauma are among the most severe impacts. Case studies such as the Syrian Civil War, the Rwandan genocide, and the Yemen crisis highlight the scale and urgency of these challenges. Addressing the humanitarian toll of wars requires global cooperation, increased humanitarian aid, and long-term strategies for peacebuilding and development to ensure that affected populations can recover and rebuild their lives.

19.4 Pathways to Peace and Conflict Resolution

Pathways to peace and conflict resolution involve a multifaceted approach that addresses the root causes of conflicts while fostering diplomacy, peacebuilding, and international cooperation. Effective conflict resolution strategies focus not only on ending violence but also on creating conditions for sustainable peace. This requires the involvement of international organizations, regional bodies, and non-governmental organizations (NGOs) that play pivotal roles in mediation, peacekeeping, and post-conflict reconstruction.

Diplomacy and mediation are fundamental tools for preventing and resolving conflicts. International organizations like the United Nations (UN) have been at the forefront of diplomatic efforts to bring warring parties to the negotiation table. The UN Charter emphasizes the importance of peaceful conflict resolution through dialogue and mediation (United Nations, 1945). The UN's peacekeeping missions, facilitated by the Department of Peace Operations, have contributed significantly to stabilizing conflict-prone regions. For instance, the UN Peacekeeping Mission in Liberia (UNMIL) helped the country transition from years of civil war to a more stable democratic society (Adebajo, 2014). Similarly, regional organizations such as the African Union (AU) have played a critical role in mediating conflicts on the continent, including the peace processes in Sudan and Somalia (Williams, 2017).

Inclusive governance and economic development are essential components of sustainable peace. Conflicts often arise from political exclusion, economic inequality, and lack of representation for marginalized groups. Inclusive governance ensures that all segments of society, including women, minorities, and youth, have a voice in decision-making processes. For example, the Good Friday Agreement of 1998, which ended decades of violence in Northern Ireland, was built on principles of power-sharing and political inclusion (Darby & Mac Ginty, 2003). By addressing historical grievances and fostering collaboration between conflicting parties, the agreement became a model for peacebuilding. Additionally, economic development plays a crucial role in conflict prevention and resolution. Poverty, unemployment, and competition over resources can exacerbate tensions, making economic growth and equitable resource distribution essential for long-term stability (Collier, 2007).

Post-conflict reconstruction is another critical pathway to peace. Societies emerging from war require significant investments in rebuilding infrastructure, institutions, and social trust. Reconstruction efforts must prioritize education, healthcare, and economic opportunities to prevent the recurrence of violence. The Truth and Reconciliation Commission (TRC) in South Africa serves as an exemplary model of post-conflict healing and reconciliation. Established after the end of apartheid, the TRC allowed victims and perpetrators to share their experiences, fostering accountability and collective healing (Tutu, 1999). By addressing past injustices and promoting forgiveness, the TRC contributed to rebuilding a fractured society.

The role of non-governmental organizations (NGOs) in peacebuilding cannot be overstated. NGOs such as International Crisis Group, Mediation Support Network, and Doctors Without Borders contribute to conflict resolution by providing humanitarian aid, facilitating dialogue, and supporting grassroots peace initiatives. NGOs often work in areas where state actors and international organizations have limited access, helping to rebuild communities and promote reconciliation. For instance, local peacebuilding initiatives in Rwanda following the 1994 genocide have focused on community-based reconciliation, empowering survivors and fostering coexistence (Clark, 2010).

International cooperation remains a cornerstone of conflict resolution. Multilateral agreements and partnerships between nations are essential for addressing transnational conflicts and fostering global stability. Initiatives such as the Sustainable Development Goals (SDGs) highlight the interconnectedness of peace, development, and human security. SDG 16 specifically calls for promoting peace, justice, and strong institutions as a foundation for sustainable development (United Nations, 2015).

In conclusion, pathways to peace and conflict resolution require a comprehensive approach that integrates diplomacy, inclusive governance, economic development, and post-conflict reconstruction. International organizations, regional bodies, and NGOs play vital roles in mediating conflicts and supporting peace processes. Successful examples, such as the Good Friday Agreement and South Africa's Truth and Reconciliation Commission, demonstrate that sustainable peace is achievable through cooperation, dialogue, and a commitment to justice. By addressing the root causes of conflicts and promoting reconciliation, societies can move toward a more peaceful and equitable future.

References

Adebajo, A. (2014). *UN peacekeeping in Africa: From the Suez crisis to the Sudan conflicts*. Lynne Rienner Publishers.

Bacevich, A. J. (2016). *America's War for the Greater Middle East: A Military History*. Random House.
Betancourt, T. S., Borisova, I., Williams, T. P., Brennan, R. T., Whitfield, T. H., De La Soudiere, M., ... & Gilman, S. E. (2013). Psychosocial adjustment and mental health in former child soldiers: Systematic review of the literature and recommendations. *Journal of Child Psychology and Psychiatry*, 54(1), 17–36. https://doi.org/10.1111/j.1469-7610.2012.02620.x
Boutros-Ghali, B. (1992). *An Agenda for Peace: Preventive Diplomacy, Peacemaking, and Peacekeeping*. United Nations.
Brown, O. (2013). Climate change and conflict. *International Affairs*, 89(5), 1123–1141. https://doi.org/10.1111/1468-2346.12068
Clark, P. (2010). *The Gacaca courts, post-genocide justice and reconciliation in Rwanda: Justice without lawyers*. Cambridge University Press.
Collier, P. (2007). *The bottom billion: Why the poorest countries are failing and what can be done about it*. Oxford University Press.
Darby, J., & Mac Ginty, R. (2003). *Contemporary peace making: Conflict, violence and peace processes*. Palgrave Macmillan.
Des Forges, A. (1999). *Leave None to Tell the Story: Genocide in Rwanda*. Human Rights Watch.
Ferris, E., & Kirisci, K. (2016). *The consequences of chaos: Syria's humanitarian crisis and the failure to protect*. Brookings Institution Press.
Gaddis, J. L. (2005). *The Cold War: A New History*. Penguin Books.
Gaddis, J. L. (2005). *The Cold War: A New History*. Penguin Press.
Gause, F. G. (2014). Beyond sectarianism: The new Middle East cold war. *Brookings Doha Center Analysis Paper*, 11.
Hobsbawm, E. J. (1994). *Age of Extremes: The Short Twentieth Century, 1914–1991*. Vintage Books.
Kaldor, M. (2013). *New and Old Wars: Organized Violence in a Global Era* (3rd ed.). Polity Press.
Karnow, S. (1997). *Vietnam: A History*. Penguin Books.
Keegan, J. (2000). *The First World War*. Vintage Books.
Kelley, C. P., Mohtadi, S., Cane, M. A., Seager, R., & Kushnir, Y. (2015). Climate change in the Fertile Crescent and implications of the recent Syrian drought. *Proceedings of the National Academy of Sciences*, 112(11), 3241–3246. https://doi.org/10.1073/pnas.1421533112
Klare, M. T. (2012). *The Race for What's Left: The Global Scramble for the World's Last Resources*. Metropolitan Books.
Lynch, M. (2016). *The new Arab wars: Uprisings and anarchy in the Middle East*. PublicAffairs.

Mamdani, M. (2001). *When victims become killers: Colonialism, nativism, and the genocide in Rwanda*. Princeton University Press.

Mundy, M. (2018). The strategies of the coalition in the Yemen war: Aerial bombardment and food war. *International Journal of Middle East Studies*, 50(4), 725–738. https://doi.org/10.1017/S0020743818000900

Obi, C. I. (2010). Oil extraction, dispossession, and resistance in Nigeria's oil-rich Niger Delta. *Canadian Journal of Development Studies*, 30(1–2), 219–236. https://doi.org/10.1080/02255189.2010.9669289

Overy, R. (2014). *The Bombing War: Europe 1939–1945*. Penguin Books.

Slim, H. (2015). *Humanitarian ethics: A guide to the morality of aid in war and disaster*. Oxford University Press.

Stueck, W. (1995). *The Korean War: An International History*. Princeton University Press.

Tutu, D. (1999). *No future without forgiveness*. Image Books.

UNHCR. (2022). *Global Trends: Forced Displacement in 2022*. United Nations High Commissioner for Refugees. https://www.unhcr.org

UNHCR. (2023). Global trends: Forced displacement in 2022. United Nations High Commissioner for Refugees. Retrieved from https://www.unhcr.org

United Nations. (1945). *Charter of the United Nations*. Retrieved from https://www.un.org

United Nations. (2015). *Transforming our world: The 2030 Agenda for Sustainable Development*. Retrieved from https://sdgs.un.org

Weinberg, G. L. (2005). *A World at Arms: A Global History of World War II*. Cambridge University Press.

Weinberg, G. L. (2005). *A World at Arms: A Global History of World War II* (2nd ed.). Cambridge University Press.

Williams, P. D. (2017). *War and conflict in Africa*. Polity Press.

World Bank. (2017). *The toll of war: The economic and social consequences of the conflict in Syria*. World Bank Publications.

Yassin-Kassab, R., & Al-Shami, L. (2016). *Burning Country: Syrians in Revolution and War*. Pluto Press.

20

The Crisis of Meaning and Purpose

The crisis of meaning and purpose is a profound existential issue that has gained increasing attention in contemporary society. It refers to the widespread sense of disconnection, confusion, and loss of direction that many individuals experience in their personal lives and within the broader social context. This crisis can manifest in various forms, from feelings of nihilism and existential despair to the search for fulfillment through materialism or superficial goals. The roots of this crisis are complex, arising from a combination of cultural, psychological, and societal factors that have evolved in the modern era. These factors include the breakdown of traditional belief systems, the erosion of community ties, the rapid pace of technological change, and the pressures of a consumer-driven society.

One significant contributor to the crisis of meaning and purpose is the decline of traditional religious and cultural frameworks that once provided individuals with a clear sense of identity and purpose. In pre-modern societies, religious institutions and communal practices played a central role in shaping people's understanding of their lives and their place in the world (Taylor, 2007). However, the process of secularization, which has accelerated in the modern era, has led to the disintegration of these shared belief systems. As a result, many individuals find themselves adrift, lacking a coherent framework for understanding their existence. The decline of religious belief and the weakening of community bonds have created a vacuum, where individuals are left to navigate the complexities of life without the guidance that was once provided by collective ideologies and values (Zinnbauer et al., 1997). This shift has contributed to a widespread sense of existential uncertainty, where people struggle to find meaning in a world that seems increasingly fragmented and disenchanted.

In addition to the erosion of traditional structures, the rapid pace of technological and societal change has further exacerbated the crisis of meaning. The advent of globalization, digital technology, and the information age

DOI: 10.1201/9788743808787-25

has transformed the way individuals interact with the world. While these changes have brought about significant advancements in communication, access to information, and global interconnectedness, they have also led to a sense of disorientation and alienation. The constant barrage of information and the pressure to keep up with technological innovations can overwhelm individuals, leaving them feeling disconnected from their deeper values and long-term goals (Turkle, 2011). The rise of social media, for example, has created an environment where individuals are constantly comparing themselves to others, leading to feelings of inadequacy and a distorted sense of self-worth. In this context, people may seek meaning in transient and superficial pursuits, such as accumulating wealth, status, or material possessions, rather than engaging in deeper, more fulfilling sources of purpose (Schwartz, 2016).

Psychologically, the crisis of meaning and purpose is closely linked to a phenomenon known as "existential anxiety," which refers to the fear and discomfort that arise from confronting the inherent uncertainty and finitude of life (Yalom, 1980). This anxiety can be triggered by existential questions such as "What is the purpose of my life?" or "What happens after I die?" While these questions are a natural part of the human experience, they can become overwhelming when individuals lack a clear sense of direction or purpose. Existential anxiety is often compounded by the pressures of modern life, including the demands of work, family, and social expectations. In such an environment, people may resort to coping mechanisms such as escapism, consumerism, or substance use, which provide temporary relief but fail to address the underlying existential void. The inability to find meaning in one's life can lead to feelings of depression, isolation, and despair, which are increasingly prevalent in modern society (Frankl, 1946/2006).

The crisis of meaning and purpose also has profound implications for mental health and well-being. Studies have shown that individuals who have a strong sense of purpose are more likely to experience better mental and physical health, greater resilience in the face of adversity, and a higher quality of life (Ryff, 1989). Conversely, those who struggle with a lack of meaning in their lives are at greater risk of developing mental health issues, such as depression and anxiety (Steger et al., 2006). The search for meaning is not merely an abstract philosophical pursuit but is intimately connected to psychological well-being. The absence of purpose can lead to a sense of hopelessness, while the pursuit of meaningful goals can provide a sense of fulfillment and direction. In this context, the crisis of meaning is not just a societal issue but a deeply personal one, affecting individuals' mental health, relationships, and overall life satisfaction.

To address the crisis of meaning and purpose, scholars and practitioners have emphasized the importance of finding and cultivating sources of meaning that are grounded in personal values, relationships, and contributions to the greater good. Viktor Frankl (1946–2006), a psychiatrist and Holocaust survivor, argued that the search for meaning is a fundamental human drive and that individuals can find purpose even in the most challenging circumstances. Frankl's concept of "logotherapy" emphasizes the importance of discovering meaning through work, love, and suffering, suggesting that individuals can transcend their circumstances by focusing on something greater than themselves. Similarly, positive psychology emphasizes the importance of pursuing meaningful goals that align with one's core values, fostering a sense of purpose through engagement in activities that promote personal growth, social connection, and contribution to society (Seligman, 2011). By fostering a sense of meaning, individuals can navigate the challenges of modern life and find fulfillment, even in the face of uncertainty.

The crisis of meaning and purpose is a multifaceted issue that arises from the erosion of traditional belief systems, the rapid pace of technological and societal change, and the existential anxiety that accompanies the search for identity and direction. This crisis has significant implications for mental health and well-being, as individuals who lack a sense of purpose are more vulnerable to depression, anxiety, and existential despair. Addressing this crisis requires a shift toward cultivating sources of meaning that are grounded in personal values, relationships, and contributions to the greater good. By doing so, individuals can navigate the complexities of modern life and find a sense of purpose that promotes psychological resilience and fulfillment.

20.1 Materialism and Existential Challenges

Materialism, the belief that material wealth and possessions are the primary sources of happiness and well-being, has become a dominant cultural force in many contemporary societies. Rooted in capitalist frameworks, materialism often equates success with the accumulation of wealth, consumer goods, and status. This worldview, however, presents significant existential challenges, as it tends to overshadow more intrinsic values such as personal growth, relationships, and the search for meaning. While material possessions can provide temporary satisfaction, they often fail to fulfill deeper psychological needs, leading to a paradox of increased consumption and decreasing life satisfaction.

One of the key existential challenges posed by materialism is the erosion of intrinsic values in favor of extrinsic goals. Materialistic individuals often define their self-worth based on external markers such as income, status, and possessions. This focus on material success can overshadow more meaningful pursuits, such as cultivating deep relationships, engaging in creative expression, and fostering personal growth (Kasser, 2002). In societies that emphasize material wealth, individuals may prioritize short-term gratification and the acquisition of goods over long-term well-being, contributing to a sense of emptiness and dissatisfaction. The "hedonic treadmill" theory suggests that as people acquire more possessions or achieve higher status, they quickly adapt to these new circumstances, leading to a continual desire for more, without lasting happiness (Brickman & Campbell, 1971). This cycle of desire and consumption often leaves individuals feeling unfulfilled, despite their material wealth.

Psychologically, materialism is linked to a range of negative outcomes, including lower life satisfaction, higher anxiety, and increased rates of depression (Kasser & Ryan, 1993). Materialistic individuals often struggle to form deep, meaningful relationships, as they may prioritize career success or material acquisition over emotional connection (Simmons, 2002). This can lead to feelings of isolation, as individuals may become disconnected from others who do not share the same values. Furthermore, materialism fosters a sense of alienation, where individuals feel disconnected from their true selves or the broader community. In a society that prioritizes material wealth, people may feel that their worth is measured by what they own rather than who they are, which can exacerbate feelings of loneliness and existential anxiety (Twenge, 2017).

In addition to psychological distress, materialism can contribute to broader existential challenges related to the search for meaning and purpose. In a materialistic culture, success is often defined by wealth and possessions, leaving little room for individuals to explore deeper questions of purpose, identity, and the meaning of life. This focus on material accumulation can overshadow more existential concerns, such as one's role in the world, the pursuit of personal values, and the cultivation of a meaningful life. Viktor Frankl's (1946–2006) existential philosophy emphasizes the importance of finding meaning in life, even in the face of suffering. According to Frankl, the search for meaning is central to human existence, and it cannot be fulfilled by external possessions or achievements. Materialism, however, often distracts individuals from this deeper search for meaning, leading to a sense of spiritual emptiness and disillusionment.

The existential crisis induced by materialism is not only personal but also societal. In societies where consumerism is dominant, the relentless pursuit of material wealth can contribute to environmental degradation, social inequality, and a lack of collective responsibility. The focus on economic growth and material consumption often comes at the expense of ecological sustainability, leading to issues such as climate change, resource depletion, and biodiversity loss (Kasser, 2002). These global challenges present existential threats to humanity's survival and highlight the limitations of a materialistic worldview that prioritizes short-term economic gain over long-term ecological health. As individuals and communities grapple with the consequences of their consumption patterns, they may experience a profound sense of disconnection from nature and a crisis of purpose.

Despite the pervasive influence of materialism, there is a growing recognition of the need for alternative approaches to life that prioritize intrinsic values and well-being. Scholars in the fields of positive psychology, existential philosophy, and spiritual studies have emphasized the importance of personal growth, self-actualization, and social connection as sources of fulfillment (Ryff, 1989). These intrinsic values, in contrast to material wealth, contribute to a deeper sense of meaning and satisfaction in life. Research has shown that individuals, who prioritize intrinsic goals, such as developing meaningful relationships, contributing to society, and engaging in personal growth, tend to experience greater well-being and life satisfaction (Deci & Ryan, 2000). This shift toward intrinsic values can help individuals navigate the existential challenges posed by materialism and foster a more fulfilling and meaningful life.

Moreover, movements such as minimalism, voluntary simplicity, and degrowth have emerged as responses to the existential crisis of materialism. These movements advocate for a reduction in consumption and a focus on non-material sources of fulfillment, such as community engagement, creativity, and spiritual growth (Alexander, 2013). By embracing these alternative values, individuals can find meaning and purpose in life that is not dependent on the accumulation of material wealth. These movements challenge the dominant narrative of endless consumption and promote a more sustainable and equitable approach to living. By prioritizing well-being over material accumulation, individuals can cultivate a sense of meaning and purpose that transcends the existential emptiness often associated with materialism.

Materialism presents significant existential challenges by fostering a focus on extrinsic goals such as wealth and possessions, which often fail to provide lasting fulfillment. This focus on material success can contribute to psychological distress, social alienation, and a lack of meaning in life. Moreover, materialism has broader societal implications, contributing to environmental degradation and social inequality. To address these existential challenges, individuals and communities must explore alternative pathways to meaning and fulfillment, focusing on intrinsic values such as personal growth, relationships, and social responsibility. By shifting away from materialism and embracing these deeper values, individuals can cultivate a more sustainable, meaningful, and fulfilling life.

20.2 Decline of Traditional Belief Systems

The decline of traditional belief systems, including religion, cultural norms, and social institutions, is a notable trend in many modern societies. This shift has been driven by a variety of factors, including the rise of secularism, globalization, scientific advancements, and the increasing influence of individualism. As societies become more interconnected and technologically advanced, traditional belief systems that once provided structure, meaning, and a sense of community are increasingly being questioned or abandoned. This decline has profound implications for identity, social cohesion, and the moral frameworks that guide human behavior.

One of the primary reasons for the decline of traditional belief systems is the rise of secularism. Secularization theory, which was popularized by sociologists such as Max Weber and Émile Durkheim, suggests that as societies modernize, religious institutions and beliefs lose their influence over social and cultural life (Bruce, 2011). The expansion of scientific knowledge and rational thought has led many individuals to question religious doctrines and traditional worldviews that were once accepted as absolute truths. The increasing emphasis on empirical evidence, technological progress, and the scientific method has provided alternative explanations for phenomena that were previously attributed to divine or supernatural forces. As a result, traditional religious beliefs, particularly in Western societies, have seen a significant decline in both adherence and practice (Wilson, 2002). This secular shift has been particularly evident in Europe, where church attendance and religious affiliation have drastically decreased over the past century (Davie, 2007).

Globalization has also played a significant role in the decline of traditional belief systems. The interconnectedness of the world through technology, trade, and migration has exposed individuals to diverse cultures, beliefs, and worldviews. In this globalized environment, traditional belief systems are increasingly being challenged by the exposure to alternative perspectives and the blending of cultures. As people come into contact with different religions, philosophies, and lifestyles, they may begin to question the validity of their own traditions and beliefs. The spread of global media, including social media, has further facilitated this process by allowing individuals to access a wide range of ideas and perspectives, often leading to a more relativistic view of truth and belief (Giddens, 2002). This cultural exchange, while enriching, can also erode the dominance of traditional belief systems, as individuals may adopt more cosmopolitan or pluralistic views that are less tied to local customs and traditions.

The increasing emphasis on individualism and personal autonomy in modern societies has also contributed to the decline of traditional belief systems. In many cultures, particularly in the West, there has been a shift from collective identities rooted in family, community, and religious affiliation to more individualistic identities based on personal choice and self-expression (Putnam, 2000). This shift is reflected in the growing importance of personal freedom and self-determination, which often leads individuals to question or reject traditional authority figures, such as religious leaders, elders, or political institutions. The rise of consumer culture, which emphasizes personal desires and material success, further reinforces individualism by encouraging people to prioritize their own needs and preferences over collective values and traditions. This individualistic mindset can undermine the communal and intergenerational transmission of traditional beliefs, as younger generations are less likely to adopt the values and practices of their elders (Bellah et al., 1985).

Scientific advancements and technological progress have also played a significant role in the decline of traditional belief systems. As science has provided more accurate and comprehensive explanations for natural phenomena, traditional religious and mythological explanations have become less compelling. For example, the development of evolutionary theory, the discovery of the heliocentric model of the solar system, and the advancement of medical science have all challenged traditional religious narratives about the origins of life, the structure of the universe, and the nature of the human body (Dawkins, 2006). The success of science in providing practical solutions to human problems, such as curing diseases, improving living standards, and extending life expectancy, has further diminished the perceived need for religious or supernatural explanations. In many ways, science has replaced traditional belief systems as the primary source of authority and knowledge in modern societies.

However, the decline of traditional belief systems does not necessarily mean the end of meaning-making or moral frameworks. While many individuals have moved away from organized religion, they continue to seek meaning and purpose in other areas of life. This search for meaning is often expressed through secular ideologies, such as humanism, existentialism, or environmentalism, which provide alternative frameworks for understanding the world and guiding ethical behavior (Nagel, 2012). Moreover, the decline of traditional belief systems has led to the rise of new forms of spirituality and belief, including New Age movements, mindfulness practices, and interest in Eastern philosophies. These new belief systems often blend elements of traditional religions with contemporary values, offering individuals a more flexible and personalized approach to spirituality (Heelas, 1996).

The decline of traditional belief systems has also raised concerns about social cohesion and moral values. Many critics argue that the erosion of traditional values has led to a decline in social trust, community engagement, and ethical behavior. Without a shared belief system to unite individuals, societies may experience increased fragmentation and polarization, as people struggle to find common ground on moral and political issues (Putnam, 2000). This decline in social cohesion can manifest in various forms, including political polarization, social alienation, and the weakening of communal institutions. However, others argue that the decline of traditional belief systems provides an opportunity for the development of more inclusive, pluralistic, and flexible moral frameworks that can accommodate the diverse values and beliefs of a globalized world (Giddens, 2002).

The decline of traditional belief systems is a complex and multifaceted phenomenon driven by secularization, globalization, individualism, and scientific progress. While this decline has led to the erosion of long-established norms and values, it has also created space for the development of new belief systems and moral frameworks. The decline of traditional beliefs raises important questions about the future of social cohesion, identity, and meaning in an increasingly diverse and interconnected world. As societies continue to evolve, individuals will likely continue to seek new ways to navigate the challenges of modern life, while grappling with the implications of the decline of traditional belief systems.

20.3 The Search for New Philosophies of Life

In the face of rapid societal changes, environmental crises, and growing existential concerns, many individuals are increasingly seeking new philosophies of life that can provide meaning, purpose, and guidance. Traditional belief systems, including religion and cultural norms, which have long provided frameworks for understanding life's purpose, have become less dominant in many parts of the world. This decline in traditional worldviews, exacerbated by secularization, scientific progress, and globalization, has prompted a widespread search for new ways to make sense of the complexities of modern existence. As a result, new philosophical approaches, often drawing from diverse traditions and offering more individualized frameworks, have emerged to address the challenges of contemporary life.

One significant factor driving the search for new philosophies of life is the growing sense of existential uncertainty. With the rapid pace of technological change, environmental degradation, and political instability, many individuals experience a crisis of meaning. This sense of disillusionment, exacerbated by the erosion of traditional belief systems, has led to a search for alternative ways of understanding one's place in the world. Existentialism, for instance, offers a philosophical approach that focuses on individual freedom, responsibility, and the search for meaning in an often indifferent or absurd universe. Prominent existentialist thinkers like Jean-Paul Sartre and Albert Camus argued that life has no inherent meaning, and it is up to individuals to create their own purpose through authentic choices and actions (Sartre, 1943; Camus, 1942). This philosophy resonates with many contemporary individuals who feel disillusioned by traditional religious or cultural structures and seek a more personal, self-determined approach to life's meaning.

In addition to existentialism, humanism has gained prominence as a philosophy that seeks to promote human flourishing, autonomy, and dignity in the absence of traditional religious frameworks. Humanism emphasizes the importance of reason, compassion, and ethical responsibility, arguing that individuals can lead fulfilling lives through the pursuit of knowledge, creativity, and social responsibility (Rosenstand, 2013). As people turn away from religious explanations of existence, humanism provides a secular alternative that emphasizes the potential for human growth and development. This philosophy is particularly appealing in a globalized world where individuals are increasingly exposed to diverse cultural perspectives and values, and where the emphasis on personal choice and self-expression is paramount.

Another emerging philosophical framework is deep ecology, which challenges traditional anthropocentric views and advocates for a more holistic understanding of humanity's relationship with the natural world. Deep ecology, as articulated by thinkers such as Arne Naess and Bill Devall, argues that human beings are not separate from nature but are deeply interconnected with all living and non-living entities. This philosophy calls for a radical shift in how humans perceive and interact with the environment, emphasizing the intrinsic value of nature and the need for ecological sustainability (Naess, 1973; Devall & Sessions, 1985). As concerns over environmental degradation, climate change, and biodiversity loss intensify, deep ecology offers a framework for rethinking human existence with the Earth, encouraging a deeper sense of responsibility and respect for the natural world.

In addition to philosophical traditions such as existentialism, humanism, and deep ecology, the search for new philosophies of life is also reflected in the rise of spiritual but non-religious movements. Many individuals are increasingly turning to practices such as mindfulness, meditation, and yoga, which offer a sense of inner peace, self-awareness, and connection to the present moment. These practices, often rooted in Eastern traditions like Buddhism and Hinduism, have been integrated into Western contexts and are seen as ways to cope with the stresses of modern life. Mindfulness, in particular, has gained widespread popularity due to its focus on cultivating awareness and presence, helping individuals navigate the challenges of living in a fast-paced, consumer-driven society (Kabat-Zinn, 2003). These spiritual practices, while not necessarily tied to a specific religious belief system, provide individuals with tools for self-reflection and personal growth, addressing the existential concerns that arise in a rapidly changing world.

The search for new philosophies of life is also influenced by the increasing focus on social justice, equity, and collective well-being. As global inequalities, environmental crises, and social divisions continue to grow, there is a growing recognition that individual fulfillment cannot be achieved without addressing broader systemic issues. Philosophies such as critical theory, postmodernism, and ecofeminism have emerged as frameworks for understanding and addressing the complexities of power, identity, and social justice. These approaches challenge traditional notions of authority, truth, and progress, advocating for a more inclusive, pluralistic, and egalitarian society. For example, ecofeminism, which combines environmentalism and feminism, emphasizes the interconnectedness of social, environmental, and gender justice, advocating for a holistic approach to addressing issues such as climate change, resource exploitation, and gender inequality (Plumwood, 1993). Such philosophies offer a way forward in a world marked by social fragmentation and ecological degradation, emphasizing the need for collective action and solidarity.

While the search for new philosophies of life reflects the complexities of modern existence, it also highlights a desire for deeper meaning, purpose, and connection in a world that can often feel fragmented and uncertain. As individuals increasingly move away from traditional belief systems, they are seeking alternative ways to make sense of their lives and the world around them. New philosophical frameworks, whether secular or spiritual, individualistic or collective, offer avenues for personal growth, social responsibility, and ecological sustainability. These emerging philosophies reflect a broader cultural shift toward greater self-awareness, ethical responsibility, and interconnectedness, offering hope for a more meaningful and sustainable future.

References

Alexander, S. (2013). *The economics of happiness: Building genuine wealth*. University of Otago Press.

Bellah, R. N., Madsen, R., Sullivan, W. M., Swidler, A., & Tipton, S. M. (1985). *Habits of the heart: Individualism and commitment in American life*. University of California Press.

Brickman, P., & Campbell, D. T. (1971). Hedonic relativism and planning the good society. In M. H. Appley (Ed.), *Adaptation-level theory: A symposium* (pp. 287–302). Academic Press.

Bruce, S. (2011). *Secularization: In defence of an idea*. Oxford University Press.

Camus, A. (1942). *The myth of Sisyphus*. Gallimard.

Davie, G. (2007). *The sociology of religion: A critical agenda*. Sage Publications.

Dawkins, R. (2006). *The God delusion*. Houghton Mifflin Harcourt.

Deci, E. L., & Ryan, R. M. (2000). The "what" and "why" of goal pursuits: Human needs and the self-determination of behavior. *Psychological Inquiry, 11*(4), 227–268. https://doi.org/10.1207/S15327965PLI1104_01

Devall, B., & Sessions, G. (1985). *Deep ecology: Living as if nature mattered*. Gibbs M. Smith.

Frankl, V. E. (2006). *Man's search for meaning*. Beacon Press. (Original work published 1946)

Giddens, A. (2002). *Runaway world: How globalization is reshaping our lives*. Routledge.

Heelas, P. (1996). *The new age movement: The celebration of the self and the sacralization of modernity*. Blackwell.

Kabat-Zinn, J. (2003). *Full catastrophe living: Using the wisdom of your body and mind to face stress, pain, and illness*. Delta Trade Paperbacks.

Kasser, T. (2002). *The high price of materialism*. MIT Press.

Kasser, T., & Ryan, R. M. (1993). A dark side of the American dream: Correlates of financial success as a central life aspiration. *Journal of Personality and Social Psychology, 65*(2), 410–422. https://doi.org/10.1037/0022-3514.65.2.410

Naess, A. (1973). *The shallow and the deep, long-range ecology movement: A summary*. Inquiry, 16(1–4), 95–100. https://doi.org/10.1080/00201747308601682

Nagel, T. (2012). *Mind and cosmos: Why the materialist neo-Darwinian conception of nature is almost certainly false*. Oxford University Press.

Plumwood, V. (1993). *Feminism and the mastery of nature*. Routledge.

Putnam, R. D. (2000). *Bowling alone: The collapse and revival of American community*. Simon & Schuster.

Rosenstand, N. (2013). *The moral of the story: An introduction to ethics* (8th ed.). McGraw-Hill.

Ryff, C. D. (1989). Happiness is everything, or is it? *Explorations on the meaning of psychological well-being*. *Journal of Personality and Social Psychology, 57*(6), 1069–1081. https://doi.org/10.1037/0022-3514.57.6.1069

Sartre, J. P. (1943). *Being and nothingness: An essay on phenomenological ontology*. Gallimard.

Schwartz, B. (2016). *The paradox of choice: Why more is less*. HarperCollins.

Seligman, M. E. P. (2011). *Flourish: A visionary new understanding of happiness and well-being*. Free Press.

Simmons, A. L. (2002). *The materialist self: Identity and the pursuit of happiness*. University of California Press.

Steger, M. F., Frazier, P., Oishi, S., & Kaler, M. (2006). The meaning in life questionnaire: Assessing the presence of and search for meaning in life. *Journal of Counseling Psychology, 53*(1), 80–93. https://doi.org/10.1037/0022-0167.53.1.80

Taylor, C. (2007). *A secular age*. Belknap Press of Harvard University Press.

Turkle, S. (2011). *Alone together: Why we expect more from technology and less from each other*. Basic Books.

Twenge, J. M. (2017). *Igen: Why today's super-connected kids are growing up less rebellious, more tolerant, less happy—and completely unprepared for adulthood*. Atria Books.

Wilson, B. R. (2002). *Religion in a secular society: A sociological view*. Routledge.

Yalom, I. D. (1980). *Existential psychotherapy*. Basic Books.

Zinnbauer, B. J., Pargament, K. I., Cole, B., Rye, M. S., Butter, E. M., & Belavich, T. G. (1997). Religion and spirituality: Unfuzzying the fuzzy. *Journal for the Scientific Study of Religion, 36*(4), 549–564. https://doi.org/10.2307/1387689

PART VI

Pathways to Solutions

21

Sustainability and Resilience

In addressing the complex challenges of the modern world, ranging from environmental degradation to social inequality, it is essential to explore viable pathways to solutions that integrate interdisciplinary approaches, sustainable practices, and transformative systems. These solutions require a holistic perspective, recognizing the interconnectedness of global issues and the need for collaborative action across multiple sectors. Pathways to solutions involve both short-term and long-term strategies that address immediate concerns and lay the groundwork for sustainable, equitable, and resilient futures.

One critical pathway to solutions is the adoption of sustainable development practices that prioritize environmental protection, social equity, and economic stability. The United Nations Sustainable Development Goals (SDGs) provide a comprehensive framework for addressing global challenges, emphasizing the need for integrated approaches to poverty reduction, gender equality, climate action, and sustainable resource use (United Nations, 2015). Achieving these goals requires collaboration between governments, businesses, civil society, and individuals to create policies and practices that promote environmental stewardship, social inclusion, and economic opportunity. For instance, transitioning to a circular economy, which focuses on reducing waste and promoting the reuse and recycling of materials, is a key strategy for mitigating environmental impacts while fostering economic resilience (Geissdoerfer et al., 2017). By shifting away from linear models of production and consumption, a circular economy offers a pathway to decoupling economic growth from environmental degradation.

Another important pathway to solutions is the emphasis on transformative education and lifelong learning. Education plays a pivotal role in empowering individuals and communities to understand and address global challenges. By fostering critical thinking, creativity, and problem-solving skills, education can equip individuals with the tools necessary to navigate

DOI: 10.1201/9788743808787-27

the complexities of a rapidly changing world. Moreover, transformative education encourages active participation in social, environmental, and economic issues, promoting a sense of responsibility and agency in addressing global challenges (Mezirow, 2000). Lifelong learning, particularly in the context of rapidly advancing technologies and shifting labor markets, is essential for individuals to remain adaptable and resilient in the face of change. By integrating sustainability and social justice into educational curricula, societies can cultivate a generation of informed, engaged citizens who are equipped to contribute to solutions for pressing global issues.

In addition to education, fostering innovation and technological advancements plays a significant role in finding pathways to solutions. Technologies such as renewable energy, artificial intelligence, and biotechnology have the potential to address critical issues such as climate change, healthcare, and food security. However, the development and deployment of these technologies must be done ethically and inclusively to ensure that they benefit all members of society. For example, renewable energy technologies such as solar and wind power offer sustainable alternatives to fossil fuels, helping to mitigate climate change while creating new economic opportunities (Sovacool et al., 2016). Similarly, advancements in biotechnology, such as genetically modified crops, can contribute to food security by increasing agricultural productivity and resilience to climate-related disruptions. However, these innovations must be carefully managed to avoid unintended consequences, such as environmental harm or exacerbating social inequalities.

Addressing social and economic inequalities is another critical pathway to solutions. Inequality, whether in terms of income, education, or access to healthcare, exacerbates many of the global challenges we face. A key strategy for addressing inequality is promoting inclusive economic growth that ensures the benefits of development are shared equitably across society. This can be achieved through policies that support fair wages, access to quality education, and social protection systems that reduce poverty and social exclusion (Sen, 1999). Furthermore, efforts to address systemic discrimination based on race, gender, and other forms of social stratification are essential for creating more just and cohesive societies. By addressing the root causes of inequality, societies can foster greater social cohesion, stability, and resilience in the face of global challenges.

Finally, global cooperation and governance are essential pathways to solutions. Many of the most pressing issues, such as climate change, pandemics, and global migration, transcend national borders and require coordinated action at the international level. Strengthening global governance structures, such as the United Nations and regional organizations, is crucial for fostering collaboration and ensuring that solutions are equitable and effective. Additionally, promoting multilateralism and diplomacy can help resolve conflicts and build trust among nations, facilitating collective action on shared challenges (Keohane, 2005). Global cooperation also requires the active involvement of non-state actors, including civil society organizations, the private sector, and local communities, to ensure that solutions are inclusive and reflect the needs and aspirations of all people.

Pathways to solutions for the complex challenges of the modern world require a multifaceted approach that integrates sustainable practices, transformative education, technological innovation, social equity, and global cooperation. By addressing these issues in an interconnected and holistic manner, societies can create the conditions for a more sustainable, equitable, and resilient future. However, achieving these solutions requires the active participation of all sectors of society and a commitment to long-term, systemic change. Only through collective action and shared responsibility can we hope to address the pressing challenges of our time.

Sustainability and resilience are two interrelated concepts that have gained significant attention in the context of addressing global environmental, social, and economic challenges. While sustainability focuses on ensuring that resources are used in ways that meet present needs without compromising the ability of future generations to meet their own needs, resilience emphasizes the capacity of systems – whether ecological, social, or economic – to adapt to and recover from disruptions (Folke, 2006). Together, these concepts provide a framework for understanding how societies can navigate the complexities of climate change, biodiversity loss, and social inequality while maintaining long-term well-being.

Sustainability is often framed within the context of the triple bottom line, which includes environmental, social, and economic dimensions (Elkington, 1997). Environmental sustainability involves the responsible use and conservation of natural resources, ensuring that ecosystems can continue to provide the services that support life, such as clean air, water, and biodiversity. Social sustainability, on the other hand, focuses on creating equitable societies where all individuals have access to basic needs, opportunities for development, and the ability to participate in decision-making processes. Economic sustainability aims to create stable economies that can provide livelihoods, reduce poverty, and foster long-term prosperity without depleting natural resources or exacerbating inequality. The intersection of these three pillars forms the foundation for sustainable development, which seeks to balance the needs of the present with the preservation of the future (World Commission on Environment and Development, 1987).

Resilience, as a complementary concept, is concerned with the ability of systems to absorb shocks, adapt to changes, and recover from disturbances. In the context of climate change, resilience refers to the capacity of both human and natural systems to withstand extreme weather events, such as floods, droughts, and storms, and to adapt to long-term shifts in climate patterns. For example, communities that rely on agriculture may build resilience by diversifying their crops, implementing water-saving technologies, or adopting new farming practices that are more resistant to changing climate conditions (Adger, 2003). Similarly, ecosystems can demonstrate resilience by maintaining their functions and services in the face of disturbances, such as forest fires or coral bleaching events, through mechanisms like regeneration and adaptation (Folke et al., 2002). Building resilience, therefore, is not just about recovering from crises but also about preparing for future challenges and ensuring that systems can continue to thrive despite uncertainty.

The relationship between sustainability and resilience is crucial for addressing the global challenges of the 21st century. While sustainability focuses on creating systems that are capable of long-term stability, resilience emphasizes the ability to cope with and recover from unexpected disruptions. In practice, these two concepts often overlap, as systems that are sustainable are typically more resilient to change, and resilient systems are more likely to be sustainable in the long term. For example, a city that invests in renewable energy, efficient public transportation, and green infrastructure is not only promoting sustainability by reducing its environmental footprint but is also building resilience by diversifying its energy sources and reducing its dependence on vulnerable infrastructure (Reed et al., 2013). Similarly, communities that prioritize social equity, such as providing access to education, healthcare, and social safety nets, are better equipped to withstand economic shocks and social unrest, thereby enhancing their resilience (Cutter et al., 2008).

At the global level, the interconnectedness of environmental, social, and economic systems means that sustainability and resilience must be approached in a holistic manner. Climate change, for instance, is a global challenge that requires coordinated efforts across countries and sectors to reduce greenhouse gas emissions, protect vulnerable populations, and promote sustainable development. The Paris Agreement, adopted in 2015, is an example of a global effort to enhance both sustainability and resilience by aiming to limit global warming to well below 2°C and by promoting adaptive strategies for countries that are already experiencing the impacts of climate change (UNFCCC, 2015). Similarly, the 2030 Agenda for Sustainable Development, with its 17 Sustainable Development Goals (SDGs), provides a comprehensive framework for addressing the interconnected challenges of poverty, inequality, climate change, and environmental degradation, while promoting resilience through inclusive and sustainable development (United Nations, 2015).

Sustainability and resilience are essential concepts for addressing the complex and interconnected challenges facing the world today. By fostering sustainable practices that promote environmental conservation, social equity, and economic stability, societies can create systems that are capable of enduring long-term change. Simultaneously, by building resilience into these systems, societies can ensure that they are able to adapt to and recover from unforeseen shocks, whether they arise from climate change, economic crises, or social unrest. Ultimately, the integration of sustainability and resilience provides a pathway toward a more equitable, stable, and thriving future for all.

21.1 Renewable Energy and Climate Adaptation

The dual challenges of climate change and energy sustainability are among the most pressing global issues of the 21st century. Renewable energy and climate adaptation are two critical areas where the intersection of environmental, social, and technological strategies can significantly mitigate the impacts of climate change while promoting sustainable development. Renewable energy, which includes sources such as solar, wind, hydro, and geothermal, is central to reducing greenhouse gas emissions and transitioning away from fossil fuels. Meanwhile, climate adaptation refers to the adjustments that societies, ecosystems, and economies must make to cope with the unavoidable impacts of climate change. Together, renewable energy and climate adaptation represent complementary strategies for fostering resilience and sustainability in a changing climate.

Renewable energy plays a crucial role in climate mitigation by reducing reliance on fossil fuels, which are the primary contributors to global warming. The transition to renewable energy systems can significantly lower carbon emissions, thus helping to limit global temperature rise and reduce the severity of climate impacts. According to the Intergovernmental Panel on Climate Change (IPCC), achieving a sustainable energy future requires scaling up the deployment of renewable energy technologies to meet global energy demand while minimizing environmental harm (IPCC, 2018). Solar, wind, and hydropower, in particular, have emerged as the most widely adopted renewable energy sources due to their scalability, decreasing costs, and ability to provide clean electricity. For instance, solar energy has seen dramatic cost reductions over the past decade, making it a viable option for both developed and developing countries (IRENA, 2020). Similarly, wind energy is expanding rapidly, with both onshore and offshore wind farms providing substantial contributions to national energy grids in countries such as Germany, China, and the United States (Global Wind Energy Council, 2020).

The role of renewable energy in climate adaptation, however, extends beyond mitigation. In many regions, especially in developing countries and small island states, renewable energy can serve as a vital component of climate adaptation strategies. These regions are often the most vulnerable to the impacts of climate change, including rising sea levels, extreme weather events, and disruptions to food and water security. Renewable energy systems, particularly decentralized and off-grid solutions, offer an opportunity to build resilience by providing reliable and affordable energy in areas that are not connected to national power grids. For example, solar-powered microgrids can supply electricity to remote communities, reducing their dependence on fossil fuels and enhancing their ability to cope with climate-induced disruptions, such as power outages caused by storms or floods (Sovacool et al., 2018). Moreover, renewable energy technologies can be integrated with other climate adaptation measures, such as water pumping for irrigation, refrigeration for food storage, and cooling systems for healthcare facilities, all of which are essential for maintaining well-being in the face of climate change.

The integration of renewable energy into climate adaptation efforts also offers a pathway for addressing the vulnerabilities of key sectors, such as agriculture, water resources, and health. For example, renewable energy can support climate-resilient agricultural practices by powering irrigation systems, greenhouse farming, and precision agriculture technologies that optimize water and energy use (Liu et al., 2019). In the water sector, renewable energy can enhance water access and management in regions facing droughts or reduced water availability. Solar-powered water pumps, for instance, can be used to extract groundwater or pump water for irrigation, reducing reliance on traditional energy sources that may be unreliable or costly (Awerbuch & Berger, 2003). Additionally, renewable energy can contribute to climate-resilient health systems by providing reliable electricity for hospitals, clinics, and refrigeration of vaccines, which is particularly critical in regions vulnerable to climate-related health risks, such as the spread of infectious diseases and heatwaves (Baker et al., 2020).

Despite the potential benefits, the deployment of renewable energy for climate adaptation faces several challenges, particularly in terms of financing, infrastructure, and policy frameworks. The initial capital costs of renewable energy technologies, while decreasing, remain a barrier for many low-income countries and communities. Furthermore, the integration of renewable energy into existing energy systems requires significant investments in infrastructure, such as grid expansion, storage solutions, and energy management systems. To address these challenges, international financial institutions, governments, and the private sector must collaborate to provide funding, technical support, and capacity-building to enable the widespread adoption of renewable energy for climate adaptation (UNFCCC, 2019). Policy frameworks that prioritize renewable energy deployment and climate adaptation must also be developed to ensure that these solutions are accessible, equitable, and aligned with broader climate goals, such as those outlined in the Paris Agreement.

Renewable energy and climate adaptation are essential components of a comprehensive strategy to address the impacts of climate change. By transitioning to renewable energy systems, societies can reduce their carbon footprint and enhance resilience to climate-related disruptions. Furthermore, renewable energy technologies offer a range of opportunities for climate adaptation, particularly in vulnerable regions, by providing reliable and sustainable energy solutions that support key sectors such as agriculture, water management, and health. However, overcoming the financial, infrastructural, and policy challenges associated with renewable energy deployment will require coordinated efforts at the local, national, and international levels. Ultimately, the integration of renewable energy into climate adaptation strategies offers a promising pathway to a more sustainable and resilient future.

21.2 Circular Economies and Sustainable Practices

Circular economies and sustainable practices are increasingly recognized as essential frameworks for addressing the global challenges of resource depletion, waste accumulation, and environmental degradation. Unlike the traditional linear economy, which follows a "take-make-dispose" model, the circular economy promotes a regenerative approach to production and consumption. It aims to minimize waste, keep products and materials in use for as long as possible, and reduce the environmental impact of industrial processes (Geissdoerfer et al., 2017). The transition to a circular economy involves redesigning business models, supply chains, and consumer behavior to prioritize sustainability, resource efficiency, and the reuse of materials.

One of the core principles of a circular economy is the concept of "closing the loop" of product lifecycles. This involves designing products that are durable, repairable, and recyclable, thereby reducing the need for virgin materials and minimizing waste. Circular business models focus on extending the life of products through practices such as remanufacturing, refurbishing, and upcycling (Bocken et al., 2016). For example, companies in the electronics and fashion industries are increasingly adopting circular practices by offering repair services, using recycled materials in their products, and establishing take-back schemes to recover old products for reuse. These practices help reduce the environmental impact of production and consumption by decreasing the demand for raw materials and reducing waste sent to landfills.

In addition to product design and manufacturing, the circular economy also emphasizes sustainable practices across the entire supply chain. This includes sourcing materials responsibly, reducing energy consumption, and minimizing emissions. For instance, companies are exploring the use of renewable energy sources in their production processes, such as solar and wind power, to reduce their carbon footprint (Lacy & Rutqvist, 2015). Similarly, the circular economy encourages businesses to adopt sustainable practices in logistics, such as optimizing transportation routes to reduce fuel consumption and using biodegradable or recyclable packaging materials. By integrating sustainability into supply chain management, businesses can reduce their environmental impact while also creating economic value through increased efficiency and reduced resource costs.

The role of consumers in the circular economy is also critical, as their purchasing decisions and behaviors can drive demand for sustainable products and services. Consumer awareness and engagement are essential for promoting the adoption of circular practices. For example, consumers can choose to buy products made from recycled materials, support brands that prioritize sustainability, and participate in recycling programs. Furthermore, the sharing economy, which is a key component of the circular economy, encourages consumers to rent, share, or exchange goods and services rather than purchase new items. This reduces the need for new production and promotes the efficient use of resources. Sharing platforms for items such as cars, tools, and clothing have gained popularity as consumers seek to reduce their environmental impact while meeting their needs (Botsman & Rogers, 2010).

Circular economies are also closely linked to the concept of sustainable development, as they contribute to the achievement of several of the United Nations Sustainable Development Goals (SDGs), particularly those related to responsible consumption and production, climate action, and the protection of ecosystems. By promoting the reuse and recycling of materials, the circular economy helps reduce the extraction of finite resources and minimizes the environmental degradation associated with mining and deforestation. Additionally, circular practices contribute to climate change mitigation by reducing greenhouse gas emissions through energy efficiency and waste reduction (Murray et al., 2017). For example, the recycling of metals, plastics, and paper reduces the need for energy-intensive extraction and processing of raw materials, which is a significant source of carbon emissions.

Despite the numerous benefits of circular economies, there are challenges to their widespread implementation. One of the primary obstacles is the need for systemic change across industries, governments, and consumers. For businesses, transitioning to a circular model requires significant investments in research and development, infrastructure, and supply chain redesign. Governments play a critical role in supporting this transition by implementing policies that incentivize circular practices, such as extended producer responsibility laws, recycling mandates, and tax breaks for companies that adopt sustainable practices (Ellen MacArthur Foundation, 2019). Additionally, there is a need for greater collaboration between stakeholders, including businesses, policymakers, consumers, and waste management organizations, to create a circular economy that is inclusive, efficient, and scalable.

Education and awareness-raising are also essential for overcoming barriers to the circular economy. By increasing understanding of the benefits of circular practices and the importance of sustainability, individuals and businesses can make more informed decisions. Educational initiatives that promote the principles of circularity and sustainability at all levels, from schools to industries, can help foster a culture of sustainability and innovation. Moreover, public and private sector collaboration is necessary to create the infrastructure and policy frameworks that support the circular economy, such as recycling systems, sustainable product standards, and green innovation incentives.

The circular economy offers a transformative approach to achieving sustainability by reducing waste, conserving resources, and minimizing environmental impact. By redesigning products, supply chains, and consumption patterns, businesses and consumers can contribute to a more sustainable future. While challenges remain in terms of implementation and scaling, the circular economy holds great potential for addressing global environmental challenges and advancing the United Nations SDGs. As awareness grows and policies evolve, the transition to a circular economy can create a more resilient, equitable, and sustainable world.

21.3 Building Resilient Communities

Building resilient communities is crucial for addressing the complex challenges posed by climate change, economic instability, social inequality, and other global crises. Resilience, in the context of communities, refers to the ability to adapt to, recover from, and thrive in the face of adversity, ensuring long-term sustainability and well-being for all members of society. The concept of resilience emphasizes not only the capacity to bounce back from shocks but also the proactive measures taken to strengthen social, economic, and environmental systems in anticipation of future challenges (Walker et al., 2004). Building resilient communities requires a multifaceted approach that integrates social, environmental, and economic dimensions, with a focus on inclusivity, equity, and sustainability.

One of the key aspects of building resilient communities is fostering social capital, which refers to the networks, relationships, and trust within a community. Social capital plays a vital role in enhancing a community's ability to respond to crises and adapt to changing circumstances. Strong social networks facilitate communication, cooperation, and collective action, enabling communities to mobilize resources and support one another in times of need (Aldrich & Meyer, 2015). For example, communities with high levels of social capital are more likely to engage in collaborative efforts to address local challenges, such as disaster preparedness, public health initiatives, and environmental conservation. In contrast, communities with weak social networks may struggle to coordinate responses and may be more vulnerable to external shocks, such as economic downturns or natural disasters.

In addition to social capital, community resilience is closely linked to economic sustainability and the ability to provide for the needs of its members. Building resilient communities requires creating diverse, inclusive, and equitable economic systems that promote local economic development, job creation, and access to essential services. Economic resilience involves ensuring that communities can withstand economic shocks, such as recessions, job losses, or disruptions in supply chains, by diversifying their sources of income and fostering economic self-sufficiency (Martin et al., 2016). For example, communities that rely on a single industry, such as mining or agriculture, may be more vulnerable to fluctuations in global markets or environmental changes. By diversifying their economies and investing in sustainable industries, such as renewable energy, green technologies, and sustainable agriculture, communities can reduce their dependency on external factors and build long-term economic resilience.

Environmental sustainability is another critical component of community resilience. Communities must be able to adapt to the changing climate and mitigate the impacts of environmental degradation, such as extreme weather events, rising sea levels, and resource scarcity. Building resilient communities requires adopting sustainable practices in land use, agriculture, energy, and waste management to reduce the environmental footprint and enhance the community's capacity to cope with environmental shocks (Folke et al., 2010). For example, communities that implement sustainable water management practices, such as rainwater harvesting and water recycling, can reduce their vulnerability to droughts and water shortages. Similarly, communities that invest in green infrastructure, such as urban parks and flood barriers, can reduce the risks associated with flooding and heat waves. These proactive measures not only protect the environment but also improve the quality of life for community members by providing cleaner air, safer living conditions, and greater access to natural resources.

An essential aspect of building resilient communities is ensuring social equity and inclusivity. Resilience is not only about protecting the community as a whole but also about ensuring that vulnerable groups, such as low-income households, marginalized populations, and people with disabilities, have access to the resources and support they need to thrive. Social equity in community resilience involves addressing disparities in access to education, healthcare, housing, and employment opportunities, ensuring that all members of society can participate in decision-making processes and benefit from community development initiatives (Norris et al., 2008). Inclusive resilience strategies prioritize the needs of the most vulnerable, recognizing that the strength of a community lies in its ability to uplift all its members. For example, disaster preparedness plans should account for the specific needs of vulnerable populations, such as providing accessible evacuation routes for people with disabilities or ensuring that low-income households have access to emergency financial assistance.

Education and capacity building are also integral to fostering resilience within communities. Empowering individuals with the knowledge, skills, and resources to adapt to change is essential for building a resilient society. Education systems that promote critical thinking, problem-solving, and collaboration can help individuals and communities better navigate challenges and develop innovative solutions to emerging problems (UNESCO, 2017). Furthermore, capacity-building initiatives that focus on developing leadership skills, fostering community engagement, and strengthening local governance can enhance the community's ability to make informed decisions, mobilize resources, and respond effectively to crises. By investing in education and capacity building, communities can create a culture of resilience that empowers individuals to take ownership of their own well-being and contribute to the collective good.

Finally, building resilient communities requires a holistic approach that integrates both top-down and bottom-up strategies. Governments, NGOs, and international organizations play a crucial role in providing the policy frameworks, financial support, and technical expertise needed to support community resilience. However, local communities themselves must be actively involved in the planning, decision-making, and implementation of resilience strategies. Bottom-up approaches that engage community members in identifying their needs, priorities, and solutions can ensure that resilience strategies are context-specific, culturally appropriate, and more likely to be effective in the long term (Chandler, 2017). By fostering collaboration between local, regional, and global stakeholders, communities can build resilience that is both locally grounded and globally informed.

In conclusion, building resilient communities requires a comprehensive approach that integrates social, economic, environmental, and governance dimensions. By fostering social capital, diversifying economies, promoting environmental sustainability, ensuring social equity, and investing in education and capacity building, communities can enhance their ability to adapt to and recover from challenges. A focus on inclusivity, collaboration, and local empowerment is essential for creating resilient communities that can thrive in the face of uncertainty and contribute to a more sustainable and equitable future.

References

Adger, W. N. (2003). Social capital, collective action, and adaptation to climate change. *Economic Geography, 79*(4), 387–404. https://doi.org/10.1111/j.1944-8287.2003.tb00220.x

Aldrich, D. P., & Meyer, M. A. (2015). Social capital and community resilience. *American Behavioral Scientist, 59*(2), 254–269. https://doi.org/10.1177/0002764214550299

Awerbuch, S., & Berger, A. (2003). *Renewable energy technologies and climate change: Energy for sustainable development in developing countries*. UNDP.

Baker, M. G., Lutz, A., & French, N. P. (2020). Renewable energy and public health: A review of the role of renewable energy in climate change adaptation. *Environmental Health Perspectives, 128*(6), 67002. https://doi.org/10.1289/EHP6979

Bocken, N. M. P., Short, S. W., Rana, P., & Evans, S. (2016). A literature and practice review to develop sustainable business model archetypes. *Journal of Cleaner Production, 65*, 42–56. https://doi.org/10.1016/j.jclepro.2013.11.039

Botsman, R., & Rogers, R. (2010). *What's mine is yours: The rise of collaborative consumption*. HarperBusiness.

Chandler, D. (2017). Resilience: The governance of complexity. *Routledge.*

Cutter, S. L., Boruff, B. J., & Shirley, W. L. (2008). Social vulnerability to environmental hazards. *Social Science Quarterly, 84*(2), 242–261. https://doi.org/10.1111/j.1540-6237.2003.00198.x

Elkington, J. (1997). *Cannibals with forks: The triple bottom line of 21st century business.* Capstone Publishing.

Ellen MacArthur Foundation. (2019). *Completing the picture: How the circular economy tackles climate change.* Ellen MacArthur Foundation.

Folke, C. (2006). Resilience: The emergence of a perspective for social-ecological systems analyses. *Global Environmental Change, 16*(3), 253–267. https://doi.org/10.1016/j.gloenvcha.2006.04.002

Folke, C., Carpenter, S. R., Walker, B., Scheffer, M., Chapin, T., & Rockström, J. (2010). Resilience thinking: Integrating resilience, adaptability and transformability. *Ecology and Society, 15*(4), 20. https://doi.org/10.5751/ES-03610-150420

Folke, C., Carpenter, S. R., Walker, B., Scheffer, M., Elmqvist, T., Gunderson, L., & Holling, C. S. (2010). Resilience thinking: Integrating resilience, adaptability and transformability. *Ecology and Society, 15*(4), 20. https://doi.org/10.5751/ES-03610-150420

Geissdoerfer, M., Savaget, P., Bocken, N. M. P., & Hultink, E. J. (2017). The Circular Economy – A new sustainability paradigm? *Journal of Cleaner Production, 143*, 757–768. https://doi.org/10.1016/j.jclepro.2016.12.048

Global Wind Energy Council. (2020). *Global wind report 2020.* GWEC.

Intergovernmental Panel on Climate Change (IPCC). (2018). *Global warming of 1.5°C: An IPCC special report on the impacts of global warming of 1.5°C above pre-industrial levels and related global greenhouse gas emission pathways.* IPCC.

International Renewable Energy Agency (IRENA). (2020). *Renewable capacity statistics 2020.* IRENA.

Keohane, R. O. (2005). *After hegemony: Cooperation and discord in the world political economy* (2nd ed.). Princeton University Press.

Lacy, P., & Rutqvist, J. (2015). *Waste to wealth: The circular economy advantage.* Palgrave Macmillan.

Liu, X., Zhang, X., & Wang, S. (2019). Solar-powered irrigation systems for climate-resilient agriculture: A review. *Renewable and Sustainable Energy Reviews, 113*, 109274. https://doi.org/10.1016/j.rser.2019.109274

Martin, R., Sunley, P., & Turner, D. (2016). Local economic resilience: A theory and evidence from the UK. *Cambridge Journal of Regions,*

Economy and Society, 9(2), 265–285. https://doi.org/10.1093/cjres/rsw013

Mezirow, J. (2000). *Learning as transformation: Critical perspectives on a theory in progress*. Jossey-Bass.

Murray, A., Skene, K., & Haynes, K. (2017). The circular economy: An interdisciplinary exploration of the concept and application in a global context. *Journal of Business Ethics, 140*(3), 369–380. https://doi.org/10.1007/s10551-016-3003-4

Norris, F. H., Stevens, S. P., Pfefferbaum, B., Wyche, K. F., & Pfefferbaum, R. L. (2008). Community resilience as a metaphor, theory, set of capacities, and strategy for disaster readiness. *American Journal of Community Psychology, 41*(1–2), 127–150. https://doi.org/10.1007/s10464-007-9156-6

Reed, M. S., Fraser, E. D. G., & Dougill, A. J. (2013). An adaptive learning process for developing and applying sustainability indicators with local communities. *Ecology and Society, 18*(3), 11. https://doi.org/10.5751/ES-05475-180311

Sen, A. (1999). *Development as freedom*. Oxford University Press.

Sovacool, B. K., Kester, J., & Mitchell, C. (2018). The role of renewable energy in climate adaptation. *Nature Sustainability, 1*(5), 220–228. https://doi.org/10.1038/s41893-018-0037-2

Sovacool, B. K., Kivimaa, P., & Martiskainen, M. (2016). The new political economy of energy transitions: The case of solar power in the UK. *Energy Research & Social Science, 12*, 1–9. https://doi.org/10.1016/j.erss.2015.11.003

UNESCO. (2017). *Education for sustainable development goals: Learning objectives*. United Nations Educational, Scientific and Cultural Organization.

UNFCCC. (2015). *Paris Agreement*. United Nations Framework Convention on Climate Change.

United Nations Framework Convention on Climate Change (UNFCCC). (2019). *The Paris Agreement and the role of renewable energy in climate change adaptation*. UNFCCC.

United Nations. (2015). *Transforming our world: The 2030 agenda for sustainable development*. United Nations.

22

Global Cooperation and Solidarity

Global cooperation and solidarity are fundamental for addressing the interconnected challenges of the 21st century, including climate change, inequality, global health crises, and geopolitical conflicts. These challenges transcend national borders and require collaborative efforts that involve governments, international organizations, civil society, and the private sector. In this context, global cooperation refers to the collaborative actions taken by different nations and entities to achieve common goals. At the same time, solidarity emphasizes mutual support and shared responsibility for the well-being of all people, particularly marginalized and vulnerable populations. These concepts form the foundation for building a more just, sustainable, and peaceful global society.

One of the primary drivers of global cooperation is the recognition of shared global challenges that require collective action. Issues such as climate change, biodiversity loss, and pandemics do not respect national borders and demand coordinated responses. The Paris Agreement on climate change, adopted in 2015, is a prime example of global cooperation, where nearly every nation committed to reducing greenhouse gas emissions and limiting global warming to well below 2°C above pre-industrial levels. This agreement reflects the understanding that climate change is a global issue that requires concerted action from all nations, regardless of their level of development or historical contribution to the problem (UNFCCC, 2015). Similarly, the COVID-19 pandemic highlighted the need for global cooperation regarding public health responses, vaccine distribution, and the sharing of scientific knowledge to mitigate the spread of the virus and save lives.

Solidarity plays a critical role in global cooperation, particularly in addressing the disparities that exist between nations and communities. The principle of solidarity emphasizes the importance of mutual support, particularly for those who are most vulnerable or disadvantaged. In the context of global challenges, solidarity manifests in efforts to provide assistance

DOI: 10.1201/9788743808787-28

to countries and populations that are disproportionately affected by crises, such as developing countries facing the worst impacts of climate change or low-income communities experiencing the economic fallout of global health emergencies. Solidarity is also expressed through international aid, debt relief, and the provision of resources for development. For instance, the Global Fund to Fight AIDS, Tuberculosis, and Malaria is an example of international solidarity in action, where high-income countries contribute financial resources to support the fight against these diseases in low- and middle-income countries (The Global Fund, 2021).

Moreover, global cooperation and solidarity are essential for achieving the United Nations Sustainable Development Goals (SDGs), which are a set of 17 global objectives aimed at addressing poverty, inequality, environmental sustainability, and peace by 2030. The SDGs provide a framework for international cooperation, with the understanding that the challenges facing humanity cannot be solved by individual countries acting alone. Achieving these goals requires a collective effort that brings together governments, international organizations, the private sector, and civil society. The SDGs emphasize the importance of inclusivity, equity, and shared responsibility, with a particular focus on leaving no one behind (United Nations, 2015). For example, SDG 13 on climate action highlights the need for global solidarity in addressing the climate crisis, recognizing that climate change disproportionately affects the most vulnerable populations, including small island nations, low-income communities, and future generations.

In addition to addressing global challenges, global cooperation and solidarity are also crucial for promoting peace and security in an increasingly interconnected world. Conflicts, whether they are armed conflicts, political instability, or economic crises, often have far-reaching consequences that extend beyond national borders. International peacekeeping efforts, diplomatic negotiations, and conflict resolution mechanisms are essential for maintaining global stability. Organizations such as the United Nations (UN) and regional bodies like the European Union (EU) play pivotal roles in fostering cooperation and solidarity to prevent and resolve conflicts. The UN's peacekeeping missions, for example, provide critical support to countries recovering from conflict by ensuring security, facilitating the delivery of humanitarian aid, and supporting the rebuilding of institutions (UN Peacekeeping, 2021). These efforts reflect the understanding that peace and security are global public goods that require collective action to sustain.

However, despite the importance of global cooperation and solidarity, several challenges hinder their full realization. Geopolitical tensions, economic disparities, and competing national interests often complicate efforts to work together on global issues. For instance, during the COVID-19 pandemic, the unequal distribution of vaccines highlighted the disparities between wealthy and developing nations, with some countries having access to vaccines far earlier than others. This situation underscored the need for stronger mechanisms to ensure equitable access to essential resources in times of global crises. Furthermore, rising nationalism and protectionism in some countries pose significant barriers to international cooperation, as they prioritize national interests over global solidarity. The retreat from multilateralism, as seen in the United States' withdrawal from the Paris Agreement under the Trump administration, exemplifies the challenges of maintaining global cooperation in a world marked by competing political ideologies (Biermann et al., 2020).

To overcome these challenges, it is essential to strengthen the institutions and mechanisms that facilitate global cooperation and solidarity. This includes reforming international organizations, improving global governance, and ensuring that the voices of marginalized communities are heard in decision-making processes. The inclusion of diverse perspectives, particularly from the Global South, is crucial for ensuring that global solutions are equitable and just. Additionally, fostering a culture of global citizenship, where individuals recognize their shared responsibility for the well-being of all people and the planet, is essential for building solidarity at the global level. Education, public awareness, and cross-cultural dialogue can help cultivate this sense of interconnectedness and mutual responsibility, which is critical for addressing the complex challenges of our time.

In conclusion, global cooperation and solidarity are essential for addressing the shared challenges of the 21st century, including climate change, health crises, and geopolitical conflicts. These concepts emphasize the importance of collective action, mutual support, and shared responsibility for the well-being of all people. While challenges remain in realizing global cooperation and solidarity, strengthening international institutions, promoting inclusivity, and fostering a sense of global citizenship can help overcome these barriers and build a more just, sustainable, and peaceful world.

22.1 The Role of International Organizations

International organizations play a pivotal role in shaping global governance, addressing transnational challenges, and promoting international cooperation. These entities, whether intergovernmental or non-governmental, provide frameworks for collaboration, policy coordination, and the development of norms and standards that transcend national borders. Their influence is especially crucial in a world increasingly marked by globalization, where issues such as climate change, public health, economic instability, and conflict require collective action. This paper explores the roles and functions of international organizations, highlighting their contributions to peace, security, development, and human rights, while also addressing the challenges they face in an evolving global landscape.

One of the primary functions of international organizations is to facilitate cooperation among states on issues that cannot be effectively addressed by individual countries alone. The United Nations (UN), established in 1945, is perhaps the most prominent example of such an organization. With its broad mandate to maintain international peace and security, promote human rights, foster social and economic development, and protect the environment, the UN plays a central role in global governance (United Nations, 2020). Through its specialized agencies, such as the World Health Organization (WHO), the United Nations Environment Programme (UNEP), and the United Nations Educational, Scientific and Cultural Organization (UNESCO), the UN coordinates international efforts to address a wide range of global challenges. For instance, the WHO's leadership during the COVID-19 pandemic demonstrated the organization's critical role in coordinating global public health responses, providing technical guidance, and supporting the equitable distribution of vaccines (World Health Organization, 2020).

In addition to peace and security, international organizations are instrumental in promoting economic development and addressing global inequalities. The World Bank and the International Monetary Fund (IMF) are key institutions in the global economic system, providing financial assistance, technical support, and policy advice to countries in need. These organizations help developing nations build infrastructure, reduce poverty, and strengthen their economies. The World Trade Organization (WTO), another significant international organization, regulates global trade by establishing rules for international commerce and resolving trade disputes. By fostering economic cooperation and ensuring that global trade operates within a set of agreed-upon rules, the WTO contributes to the stability and predictability of the global economy (Bown, 2019).

International organizations also play a critical role in addressing environmental challenges that transcend national borders. The Paris Agreement, adopted under the United Nations Framework Convention on Climate Change (UNFCCC) in 2015, is an example of a multilateral effort to combat climate change. This agreement, signed by nearly every nation in the world, sets global targets for reducing greenhouse gas emissions and limiting global warming to well below 2°C above pre-industrial levels. The UNFCCC, alongside other organizations such as the World Bank and the Global Environment Facility, coordinates international financial mechanisms and technical support to help countries implement climate change mitigation and adaptation strategies (Biermann et al., 2020). Similarly, the Convention on Biological Diversity (CBD) works to conserve global biodiversity, addressing the alarming rates of species loss and habitat destruction that threaten ecosystems and human livelihoods.

Human rights are another area where international organizations have made significant contributions. The United Nations Human Rights Council (UNHRC) and the Office of the High Commissioner for Human Rights (OHCHR) are tasked with promoting and protecting human rights worldwide. Through conventions, treaties, and monitoring mechanisms, these organizations work to uphold international human rights standards and hold violators accountable. The Universal Declaration of Human Rights, adopted by the UN in 1948, remains a cornerstone document in the global human rights framework, providing a foundation for subsequent treaties and agreements aimed at safeguarding the dignity and rights of individuals (United Nations, 1948). Moreover, international human rights organizations such as Amnesty International and Human Rights Watch play a critical role in advocacy, documenting abuses, and pressuring governments to uphold human rights standards.

Despite their significant contributions, international organizations face a range of challenges that can hinder their effectiveness. One of the most significant challenges is the issue of state sovereignty. Many countries are reluctant to cede authority to international bodies, especially when it comes to sensitive issues such as human rights, environmental regulation, and economic policies. The ability of international organizations to enforce their decisions is often limited, relying on the cooperation of member states to implement resolutions or sanctions. For example, the UN Security Council's inability to intervene effectively in the Syrian civil war highlights the challenges of achieving consensus among powerful nations, particularly when their national interests are at stake (Mackay, 2020).

Another challenge facing international organizations is the growing trend of nationalism and populism in various parts of the world. As countries prioritize their own interests over international cooperation, multilateralism is increasingly under threat. The United States' withdrawal from the Paris Agreement under the Trump administration and its decision to exit the World Health Organization during the COVID-19 pandemic are examples of how nationalism can undermine global cooperation and weaken the effectiveness of international organizations (Biermann et al., 2020). This shift toward unilateralism poses significant challenges for international organizations, which rely on multilateral consensus to address global issues.

Moreover, international organizations often face criticism for their perceived lack of accountability and transparency. In some cases, these organizations have been accused of prioritizing the interests of powerful states or multinational corporations over the needs of vulnerable populations. For instance, the World Bank and IMF have been criticized for imposing austerity measures on developing countries, which some argue exacerbate poverty and inequality (Stiglitz, 2002). To address these concerns, there have been calls for reforming international organizations to ensure greater inclusivity, transparency, and accountability in their decision-making processes.

In conclusion, international organizations play a crucial role in addressing global challenges, promoting peace and security, advancing human rights, and fostering economic development. Through their ability to coordinate international efforts and provide a platform for dialogue and cooperation, these organizations are central to the functioning of global governance. However, they face significant challenges, including issues of state sovereignty, rising nationalism, and criticisms of accountability. To remain effective in the face of these challenges, international organizations must adapt to the changing global landscape, strengthen their legitimacy, and ensure that they serve the interests of all people, particularly the most vulnerable.

22.2 Multilateral Agreements for Global Challenges

Multilateral agreements are pivotal in addressing global challenges that transcend national borders, requiring collective action and cooperation among multiple countries. These agreements, negotiated and signed by numerous states, create frameworks for resolving issues such as climate change, biodiversity loss, public health, and global security. The necessity of multilateralism has become increasingly evident in the face of global crises that no single nation can tackle alone. This paper explores the importance of multilateral agreements in addressing pressing global challenges, examining key examples, their successes, and the obstacles they face in an interconnected world.

One of the most significant multilateral agreements is the Paris Agreement on climate change, adopted in 2015 under the United Nations Framework Convention on Climate Change (UNFCCC). The agreement brought together 196 countries to set global targets for reducing greenhouse gas emissions, aiming to limit global temperature rise to well below 2°C above pre-industrial levels, with efforts to limit it to 1.5°C. This agreement reflects the urgency of addressing climate change and the recognition that it is a global problem requiring collective solutions. The Paris Agreement established a framework for countries to submit nationally determined contributions (NDCs) outlining their emission reduction targets and climate action plans (UNFCCC, 2015). It also created mechanisms for financial support to developing countries to help them mitigate and adapt to climate change. Despite its landmark status, the agreement faces challenges in ensuring full implementation, with some countries, such as the United States under the Trump administration, withdrawing from the agreement or failing to meet their targets (Biermann et al., 2020). However, the re-engagement of the U.S. under President Biden has renewed optimism regarding the agreement's effectiveness in fostering global cooperation on climate action.

In addition to climate change, multilateral agreements have been critical in addressing global health crises. The World Health Organization (WHO) plays a central role in coordinating international responses to health emergencies, most notably during the COVID-19 pandemic. The International Health Regulations (IHR), a legally binding framework adopted by the WHO, aim to prevent the spread of diseases across borders and enhance global health security (WHO, 2005). The COVID-19 pandemic highlighted the importance of global cooperation in managing health crises, as countries were required to share information, coordinate responses, and ensure equitable access to vaccines and treatments. The COVAX initiative, launched by the WHO and other partners, was a multilateral effort aimed at ensuring equitable distribution of COVID-19 vaccines, especially to low- and middle-income countries (GAVI, 2020). While the initiative faced logistical and political challenges, it underscored the importance of multilateralism in tackling global health issues, emphasizing the need for solidarity and shared responsibility in public health.

Another critical area where multilateral agreements have made significant strides is biodiversity conservation. The Convention on Biological Diversity (CBD), adopted in 1992 at the Earth Summit in Rio de Janeiro, is a legally binding treaty that aims to conserve biodiversity, promote the sustainable use of natural resources, and ensure the fair and equitable sharing of benefits arising from genetic resources (CBD, 1992). The CBD has been instrumental in fostering international cooperation on biodiversity conservation, with its member states regularly meeting to review progress and adopt new strategies. The Aichi Biodiversity Targets, adopted in 2010 as part of the CBD's strategic plan, set ambitious goals for biodiversity conservation, including the protection of ecosystems, species, and genetic diversity.

However, despite these efforts, biodiversity loss continues at an alarming rate, and the targets have not been fully achieved. The Post-2020 Global Biodiversity Framework, which will replace the Aichi targets, aims to set new, more ambitious goals to halt biodiversity loss by 2030 (UNEP, 2022). The success of this framework will depend on the commitment of countries to implement effective conservation policies and mobilize the necessary resources.

Multilateral agreements are also essential in addressing global security challenges, such as the proliferation of nuclear weapons. The Treaty on the Non-Proliferation of Nuclear Weapons (NPT), signed in 1968, is one of the most significant multilateral agreements in the realm of international security. The NPT aims to prevent the spread of nuclear weapons, promote disarmament, and encourage the peaceful use of nuclear energy. While the treaty has been successful in limiting the number of nuclear-armed states and fostering disarmament discussions, it has faced challenges, particularly with countries like North Korea and Iran, which have pursued nuclear weapons programs outside the treaty framework (Kroenig, 2018). The NPT's effectiveness is contingent upon the political will of its signatories to adhere to its provisions and engage in meaningful dialogue on disarmament.

Despite the importance of multilateral agreements, several challenges hinder their effectiveness in addressing global challenges. One of the primary obstacles is the lack of enforcement mechanisms. While multilateral agreements often establish frameworks for cooperation, they typically lack binding enforcement mechanisms, relying on the goodwill and cooperation of states to implement the agreed-upon measures. This can lead to discrepancies between the commitments made in international agreements and their actual implementation. For instance, while the Paris Agreement sets emission reduction targets, there is no formal mechanism to enforce compliance, and countries are free to adjust their targets as they see fit (Biermann et al., 2020). This lack of accountability can undermine the effectiveness of multilateral agreements, especially when powerful countries are unwilling to take meaningful action.

Another challenge is the political and economic interests of states, which can impede progress in multilateral negotiations. Countries often prioritize their national interests over global cooperation, particularly when the benefits of an agreement are perceived to be uneven or when national sovereignty is at stake. For example, the failure to reach a comprehensive global agreement on climate change at the 2009 Copenhagen Summit was largely due to disagreements between developed and developing countries over financial contributions and emission reduction targets (Biermann et al.,

2020). Similarly, the reluctance of some states to commit to multilateral environmental agreements, such as the CBD, can hinder global efforts to address pressing environmental issues.

In conclusion, multilateral agreements are essential tools for addressing global challenges, from climate change and biodiversity loss to public health and security. These agreements facilitate international cooperation, provide frameworks for collective action, and help to create shared norms and standards. However, their effectiveness is often limited by the lack of enforcement mechanisms, the political and economic interests of states, and the challenges of achieving consensus among diverse actors. To enhance the impact of multilateral agreements, it is crucial to strengthen their implementation mechanisms, foster greater political will, and ensure that they are inclusive and equitable, reflecting the needs and interests of all countries, particularly the most vulnerable.

22.3 Bridging North-South Divides

The divide between the Global North and Global South is a critical issue in international development and global equity. This divide is rooted in historical patterns of colonization, economic disparity, and uneven development. The Global North, which includes wealthy industrialized nations, has historically benefited from colonial exploitation, industrialization, and access to resources, while the Global South, which encompasses many developing nations, continues to grapple with the legacies of colonialism, economic dependence, and limited access to resources. Bridging this divide is essential for fostering global solidarity, addressing systemic inequalities, and ensuring sustainable development. This paper explores the dynamics of the North-South divide, the challenges in bridging this gap, and the pathways toward greater global equity.

The North-South divide is often characterized by stark disparities in wealth, access to technology, healthcare, education, and political power. According to the United Nations Development Programme (UNDP), while the Global North represents a small fraction of the world's population, it controls a disproportionate share of global wealth, technology, and resources (UNDP, 2020). In contrast, many countries in the Global South face significant challenges related to poverty, underdevelopment, and the lack of infrastructure. These disparities are exacerbated by global economic systems that perpetuate inequality, such as trade imbalances, debt dependency, and the exploitation of natural resources. The World Systems Theory, proposed by sociologist Immanuel Wallerstein, argues that the global economic system is structured in a way that benefits core nations (Global North) while peripheral nations (Global South) remain dependent and marginalized (Wallerstein, 1974). This theory highlights the systemic nature of the North-South divide, which is reinforced by global capitalism and historical patterns of exploitation.

Efforts to bridge the North-South divide have been ongoing for decades, with varying degrees of success. One of the most significant frameworks for addressing global inequalities is the United Nations Sustainable Development Goals (SDGs), adopted in 2015. The SDGs are a universal agenda aimed at ending poverty, protecting the planet, and ensuring prosperity for all by 2030. While the SDGs are designed to apply to all countries, they recognize the need for special attention to the challenges faced by the Global South, including issues such as poverty eradication, access to education, gender equality, and climate change adaptation. The SDGs emphasize the importance of global partnerships and the need for cooperation between developed and developing nations to achieve sustainable development (United Nations, 2015). However, the implementation of the SDGs has been uneven, and many countries in the Global South continue to struggle with the resources and capacity to meet these ambitious goals. The ongoing challenges of financing development, addressing debt burdens, and ensuring equitable access to technology remain significant obstacles in bridging the North-South divide.

Another key initiative aimed at addressing global inequalities is international trade reform. The Global South has long been disadvantaged in global trade systems, with developed nations benefiting from trade policies that favor their industries and agricultural sectors. The World Trade Organization (WTO), which governs international trade, has been criticized for perpetuating this imbalance, as its rules often favor wealthier countries and multinational corporations. The Doha Development Round of trade negotiations, launched in 2001, aimed to address the concerns of developing countries by promoting fairer trade rules and reducing barriers to market access. However, progress has been slow, and many developing countries continue to face trade barriers, subsidies, and unfair practices that hinder their economic growth (Hoekman, 2010). Reforming global trade rules to ensure a more equitable system is a crucial step in bridging the North-South divide, as it would enable developing countries to access markets, diversify their economies, and reduce their dependence on foreign aid.

Climate change is another area where the North-South divide is particularly evident. The Global North has historically been the largest emitter of greenhouse gases, while the Global South is often the most vulnerable to the impacts of climate change, such as extreme weather events, rising sea levels, and droughts. The Paris Agreement on climate change, adopted in 2015, recognizes the principle of common but differentiated responsibilities, acknowledging that developed countries should take the lead in reducing emissions and providing financial and technological support to developing countries (UNFCCC, 2015). However, many countries in the Global South argue that the financial commitments made by developed nations have not been sufficient, and that more support is needed to help them mitigate and adapt to climate change. Bridging this divide requires not only greater financial contributions from the Global North but also a commitment to climate justice, which ensures that the impacts of climate change are addressed in a way that is fair and equitable for all nations, particularly those that have contributed the least to the problem.

One of the most effective ways to bridge the North-South divide is through capacity building and technology transfer. Many countries in the Global South face significant challenges in accessing the technology and knowledge necessary to drive sustainable development. This includes technologies related to renewable energy, agriculture, healthcare, and education. The Technology Facilitation Mechanism established by the United Nations in 2015 aims to support the transfer of technology and knowledge to developing countries, particularly in areas related to sustainable development (United Nations, 2015). By facilitating the exchange of knowledge and technology, developed countries can help to empower the Global South, enabling them to address their own development challenges and reduce their dependence on foreign aid. However, the effectiveness of these efforts depends on ensuring that the technology is appropriate, affordable, and accessible and that it is accompanied by the necessary capacity-building initiatives to ensure its successful implementation.

In conclusion, bridging the North-South divide is a complex and multifaceted challenge that requires coordinated efforts at the global level. While there have been significant initiatives aimed at addressing global inequalities, such as the SDGs, international trade reform, and climate change agreements, progress has been slow, and many challenges remain. To achieve a more equitable and sustainable world, it is essential to focus on strengthening global partnerships, promoting fair trade, ensuring climate justice, and supporting capacity-building and technology transfer in the Global South. Only through collective action and a commitment to global solidarity can the North-South divide be bridged, ensuring that all nations have the opportunity to thrive and contribute to the well-being of humanity and the planet.

References

Biermann, F., Pattberg, P., van Asselt, H., & Zelli, F. (2020). The global climate regime: Taking stock and looking forward. *Global Environmental Politics, 20*(2), 1–16. https://doi.org/10.1162/glep_a_00556

Bown, C. P. (2019). The WTO and the battle for global trade. *The Brookings Institution.* https://www.brookings.edu/book/the-wto-and-the-battle-for-global-trade/

CBD. (1992). *Convention on Biological Diversity.* United Nations Environment Programme. https://www.cbd.int/convention/text/

GAVI. (2020). *COVAX: Working together to ensure equitable access to COVID-19 vaccines.* GAVI. https://www.gavi.org/covax-facility

Hoekman, B. M. (2010). *The Doha Development Agenda: The Doha Round and the global trade regime.* World Bank Publications.

Kroenig, M. (2018). *The logic of coercion: Nuclear weapons and international security.* Oxford University Press.

Mackay, R. (2020). The United Nations Security Council and the Syrian conflict. *Global Governance, 26*(3), 369–385. https://doi.org/10.1163/19426720-02603003

Stiglitz, J. E. (2002). *Globalization and its discontents.* W.W. Norton & Company.

The Global Fund. (2021). *The Global Fund to Fight AIDS, Tuberculosis, and Malaria.* https://www.theglobalfund.org/en/

UN Peacekeeping. (2021). *What we do.* United Nations. https://peacekeeping.un.org/en/what-we-do

UNDP. (2020). *Human development report 2020: The next frontier – Human development and the Anthropocene.* United Nations Development Programme. http://hdr.undp.org/en/2020-report

UNEP. (2022). *Post-2020 Global Biodiversity Framework.* United Nations Environment Programme. https://www.unep.org/resources/report/post-2020-global-biodiversity-framework

UNFCCC. (2015). *Paris Agreement.* United Nations Framework Convention on Climate Change. https://unfccc.int/process-and-meetings/the-paris-agreement/the-paris-agreement

United Nations. (1948). *Universal Declaration of Human Rights.* https://www.un.org/en/universal-declaration-human-rights/

United Nations. (2015). *Transforming our world: The 2030 Agenda for Sustainable Development.* United Nations. https://sdgs.un.org/2030agenda

United Nations. (2020). *The United Nations at a glance.* https://www.un.org/en/sections/about-un/overview/index.html

Wallerstein, I. (1974). *The modern world-system: Capitalist agriculture and the origins of the European world-economy in the sixteenth century*. Academic Press.

WHO. (2005). *International Health Regulations (2005)*. World Health Organization. https://www.who.int/publications/i/item/9789241580496

World Health Organization. (2020). *COVID-19 strategy update*. https://www.who.int/publications/i/item/covid-19-strategy-update

23

Innovations for a Better Future

Innovation has always been a driving force for societal progress, and in the context of the 21st century, it holds the potential to address some of the most pressing challenges humanity faces. From technological advancements to social and environmental innovations, these transformative changes are shaping a future that is more sustainable, equitable, and interconnected. As global issues such as climate change, inequality, and health crises intensify, the need for innovative solutions that can drive systemic change has never been more urgent. This paper explores several key areas of innovation that are critical for a better future, including technological advancements, social innovations, and environmental solutions.

One of the most significant areas of innovation in recent years is technology, particularly in fields like artificial intelligence (AI), renewable energy, and biotechnology. AI has the potential to revolutionize industries, from healthcare to agriculture, by improving efficiency, personalizing services, and solving complex problems. In healthcare, AI can enhance diagnostics, optimize treatment plans, and facilitate drug discovery (Topol, 2019). In agriculture, AI-driven precision farming techniques can help optimize resource use, reduce waste, and increase crop yields, contributing to food security (Zhang et al., 2020). Furthermore, renewable energy technologies, such as solar, wind, and geothermal, are rapidly advancing, providing more sustainable alternatives to fossil fuels. Innovations in energy storage, such as advanced batteries, are addressing the intermittency issues associated with renewable energy, making it more reliable and accessible (Luo et al., 2015). These technological innovations are crucial for reducing carbon emissions, mitigating climate change, and creating a more sustainable energy future.

In addition to technological advancements, social innovations are essential for creating a more equitable and inclusive future. Social innovations refer to new ideas, practices, and models that address societal challenges and improve the well-being of individuals and communities. One example

DOI: 10.1201/9788743808787-29

of social innovation is the sharing economy, which leverages digital platforms to enable the sharing of resources, such as housing (e.g., Airbnb) and transportation (e.g., Uber), reducing waste and promoting more efficient use of resources (Botsman, 2017). Another example is social entrepreneurship, where individuals or organizations create innovative solutions to social problems, such as poverty, education, and healthcare. Social enterprises, such as Grameen Bank, which provides micro-loans to empower impoverished individuals, particularly women, demonstrate the power of innovation to address deep-rooted social inequalities (Yunus, 2007). By fostering inclusive models that prioritize the well-being of all people, social innovations are contributing to the creation of more cohesive and resilient societies.

Environmental innovations are also critical for ensuring a sustainable future. As the world faces the dual crises of climate change and biodiversity loss, innovative solutions in environmental conservation and sustainability are essential. One promising area is circular economy practices, which aim to minimize waste and maximize the reuse of resources. In a circular economy, products are designed for longevity, repairability, and recyclability, and waste is seen as a resource rather than a byproduct. Companies like Patagonia and Interface are leading the way in implementing circular economy principles by using recycled materials in their products and developing closed-loop production systems (Ellen MacArthur Foundation, 2017). Furthermore, nature-based solutions, such as reforestation and ecosystem restoration, are gaining recognition as effective strategies for mitigating climate change and enhancing biodiversity. These solutions not only sequester carbon but also provide vital ecosystem services, such as water purification and flood protection, contributing to the resilience of communities and ecosystems (Seddon et al., 2020). By integrating nature into environmental solutions, these innovations offer a sustainable path forward for addressing both climate change and biodiversity loss.

The role of education and lifelong learning in driving innovation cannot be overstated. As the world continues to change at a rapid pace, individuals and societies must adapt to new technologies, shifting economic structures, and evolving social norms. Digital learning platforms and online education have made knowledge more accessible than ever before, democratizing education and enabling people around the world to acquire the skills needed to thrive in the modern economy (Bakhshi et al., 2017). Furthermore, fostering a culture of innovation through education systems that emphasize creativity, critical thinking, and problem-solving is essential for cultivating the next generation of innovators. By equipping individuals with the tools to think outside the box and tackle global challenges, education can play a central role in driving positive change and ensuring that innovation serves the common good.

Despite the tremendous potential of these innovations, challenges remain in ensuring that they are implemented equitably and sustainably. Global cooperation is essential for scaling up innovative solutions and addressing the systemic barriers that hinder progress. Governments, businesses, and civil society must work together to create policies that support innovation, promote research and development, and ensure that the benefits of innovation are shared across society. Additionally, ethical considerations must be at the forefront of innovation, ensuring that new technologies and practices do not exacerbate inequalities or harm vulnerable populations. For example, the rise of AI and automation raises concerns about job displacement and privacy, and these issues must be addressed through inclusive policies and regulations (Brynjolfsson & McAfee, 2014). Similarly, environmental innovations must be designed to ensure that they do not disproportionately impact marginalized communities or exacerbate existing environmental injustices.

Innovations in technology, social practices, and environmental solutions offer immense potential for creating a better future. By embracing these innovations, societies can address pressing global challenges, from climate change to inequality, and build more sustainable, resilient, and inclusive communities. However, to ensure that these innovations benefit all people, it is essential to prioritize equity, inclusivity, and ethical considerations in their development and implementation. Through global cooperation and a commitment to shared goals, innovation can be harnessed as a powerful tool for creating a better future for all.

23.1 Technological Solutions to Global Problems

Technological innovations have long been recognized as pivotal drivers of progress and solutions to some of the most pressing global challenges. In the contemporary world, where issues such as climate change, poverty, inequality, and health crises are increasingly urgent, technology holds the potential to provide scalable and sustainable solutions. From renewable energy technologies to artificial intelligence (AI) and biotechnology, technological solutions are reshaping the way we approach and address global problems. This paper explores several key technological solutions that are having a profound impact on global issues, focusing on climate change mitigation, healthcare improvements, and resource management.

One of the most critical technological solutions in addressing global challenges is the development of renewable energy technologies. The transition from fossil fuels to renewable energy sources such as solar, wind, and hydropower is essential to mitigate the impacts of climate change. According to the International Renewable Energy Agency (IRENA, 2020), renewable energy technologies have become increasingly cost-competitive, making them viable alternatives to traditional energy sources. Solar photovoltaic (PV) systems, wind turbines, and geothermal energy are already contributing significantly to global energy production, and innovations in energy storage, such as advanced batteries, are helping to address the intermittency issues associated with renewable energy (Luo et al., 2015). Furthermore, the development of smart grids and decentralized energy systems is enabling more efficient energy distribution and reducing transmission losses, making renewable energy more accessible to communities around the world (IRENA, 2020). These technological advancements not only help to reduce carbon emissions but also promote energy security and economic development, particularly in underserved regions.

In the field of healthcare, technological solutions are revolutionizing diagnostics, treatment, and disease prevention, improving health outcomes worldwide. One of the most promising areas of innovation is the use of artificial intelligence (AI) and machine learning in medical research and clinical practice. AI algorithms are now being used to analyze vast amounts of medical data, identify patterns, and make predictions that can aid in early diagnosis and personalized treatment plans. For example, AI-powered diagnostic tools have shown great promise in detecting diseases such as cancer, diabetes, and cardiovascular conditions at early stages, which can significantly improve patient outcomes (Esteva et al., 2019). Additionally, AI is playing a key role in drug discovery, helping researchers to identify potential therapeutic compounds more quickly and efficiently than traditional methods (Vamathevan et al., 2019). Another critical area of healthcare innovation is telemedicine, which has expanded access to medical services, particularly in remote or underserved regions. Through telemedicine platforms, patients can consult with healthcare providers via video calls, reducing the need for travel and improving access to essential care (Smith et al., 2020). These technological solutions are transforming healthcare systems, making them more efficient, equitable, and accessible.

Biotechnology is another area where technological innovations are making significant strides in addressing global challenges, particularly in agriculture and food security. The development of genetically modified organisms (GMOs) and precision agriculture techniques is helping to increase food production, improve crop resilience, and reduce the environmental impact of farming. GMOs, for instance, have been engineered to resist pests, tolerate drought, and enhance nutritional content, thereby addressing food security challenges in regions facing climate variability and population growth (Qaim, 2016). Precision agriculture, which involves the use of sensors, drones, and data analytics, allows farmers to optimize the use of resources such as water, fertilizers, and pesticides, reducing waste and environmental degradation (Zhang et al., 2020). These biotechnological innovations are essential for feeding the growing global population while minimizing the environmental footprint of agricultural practices.

In the realm of resource management, technological solutions are playing a crucial role in improving the efficiency of natural resource use and reducing waste. One of the most significant innovations in this area is the concept of the circular economy, which emphasizes the reuse, recycling, and repurposing of materials to create a closed-loop system where waste is minimized, and resources are kept in use for as long as possible. Technologies such as advanced recycling systems, waste-to-energy technologies, and biodegradable materials are enabling industries to reduce their reliance on virgin resources and minimize environmental pollution (Ellen MacArthur Foundation, 2017). For example, companies are developing new methods to recycle plastic waste, converting it into new products or even energy, thereby reducing the accumulation of plastic in landfills and oceans. Additionally, innovations in water purification and desalination technologies are helping to address the global water crisis by providing clean water in areas where freshwater resources are scarce (Al-Karaghouli & Kazmerski, 2013). These technological solutions are essential for sustainable resource management and ensuring the availability of essential resources for future generations.

Despite the promising potential of technological solutions, there are challenges in ensuring that these innovations are implemented equitably and sustainably. Digital divides remain a significant barrier to the widespread adoption of technology, particularly in low-income and rural areas. Access to cutting-edge technologies such as AI, renewable energy, and healthcare innovations is often limited by infrastructure, affordability, and lack of digital literacy (Hilbert, 2016). To address these disparities, it is crucial to ensure that technological solutions are accessible to all, particularly marginalized communities. Furthermore, the rapid pace of technological development raises important ethical considerations related to privacy, security, and the potential for job displacement due to automation (Brynjolfsson & McAfee, 2014). As new technologies are deployed to address global challenges, it is essential to consider their social, economic, and environmental impacts and ensure that they contribute to the well-being of all people.

In conclusion, technological innovations are playing a pivotal role in addressing some of the most pressing global challenges, from climate change and healthcare to food security and resource management. Advances in renewable energy, AI, biotechnology, and resource efficiency are reshaping the way we approach global problems, offering scalable and sustainable solutions. However, to ensure that these innovations benefit all people and contribute to a more equitable and sustainable future, it is essential to address challenges related to access, equity, and ethics. By fostering global cooperation and promoting inclusive policies, technological solutions can be harnessed as powerful tools for creating a better world for current and future generations.

23.2 Grassroots Movements and Local Initiatives

Grassroots movements and local initiatives have become central to addressing a wide range of social, environmental, and political challenges. These movements are typically characterized by bottom-up, community-driven efforts aimed at creating positive change, often in response to issues that are overlooked or inadequately addressed by mainstream institutions or governments. Whether advocating for environmental justice, human rights, or local economic development, grassroots movements play a critical role in fostering social transformation and empowering marginalized communities. This paper explores the significance of grassroots movements and local initiatives in driving change, their strategies for mobilization, and the challenges they face in achieving their goals.

One of the most notable aspects of grassroots movements is their ability to empower local communities by giving them a voice in decision-making processes. According to Tarrow (2011), grassroots movements often emerge in response to a perceived lack of representation in political and social spheres. These movements are typically led by individuals or small groups who mobilize local resources and networks to address issues that directly affect their communities. For example, the environmental justice movement in the United States arose in response to the disproportionate exposure of low-income and minority communities to environmental hazards such as toxic waste and pollution (Bullard, 2005). Grassroots organizations like the South West Organizing Project in New Mexico and the United Church of Christ's Commission for Racial Justice were instrumental in raising awareness about the link between environmental degradation and social inequality. These movements have succeeded in influencing policy changes at the local, state, and national levels, demonstrating the power of grassroots advocacy in promoting environmental and social justice.

Grassroots movements also often emphasize local solutions to global problems. In contrast to top-down approaches that may overlook the unique needs of specific communities, local initiatives are tailored to the cultural, economic, and environmental contexts in which they operate. For instance, in the context of climate change, many grassroots movements focus on community-based adaptation strategies that are more appropriate for local conditions. The "Transition Towns" movement, which began in the United Kingdom and has spread globally, is an example of a local initiative aimed at building resilience to climate change through sustainable practices such as local food production, renewable energy use, and community cooperation (Hopkins, 2008). These initiatives not only help mitigate the impacts of climate change but also foster a sense of community and solidarity, which is crucial for long-term sustainability.

The success of grassroots movements often depends on their ability to mobilize resources and build broad-based coalitions. Local initiatives typically rely on volunteer efforts, donations, and community-based organizing to fund their activities and support their goals. One of the key strategies employed by grassroots movements is networking and alliances with other organizations, including NGOs, labor unions, and academic institutions. These alliances help to amplify the movement's message and increase its visibility. For example, the global "Fridays for Future" movement, initiated by Swedish teenager Greta Thunberg, has galvanized millions of young people around the world to demand action on climate change. Through social media platforms and collaborations with local environmental groups, the movement has created a global network that advocates for policy changes while promoting sustainable living practices (Thunberg, 2019). Such movements highlight the importance of digital tools and social media in facilitating communication and coordination among grassroots organizations, enabling them to overcome geographical barriers and reach a wider audience.

Despite their potential for creating change, grassroots movements face several challenges that can hinder their effectiveness. One of the primary obstacles is resistance from powerful institutions and governments that may view these movements as a threat to the status quo. In many cases, grassroots activists face political repression, criminalization, or even violence, particularly when their efforts challenge entrenched economic or political interests. For example, in countries with authoritarian regimes, activists advocating for human rights or environmental protection may face imprisonment or harassment (Klein, 2007). Additionally, grassroots movements often struggle with issues related to sustainability and funding. While many movements begin with passionate volunteer efforts, maintaining momentum over time can be difficult without a stable source of funding or institutional support (Juris, 2008). To address these challenges, grassroots organizations must be adaptive, flexible, and resilient, often seeking innovative ways to sustain their work and protect their members.

Another challenge faced by grassroots movements is the fragmentation of efforts. While grassroots movements can be powerful, their decentralized nature can sometimes lead to a lack of coordination or unity among different groups working on similar issues. According to Diani (2000), the success of a social movement often depends on the ability to build a cohesive collective identity and a shared agenda. However, when movements are composed of multiple, diverse organizations, there can be disagreements over strategy, goals, or leadership. In such cases, the movement may struggle to present a unified front, which can undermine its impact. To overcome this, effective grassroots movements often engage in dialogue and collaboration to align their efforts and foster a sense of shared purpose.

Despite these challenges, grassroots movements and local initiatives continue to be powerful agents of change, particularly in the face of global crises such as climate change, inequality, and social injustice. Their ability to mobilize communities, advocate for local solutions, and challenge institutional power makes them indispensable in the fight for a more equitable and sustainable future. By focusing on community empowerment, collaboration, and resilience, grassroots movements are helping to shape a new paradigm of social and environmental responsibility.

23.3 The Role of Youth and Intergenerational Collaboration

The role of youth in shaping social, political, and environmental change has become increasingly significant in recent years, especially as global challenges such as climate change, inequality, and political polarization intensify. Youth movements, powered by young people's passion, creativity, and desire for a better future, have gained global recognition for their ability to mobilize action and challenge established systems. However, the true potential of youth activism can be fully realized when it is complemented by intergenerational collaboration, where older generations provide guidance, wisdom, and resources, while younger generations bring fresh perspectives and innovative solutions. This paper explores the importance of youth in global change efforts and the benefits of intergenerational collaboration in addressing complex societal issues.

Youth have historically played a pivotal role in social and political movements. From the civil rights movement in the United States to the anti-apartheid movement in South Africa, young people have been at the forefront of pushing for social justice and equality (Bennet, 2012). More recently, youth-led movements such as Fridays for Future, initiated by Swedish climate activist Greta Thunberg, and March for Our Lives, a student-led campaign advocating for gun control in the U.S., have captured global attention. These movements highlight the power of youth in demanding change and raising awareness about urgent issues. As digital natives, today's youth are particularly adept at using social media and digital platforms to organize, advocate, and mobilize on a global scale (Boulianne, 2015). The ability to connect with peers across borders and share information rapidly has transformed the way youth engage with social and political issues, amplifying their voices and impact.

Despite their energy and innovation, youth often face challenges in achieving meaningful change. One of the primary obstacles is the lack of institutional power and decision-making authority. While young people are often the most affected by global challenges such as climate change and economic inequality, they are frequently excluded from high-level policy discussions and governance processes (United Nations, 2018). As a result, their ability to influence long-term solutions can be limited. This is where intergenerational collaboration becomes crucial. By working alongside older generations – who may have more experience, networks, and access to resources – youth can bridge the gap between activism and policy change. Older generations, in turn, can benefit from the fresh ideas and energy that young people bring to the table.

Intergenerational collaboration can take many forms, ranging from mentorship programs to joint advocacy efforts. A key benefit of such collaboration is the exchange of knowledge and perspectives. Older generations can share their experiences, lessons learned, and strategies for navigating complex political and social systems, while youth can offer new ideas, technological expertise, and a sense of urgency. For example, in the context of climate change, older activists and policymakers can provide historical context and insight into past environmental movements, while youth activists can bring innovative approaches to sustainability, such as advocating for renewable energy technologies and sustainable consumption patterns (UNESCO, 2019). The combination of these perspectives can lead to more comprehensive and effective solutions.

Furthermore, intergenerational collaboration fosters mutual respect and understanding between different age groups, which is essential for creating cohesive and inclusive societies. The generational divide, particularly in the context of political and social issues, can often lead to misunderstandings and tensions. However, when youth and older generations come together to collaborate, they can break down stereotypes and build solidarity across age groups. This is especially important in addressing issues such as social inequality, where youth may face discrimination or exclusion based on age, while older generations may struggle to understand the challenges that younger people face in a rapidly changing world. By working together, these generations can find common ground and advocate for policies that benefit all members of society, regardless of age.

Intergenerational collaboration also strengthens social cohesion and promotes the development of inclusive policies. When different generations collaborate, they can advocate for policies that address the needs of both younger and older populations, ensuring that the voices of all age groups are heard in decision-making processes. For example, in the area of education, youth may call for curricula that reflect the challenges of the 21st century, such as climate change, while older generations may emphasize the importance of traditional knowledge and skills. By working together, these generations can create educational systems that equip young people with the tools they need to address global challenges while preserving cultural heritage and traditional wisdom (Berkes, 2009).

However, the success of intergenerational collaboration depends on creating spaces for dialogue and partnership. This requires intentional efforts to foster mutual understanding and respect between youth and older generations. It is also important to recognize the diversity within both age groups, as not all youth or older individuals share the same experiences or perspectives. For example, youth from different cultural, economic, and social backgrounds may have varying views on issues such as climate change or economic inequality, just as older generations may have different approaches to solutions. Creating inclusive spaces for dialogue, where all voices are heard and valued, is essential for building effective intergenerational partnerships (Lester, 2015).

Youth have an essential role to play in shaping the future, and their activism and energy are invaluable in addressing global challenges. However, their efforts can be significantly enhanced through intergenerational collaboration, where older generations provide mentorship, experience, and resources. By working together, youth and older generations can bridge the gap between activism and policy, exchange knowledge and perspectives, and advocate for more inclusive and sustainable solutions. As the world faces complex challenges, the power of youth and intergenerational collaboration will be crucial in building a more just, equitable, and sustainable future.

References

Al-Karaghouli, A., & Kazmerski, L. L. (2013). *Energy consumption and water production cost of water desalination technologies*. Desalination, 314, 1–14. https://doi.org/10.1016/j.desal.2013.03.007

Bakhshi, H., Freeman, A., & Higgs, P. (2017). *The Geography of Creativity*. Nesta. https://www.nesta.org.uk/report/the-geography-of-creativity/

Bennet, L. (2012). *Youth, political participation, and social media: A review of the literature*. Journal of Youth Studies, 15(1), 1–18. https://doi.org/10.1080/13676261.2011.596255

Berkes, F. (2009). *Evolution of co-management: Role of knowledge generation, bridging organizations, and social learning*. Journal of Environmental Management, 90(5), 1692–1702. https://doi.org/10.1016/j.jenvman.2008.12.001

Botsman, R. (2017). *Who can you trust? How technology brought us together and why it might drive us apart*. Penguin.

Boulianne, S. (2015). *Social media use and participation: A meta-analysis of current research*. Information, Communication & Society, 18(5), 524–538. https://doi.org/10.1080/1369118X.2015.1008542

Brynjolfsson, E., & McAfee, A. (2014). *The Second Machine Age: Work, Progress, and Prosperity in a Time of Brilliant Technologies*. W. W. Norton & Company.

Bullard, R. D. (2005). *The quest for environmental justice: Human rights and the politics of pollution*. Sierra Club Books.

Diani, M. (2000). *Social movements and networks: Relational approaches to collective action*. Oxford University Press.

Ellen MacArthur Foundation. (2017). *Towards the Circular Economy: Economic and business rationale for an accelerated transition*. Ellen MacArthur Foundation. https://www.ellenmacarthurfoundation.org/

Esteva, A., Kuprel, B., Novoa, R. A., et al. (2019). *Dermatologist-level classification of skin cancer with deep neural networks*. Nature, 542(7639), 115–118. https://doi.org/10.1038/nature21056

Hilbert, M. (2016). *Digital gender divide or technologically empowered women in developing countries?*. Women's Studies International Forum, 54, 66–77. https://doi.org/10.1016/j.wsif.2015.08.001

Hopkins, R. (2008). *The transition handbook: From oil dependency to local resilience*. Green Books.

International Renewable Energy Agency (IRENA). (2020). *Renewable Capacity Statistics 2020*. IRENA. https://www.irena.org/Statistics

Juris, J. S. (2008). *Networking futures: The movements against corporate globalization*. Duke University Press.

Klein, N. (2007). *The shock doctrine: The rise of disaster capitalism*. Metropolitan Books.

Lester, L. (2015). *Youth and intergenerational collaboration: A path to a more inclusive future*. Journal of Social Policy, 44(4), 1–18. https://doi.org/10.1017/S0047279415000133

Luo, X., Wang, J., & Dooner, M. (2015). *A review of energy storage technologies and their applications in renewable energy systems*. Renewable and Sustainable Energy Reviews, 40, 121–132. https://doi.org/10.1016/j.rser.2014.07.117

Qaim, M. (2016). *Genetically modified crops and agricultural development*. PNAS, 113(19), 5175–5181. https://doi.org/10.1073/pnas.1600406113

Seddon, N., Chausson, A., Berry, P., & Girardin, C. A. (2020). *Understanding the value of nature-based solutions for climate change adaptation and resilience*. Science, 369(6509), 1–9. https://doi.org/10.1126/science.aaz9530

Smith, A. C., Thomas, E., Snoswell, C. L., et al. (2020). *Telemedicine and COVID-19: The rapid adoption of virtual care*. Journal of Telemedicine and Telecare, 26(5), 309–313. https://doi.org/10.1177/1357633X20916567

Tarrow, S. (2011). *Power in movement: Social movements and contentious politics* (3rd ed.). Cambridge University Press.

Thunberg, G. (2019). *No one is too small to make a difference*. Penguin Books.

Topol, E. (2019). *Deep Medicine: How Artificial Intelligence Can Make Healthcare Human Again*. Basic Books.

UNESCO. (2019). *Intergenerational dialogue: A key to sustainable development*. United Nations Educational, Scientific and Cultural Organization. https://www.unesco.org

United Nations. (2018). *Youth and the 2030 Agenda for Sustainable Development*. United Nations Department of Economic and Social Affairs. https://www.un.org/development/desa/youth

Vamathevan, J., Clark, D., Czodrowski, P., et al. (2019). *Applications of machine learning in drug discovery and development*. Nature Reviews Drug Discovery, 18(6), 463–477. https://doi.org/10.1038/s41573-019-0024-5

Yunus, M. (2007). *Creating a World Without Poverty: Social Business and the Future of Capitalism*. PublicAffairs.

Zhang, Y., Li, Y., & Zhang, Y. (2020). *Precision agriculture and AI: A review on recent advances*. Computers and Electronics in Agriculture, 174, 105490. https://doi.org/10.1016/j.compag.2020.105490

24

Education for Transformation

Education plays a pivotal role in fostering societal change and individual growth. In the face of global challenges such as climate change, inequality, and social fragmentation, education must evolve to prepare individuals not only for the demands of the contemporary world but also for the transformation of societal structures. Education for transformation goes beyond traditional learning; it emphasizes the development of critical thinking, empathy, and the capacity to engage with complex, interconnected global issues. This approach to education seeks to empower individuals to become agents of change, equipping them with the knowledge, skills, and values necessary to create sustainable, just, and inclusive societies. This empowerment is not just a theoretical concept but a tangible reality that each individual can embrace and act upon.

At its core, education for transformation is grounded in the belief that learning should not be merely about acquiring knowledge but also cultivating the ability to think critically and act ethically. It encourages students to question existing systems, understand their roles within them, and challenge the status quo in pursuit of positive change (Mezirow, 2000). The transformation process in education involves intellectual growth and personal and social development, fostering a sense of responsibility and agency. As such, it aligns with transformative learning theory, which posits that learning is most effective when it leads to profound shifts in individuals' worldviews, helping them to reframe their understanding of themselves and their relationship to the world (Cranton, 2006). By encouraging reflection, dialogue, and active participation, transformative education empowers learners to participate actively in societal transformation. For instance, this could mean incorporating a project-based learning approach that allows students to tackle real-world issues in a high school setting. In a university, it could involve interdisciplinary courses that explore the interconnectedness of global challenges.

DOI: 10.1201/9788743808787-30

The importance of education for transformation is particularly evident in the context of global sustainability and social justice. As the world faces complex issues such as climate change, economic inequality, and political instability, it is essential that education systems not only provide knowledge but also foster an understanding of the interconnectedness of these challenges. Education for transformation emphasizes interdisciplinary learning, where students are encouraged to explore the relationships between environmental, social, economic, and political systems. For example, an education that integrates sustainability into the curriculum helps students understand the long-term impacts of their actions on the environment and society while also fostering a sense of environmental stewardship (Sterling, 2001). This holistic approach to education prepares students to think critically about global challenges and seek solutions promoting sustainability, equity, and justice.

Moreover, education for transformation requires a shift towards participatory learning. Traditional education models often prioritize passive learning, where students receive knowledge rather than actively participate in its creation. In contrast, transformative education encourages collaboration, dialogue, and problem-solving. By engaging students in real-world issues and involving them in decision-making processes, educators can cultivate a sense of ownership and responsibility in learners. This participatory approach is particularly relevant in social and political change, as it empowers students to actively shape the future. For instance, youth-led movements such as the Fridays for Future climate strikes demonstrate the power of education that fosters active citizenship and engagement in global issues (Bennett, 2012). By promoting critical thinking and collaboration, education for transformation enables individuals to become agents of change in their communities and beyond.

Another key aspect of education for transformation is its focus on social justice and inclusivity. Education systems must be designed to address and challenge systemic inequalities, ensuring that all students, regardless of their background, have access to quality education and the opportunity to thrive. This involves creating learning environments that are inclusive, equitable, and culturally responsive, recognizing and valuing diverse perspectives and experiences. As Nussbaum (2011) argues, education should be a tool for promoting human capabilities and empowering individuals to lead flourishing lives. In this sense, education for transformation is about developing academic skills and cultivating empathy, respect, and solidarity across differences. By fostering these values, education can contribute to creating more inclusive and cohesive societies where individuals are empowered to work together to address shared challenges. This emphasis on empathy and respect is not just a byproduct of education but a fundamental goal that shapes the very fabric of our society.

Furthermore, global citizenship is an essential component of education for transformation. In an increasingly interconnected world, individuals must understand their roles as global citizens and recognize the interdependence of all people and nations. Education for transformation encourages students to see themselves as part of a larger global community with responsibilities to both local and global societies. This perspective is critical in addressing issues such as climate change, where the actions of individuals in one part of the world can have far-reaching consequences for others. By promoting global citizenship, education helps individuals develop the awareness and skills needed to engage with global issues and contribute to creating a more just and sustainable world. This sense of global citizenship is not just a concept but a call to action that each individual can heed daily.

In conclusion, education for transformation is a critical approach to addressing the challenges of the 21st century. By fostering critical thinking, interdisciplinary learning, participatory engagement, social justice, and global citizenship, it equips individuals with the tools they need to navigate and transform complex global issues. As the world faces unprecedented challenges, education must evolve to prepare individuals not only to succeed in a rapidly changing world but also to actively contribute to creating a more sustainable, equitable, and inclusive future. However, it is essential to acknowledge that this approach may face challenges and criticisms, such as the potential for bias in the selection of issues to be addressed or the difficulty of measuring the impact of transformative education. In this way, education becomes a catalyst for transformation, empowering individuals to become agents of change and advocates for a better world.

24.1 Transformative and Lifelong Learning

Transformative and lifelong learning are foundational concepts in pursuing personal and societal growth, particularly in a rapidly changing world. These two approaches, while related, have distinct characteristics. Transformative learning emphasizes the need for continuous development and the capacity for profound shifts in understanding, perspectives, and behaviors throughout an individual's life. It encourages individuals to question their assumptions and beliefs, leading to a reevaluation of their worldview. On the other hand, lifelong learning is a broader concept that refers to the continuous acquisition of knowledge and skills throughout one's life. As global challenges such as climate change, social inequality, and technological disruption evolve, transformative and lifelong learning offers critical pathways for individuals to adapt, engage with complex issues, and contribute meaningfully to societal transformation. These learning paradigms go beyond the acquisition of knowledge to include the cultivation of critical thinking, self-reflection, and the ability to act ethically and responsibly in an interconnected world.

Transformative learning theory, primarily developed by Jack Mezirow, posits that learning is most effective when it significantly changes how individuals perceive themselves and the world around them (Mezirow, 1997). This process involves a shift in perspective, where learners move from a limited or distorted understanding to a more inclusive, integrated, and critically reflective viewpoint. Central to transformative learning is the idea that individuals can confront and challenge their assumptions, beliefs, and values through critical reflection, dialogue, and experience. This process allows for a deeper understanding of personal and societal issues, leading to more informed, compassionate, and active participation in the world (Cranton, 2006). Transformative learning is particularly relevant in contexts of social change, where individuals must navigate complex global challenges and reorient their values and actions to promote justice, sustainability, and well-being.

The transformative learning process is often triggered by disorienting dilemmas – events or experiences that challenge existing worldviews and provoke reflection (Mezirow, 1997). For instance, a person may experience a significant life event, such as a personal loss, a social injustice, or a global crisis, which forces them to reconsider their assumptions about the world. Through this process of reflection, learners engage in critical thinking, examining their beliefs, values, and behaviors in light of new information or experiences. This reflective process develops new perspectives, enabling individuals to make more informed decisions and take actions that align with their revised understanding of the world. This way, transformative learning fosters personal growth and societal transformation by encouraging individuals to question the status quo and engage in meaningful change.

On the other hand, lifelong learning emphasizes continuous learning throughout an individual's life. Unlike traditional education models, which often focus on formal schooling during childhood and adolescence, lifelong learning recognizes that learning is an ongoing process that extends into adulthood and throughout one's career (Jarvis, 2006). Lifelong learning encompasses many learning experiences, including formal education, informal learning, and self-directed learning. This approach is essential in a rapidly changing world, where individuals must constantly adapt to new technologies, economic shifts, and global challenges. Lifelong learning enables individuals to acquire new skills, update existing knowledge, and stay informed about evolving issues, making it a crucial element of personal and professional development.

The importance of lifelong learning is particularly evident in the context of technological advancements and the changing nature of work. As automation, artificial intelligence, and other technological innovations reshape industries, individuals must continuously update their skills to remain relevant in the workforce. Moreover, lifelong learning fosters a sense of agency and empowerment as individuals take responsibility for their learning and development. This approach encourages self-directed learning, where individuals actively seek new knowledge and skills based on their interests and goals (Knowles, 1975). By fostering a mindset of lifelong learning, individuals can better navigate the complexities of the modern world, contributing to their personal growth and societal well-being.

While transformative learning focuses on deep, reflective change in individuals' perspectives, lifelong learning emphasizes the ongoing acquisition of knowledge and skills. Together, these approaches complement each other by fostering both personal growth and the capacity for adaptation in an ever-changing world. In the context of global challenges, such as climate change and social inequality, transformative and lifelong learning are essential for fostering the knowledge, skills, and values necessary to address these issues. For example, individuals who engage in transformative learning may develop a deeper understanding of environmental sustainability and social justice. In contrast, those who embrace lifelong learning may continually update their knowledge and skills to stay informed and contribute to solutions. By integrating both transformative and lifelong learning, individuals can develop the resilience and agency needed to navigate and address society's complex issues.

Furthermore, integrating global citizenship education into transformative and lifelong learning can help individuals develop a broader understanding of global issues and their roles in addressing them. Global citizenship education emphasizes the interconnectedness of all people and the importance of working together to promote peace, sustainability, and social justice (UNESCO, 2014). By fostering a sense of global responsibility and solidarity, this type of education can enhance the effectiveness of transformative and lifelong learning, helping individuals recognize their capacity to contribute to positive change on a global scale.

In conclusion, transformative and lifelong learning are integral to personal development, societal transformation, and global progress. These approaches encourage individuals to critically reflect on their assumptions, values, and behaviors, fostering a deeper understanding of themselves and their world. In the face of global challenges, such as climate change and social inequality, transformative and lifelong learning offer pathways for individuals to adapt, engage, and contribute to positive change. By cultivating these forms of learning, individuals can develop the knowledge, skills, and values necessary to navigate complex issues, foster social justice, and promote sustainability in an increasingly interconnected world.

24.2 Empowering Communities Through Knowledge

Empowering communities through knowledge is a central principle in pursuing social justice, equity, and sustainable development. Knowledge is a powerful tool enabling individuals and communities to better understand their environments, make informed decisions, and advocate for their rights and needs. When communities are equipped with the knowledge necessary to address their challenges, they can drive social, economic, and political change from within, fostering resilience and long-term sustainability. This process is particularly important in the context of marginalized or disenfranchised communities, where access to knowledge has historically been limited or controlled by external forces. Empowering communities through knowledge involves not only the dissemination of information but also the cultivation of critical thinking, skills development, and the creation of platforms for collective action.

The concept of knowledge empowerment emphasizes the importance of democratizing knowledge and making it accessible to all members of society, regardless of their socio-economic status, geographic location, or educational background. This process involves removing barriers to knowledge access, such as language, digital literacy, and economic inequality, and ensuring that individuals and communities have the tools to both acquire and apply knowledge effectively (Foley, 1999). In many cases, empowering communities through knowledge requires addressing systemic inequalities that prevent certain groups from accessing education and information. For example, in rural or remote areas, access to educational resources and information may be limited due to geographical isolation, lack of infrastructure, or insufficient funding for educational institutions. By creating inclusive educational opportunities and making knowledge more accessible, communities can overcome these barriers and gain the capacity to address local challenges.

One important aspect of empowering communities through knowledge is the promotion of participatory learning. Participatory learning involves engaging community members in the process of knowledge creation and decision-making, rather than simply disseminating information to them. This approach recognizes that communities possess valuable local knowledge, experiences, and insights that can contribute to solutions for the challenges they face. By fostering participatory learning, individuals are not only recipients of knowledge but also active contributors to the knowledge production process. This can take many forms, such as community-based research, collaborative workshops, or local forums for discussion and problem-solving. Participatory learning helps to build a sense of ownership and agency within communities, as individuals become more invested in the outcomes of their learning and the actions that result from it (Cornwall & Jewkes, 1995).

In addition to participatory learning, capacity building is another critical element of community empowerment through knowledge. Capacity building refers to the process of enhancing the skills, competencies, and abilities of individuals and organizations to achieve their goals and address their needs. This can include training in specific areas such as leadership, entrepreneurship, healthcare, or environmental sustainability, as well as fostering critical thinking, problem-solving, and advocacy skills. Capacity building not only increases the immediate capabilities of individuals and communities but also strengthens their long-term resilience, enabling them to adapt to changing circumstances and challenges. For example, in the context of environmental conservation, capacity building can empower local communities to manage natural resources sustainably, advocate for policies that protect their environments, and develop alternative livelihoods that are less dependent on harmful practices such as deforestation or overfishing (Berkes, 2004).

Technology and digital platforms have also become powerful tools for empowering communities through knowledge. In an increasingly digital world, access to information and education is no longer limited to physical classrooms or traditional media. Online platforms, social media, and mobile technology can provide communities with new opportunities to access educational resources, engage in global conversations, and collaborate on solutions to shared problems. For example, online learning platforms offer flexible and affordable educational opportunities, while social media enables individuals to share information, organize campaigns, and advocate for change. However, it is important to recognize that digital literacy and access to technology are still significant barriers for many communities, particularly in low-income or rural areas. Efforts to empower communities through knowledge must therefore also address the digital divide and ensure that all individuals have the skills and resources necessary to participate in the digital world (Van Dijk, 2017).

Furthermore, knowledge sharing and collaboration between communities and external organizations, such as NGOs, governments, and academic institutions, can enhance the effectiveness of community empowerment efforts. These collaborations can facilitate the exchange of expertise, resources, and strategies that can help communities address their challenges. For instance, partnerships between local communities and environmental organizations can lead to the co-production of knowledge about sustainable land management practices, while collaborations with healthcare providers can improve access to medical information and services. By fostering these relationships, communities can gain access to additional resources and expertise, while external organizations can benefit from the local knowledge and perspectives that communities bring to the table (Sullivan et al., 2017).

Empowering communities through knowledge also has broader societal implications, as it can contribute to the creation of more just, equitable, and sustainable societies. When communities are equipped with the knowledge and skills to address their own challenges, they are better able to advocate for their rights, participate in decision-making processes, and contribute to the development of policies and systems that reflect their needs and aspirations. This can lead to greater social cohesion, reduced inequality, and more inclusive development outcomes. For example, communities that are empowered to address issues such as climate change, healthcare, or education are more likely to implement solutions that are culturally appropriate, locally relevant, and sustainable in the long term.

In conclusion, empowering communities through knowledge is a critical pathway for fostering social change, resilience, and sustainable development. By ensuring that all individuals and communities have access to the information, skills, and opportunities they need to thrive, we can create a more equitable and just society. This process involves promoting participatory learning, building capacity, leveraging technology, and fostering collaboration between communities and external organizations. As communities gain the knowledge and agency to address their own challenges, they can contribute to the creation of a more sustainable, inclusive, and resilient world.

24.3 Rethinking Education for the Future

Education has long been considered a cornerstone of societal development, providing individuals with the knowledge, skills, and values necessary to participate in and contribute to the world around them. However, as we face the challenges of the 21st century, including technological advancements, climate change, globalization, and shifting economic landscapes, it has become increasingly clear that traditional educational models may not be sufficient to prepare future generations for the complexities and uncertainties of the future. Rethinking education for the future requires a fundamental reimagining of how we approach learning, the role of educators, and the goals of education itself.

One of the central areas for rethinking education is the integration of technology. Digital tools and platforms have already transformed many aspects of society, and education is no exception. Technology can enhance learning experiences by providing access to a wealth of information, facilitating collaboration across geographic boundaries, and enabling personalized learning paths that cater to individual needs and learning styles (Selwyn, 2016). However, the integration of technology must be done thoughtfully and inclusively to ensure that all students, regardless of socio-economic background, have equal access to digital resources. Moreover, education systems must go beyond simply incorporating digital tools into traditional teaching methods. Instead, they should embrace technology as a means to foster new forms of learning that encourage critical thinking, creativity, and collaboration, while also addressing the ethical implications of digital technologies (Luckin et al., 2016).

Another crucial aspect of rethinking education is the shift towards lifelong learning. In a rapidly changing world, where the half-life of knowledge is shrinking and many jobs are being automated or transformed by technology, the ability to continuously learn and adapt is essential. Lifelong learning is not just about formal education but also about fostering a mindset of curiosity, adaptability, and self-directed learning throughout one's life (Fischer et al., 2014). This shift requires a rethinking of how education is structured, moving away from a model that emphasizes static knowledge acquisition towards one that values continuous personal and professional development. In this context, education systems should focus on developing skills such as problem-solving, critical thinking, emotional intelligence, and adaptability, which are essential for navigating an unpredictable future (OECD, 2018).

Equity and inclusion are also central to the rethinking of education for the future. Traditional educational systems have often perpetuated inequalities based on factors such as race, gender, socioeconomic status, and geographic location. For education to truly serve all students and prepare them for the future, it must be inclusive and equitable. This means not only ensuring access to education but also addressing systemic barriers that prevent certain groups from fully participating in and benefiting from educational opportunities (UNESCO, 2015). Rethinking education for the future involves creating learning environments that are culturally responsive, supportive of diverse learning needs, and focused on empowering marginalized communities. This can be achieved through policies that promote inclusive curricula, teacher training, and community engagement, ensuring that education serves as a tool for social mobility and equity.

In addition to addressing equity, environmental sustainability is another key area that must be integrated into future education systems. As the world faces the urgent challenges of climate change and environmental degradation, it is essential that education systems equip students with the knowledge and skills needed to address these global crises. This includes not only teaching about environmental issues but also fostering a sense of responsibility and agency in students, empowering them to become active participants in sustainability efforts (Sterling, 2001). Education for sustainability should be embedded across all subjects and levels of education, from primary school to higher education, and should emphasize the interconnectedness of social, economic, and environmental systems. This approach will help to cultivate a generation of learners who are equipped to think critically about environmental challenges and who are motivated to create sustainable solutions.

Furthermore, global citizenship is an increasingly important aspect of education for the future. In a world that is more interconnected than ever, students must be prepared to navigate cultural diversity, global challenges, and international cooperation. Education for global citizenship involves fostering an understanding of global issues such as poverty, inequality, conflict, and climate change, as well as promoting values of empathy, respect, and social responsibility (Tarrant et al., 2015). This can be achieved through curricula that emphasize global perspectives, intercultural communication, and collaborative problem-solving. By preparing students to think globally and act locally, education can play a critical role in promoting peace, justice, and sustainable development.

Finally, educational assessment must also evolve to align with the changing goals of education. Traditional assessment methods, which often focus on rote memorization and standardized testing, may not adequately measure the skills and competencies needed for the future. Instead, assessments should be designed to evaluate a broader range of abilities, including creativity, collaboration, critical thinking, and emotional intelligence (Saavedra & Opfer, 2012). This shift in assessment practices will help to ensure that education is not just about acquiring knowledge but also about developing the skills necessary to thrive in an ever-changing world.

In conclusion, rethinking education for the future requires a comprehensive approach that addresses the evolving needs of individuals, societies, and the planet. By integrating technology, promoting lifelong learning, ensuring equity and inclusion, emphasizing sustainability, fostering global citizenship, and reimagining assessment, education systems can better prepare students for the challenges and opportunities of the 21st century. Education must evolve to become a dynamic, inclusive, and transformative force that empowers individuals to navigate an uncertain future and contribute to creating a more just, sustainable, and resilient world.

References

Bennett, L. (2012). Youth, political participation, and social media: A literature review. *Journal of Youth Studies*, 15(1), 1–18. https://doi.org/10.1080/13676261.2011.596255

Berkes, F. (2004). Rethinking community-based conservation. *Conservation Biology, 18*(3),621–630.https://doi.org/10.1111/j.1523-1739.2004.00077.x

Cornwall, A., & Jewkes, R. (1995). What is participatory research? *Social Science & Medicine, 41*(12), 1667–1676. https://doi.org/10.1016/0277-9536(95)00127-S

Cranton, P. (2006). *Understanding and promoting transformative learning: A guide for educators of adults*. Jossey-Bass.

Davies, I. (2006). Global citizenship: Abstraction or reality? *Educational Review*, 58(1), 5–25. https://doi.org/10.1080/00131910500352743

Fischer, G., Dufresne, M., & Jäkel, C. (2014). Lifelong learning and learning ecosystems: A perspective on the future of education. *International Journal of Educational Technology in Higher Education, 11*(1), 1–16. https://doi.org/10.1186/s41239-014-0013-7

Foley, G. (1999). Learning in social action: A contribution to understanding informal education. *Zed Books*.

Jarvis, P. (2006). *The theory and practice of learning* (2nd ed.). Routledge.

Knowles, M. S. (1975). *Self-directed learning: A guide for learners and teachers*. Association Press.

Luckin, R., Holmes, W., Griffiths, M., & Forcier, L. B. (2016). *Intelligence unleashed: An argument for AI in education*. Pearson Education.

Mezirow, J. (1997). *Transformative learning: Theory to practice*. New Directions for Adult and Continuing Education, 1997(74), 5–12. https://doi.org/10.1002/ace.7401

Mezirow, J. (2000). *Learning as transformation: Critical perspectives on a theory in progress*. Jossey-Bass.

Nussbaum, M. C. (2011). *Creating capabilities: The human development approach*. Harvard University Press.

OECD. (2018). *The future of education and skills: Education 2030*. OECD Publishing.

Saavedra, A. R., & Opfer, V. D. (2012). Teaching and learning 21st century skills: Lessons from the learning sciences. *The Future of Children, 22*(1), 1–19. https://doi.org/10.1353/foc.2012.0000

Sterling, S. (2001). Sustainable education: Re-visioning learning and change. *Green Books.*

Sullivan, S., Chalmers, N., & Tsing, A. L. (2017). *Collaboration and co-production in environmental knowledge*. Routledge.

Tarrant, M. A., Rubin, D. L., & Stoner, L. (2015). The development of the global citizenship scale. *Journal of Social Issues, 71*(3), 379–396. https://doi.org/10.1111/josi.12118

UNESCO. (2014). *Global citizenship education: Preparing learners for the challenges of the 21st century*. UNESCO.

UNESCO. (2015). *Education for sustainable development goals: Learning objectives*. UNESCO Publishing.

Van Dijk, J. A. G. M. (2017). *The digital divide: The internet and social inequality in international perspective* (3rd ed.). Sage Publications.

25

Ethical Leadership and Governance

25.1 Redefining Leadership for a Globalized World

Ethical leadership and governance are fundamental concepts that shape the effectiveness, integrity, and sustainability of organizations, communities, and societies. Ethical leadership refers to the practice of leading by adhering to moral principles, promoting fairness, transparency, and accountability, and making decisions that prioritize the well-being of stakeholders, the community, and the environment. Governance, on the other hand, involves the structures, processes, and practices that guide decision-making and ensure that leaders are held accountable for their actions. The relationship between ethical leadership and governance is crucial, as ethical leadership can foster good governance practices that contribute to the long-term success and credibility of organizations and institutions.

One of the key aspects of ethical leadership is the ability to make decisions based on ethical principles, such as honesty, fairness, and justice. Ethical leaders are expected to model these values in their actions, influencing their followers and creating a culture of trust and respect within the organization (Brown & Treviño, 2006). In this sense, ethical leadership is not just about making morally sound decisions but also about inspiring others to adopt ethical behavior and promoting a sense of responsibility among followers. This is particularly important in contexts where leaders are tasked with making decisions that impact a wide range of stakeholders, including employees, customers, communities, and the environment.

Good governance is closely linked to ethical leadership. Effective governance systems provide mechanisms for ensuring accountability, transparency, and stakeholder participation, which are essential for fostering trust in institutions. Ethical leadership plays a critical role in governance by ensuring that decision-makers are not only competent but also committed to acting in the best interests of the public or the organization. Good governance structures, such as independent oversight bodies, transparent decision-making processes, and regular accountability checks, can help ensure that leaders

DOI: 10.1201/9788743808787-31

are held responsible for their actions and that they adhere to ethical standards (Behn, 2001). These structures are particularly important in preventing corruption, abuse of power, and other unethical practices that can undermine public trust and organizational integrity.

An essential feature of ethical leadership is accountability, which refers to the obligation of leaders to answer for their decisions and actions. Accountability is a cornerstone of good governance, as it ensures that leaders are held responsible for their behavior and that there are consequences for unethical actions. In both public and private sectors, ethical leadership and governance require mechanisms for ensuring that leaders are answerable to stakeholders and that their actions align with the values and goals of the organization or society. These mechanisms may include regular audits, public reporting, and the establishment of ethical review boards that can evaluate the decisions made by leaders and hold them accountable for their impact (Kaptein, 2011).

Transparency is another critical element of ethical leadership and governance. Transparency refers to the openness with which decisions are made, the rationale behind those decisions, and the availability of information to stakeholders. Ethical leaders are expected to communicate openly and honestly about the decisions they make, particularly when those decisions affect public resources or the well-being of others. In the context of governance, transparency helps to prevent corruption, ensure public trust, and encourage informed participation from stakeholders. By promoting transparency, ethical leaders can foster a sense of ownership and trust among those they lead, creating a more inclusive and participatory environment (Bovens, 2007).

Ethical decision-making is another key component of ethical leadership and governance. Leaders are often faced with complex situations where the right course of action is not immediately clear. In such cases, ethical leaders must navigate competing interests, weigh the potential consequences of their decisions, and choose a path that aligns with ethical principles. Ethical decision-making involves a process of reflection, empathy, and moral reasoning, which helps leaders make choices that prioritize the long-term well-being of individuals and communities over short-term gains (Rest, 1986). This process is particularly important in situations where leaders are confronted with moral dilemmas or where the interests of different stakeholders may conflict.

Furthermore, inclusive leadership is an important aspect of ethical governance. Ethical leaders are expected to promote inclusivity by ensuring that all voices are heard and that decisions are made with consideration for the diverse needs and perspectives of stakeholders. Inclusive leadership fosters a sense of belonging and ensures that the interests of marginalized or underrepresented groups are not overlooked. In the context of governance, inclusive leadership can help to create more equitable policies and practices that reflect the values of justice and fairness (Nembhard & Edmondson, 2006). By prioritizing inclusivity, ethical leaders can contribute to the creation of more just and sustainable societies.

In conclusion, ethical leadership and governance are essential for building trust, promoting accountability, and ensuring the long-term success of organizations and societies. Ethical leadership involves making decisions based on moral principles, fostering a culture of integrity, and inspiring others to act ethically. Good governance, in turn, provides the structures and processes that hold leaders accountable for their actions and ensure that they are transparent and responsive to the needs of stakeholders. By prioritizing ethical leadership and governance, organizations and societies can create a foundation for sustainable growth, social justice, and collective well-being.

25.2 Participatory Democracy and Citizen Engagement

Participatory democracy emphasizes the active involvement of citizens in the decision-making processes that affect their lives, communities, and societies. Unlike traditional representative democracy, where elected officials make decisions on behalf of the people, participatory democracy seeks to empower individuals by encouraging direct involvement in political, social, and economic matters. This model of democracy aims to deepen democracy by enhancing transparency, accountability, and inclusivity, ensuring that the voices of all citizens, particularly marginalized groups, are heard and considered in governance processes (Fung, 2006). The role of citizen engagement is central to participatory democracy, as it fosters active participation, informed debate, and collaborative decision-making.

One of the core principles of participatory democracy is the belief that individuals should have a direct role in shaping the policies that affect their lives. This involves not only voting in elections but also engaging in public discussions, consultations, and deliberations. Participatory democracy encourages citizens to take responsibility for the political and social issues that matter to them, fostering a sense of ownership and agency in the democratic process. As such, it goes beyond merely exercising voting rights and emphasizes continuous engagement through mechanisms like town hall meetings, public forums, citizen juries, and participatory budgeting (Smith, 2009). By allowing citizens to engage in meaningful dialogue and decision-making, participatory democracy aims to create more inclusive, equitable, and responsive governance structures.

Citizen engagement is an essential component of participatory democracy. It refers to the active involvement of individuals in civic and political processes, which can take many forms, from voting and petitioning to attending public meetings and volunteering for local initiatives. Engaged citizens are more likely to advocate for their interests, hold leaders accountable, and contribute to the overall well-being of their communities (Verba, Schlozman, & Brady, 1995). Citizen engagement also plays a critical role in addressing social inequalities by ensuring that marginalized voices are included in decision-making processes. Research has shown that when citizens are actively involved in governance, they are more likely to trust political institutions, participate in community activities, and work together to address collective challenges (Putnam, 2000).

The concept of deliberative democracy is closely tied to participatory democracy and citizen engagement. Deliberative democracy emphasizes the importance of reasoned, respectful dialogue and the exchange of ideas among citizens to reach collective decisions. This approach values the quality of deliberation over mere voting or decision-making, recognizing that democratic outcomes are more legitimate when they result from informed, reflective, and inclusive discussions (Gutmann & Thompson, 2004). Deliberative processes, such as citizens' assemblies and deliberative polls, provide platforms for diverse groups to discuss complex issues, share perspectives, and collectively develop solutions. By fostering deliberation, participatory democracy encourages citizens to engage with issues critically and collaboratively, leading to more thoughtful and sustainable outcomes.

Participatory governance is another critical aspect of participatory democracy. It refers to the involvement of citizens in the decision-making processes of public institutions, such as local governments, schools, and healthcare systems. Participatory governance can take various forms, including participatory budgeting, where citizens directly influence how public funds are allocated, and community-driven development, where residents play a central role in identifying and addressing their own needs (Santos, 2006). This approach to governance not only enhances transparency and accountability but also helps to build social capital by fostering trust and cooperation among citizens and between citizens and their governments. In this way, participatory governance strengthens democratic practices and contributes to more responsive and inclusive public policies.

However, challenges to participatory democracy and citizen engagement persist, particularly in societies where political apathy, inequality, and disillusionment with political institutions are prevalent. One significant barrier is the digital divide, which can limit access to information and online participation opportunities for marginalized groups (Graham & Zook, 2006). Additionally, participatory processes can be time-consuming and require significant resources, which may deter some citizens from engaging fully. Furthermore, the effectiveness of participatory democracy depends on the willingness of political elites and institutions to genuinely incorporate citizen input into decision-making. In many cases, participatory processes are merely symbolic, and citizens' contributions are not meaningfully considered or implemented (Baiocchi, Heller, & Silva, 2011).

Despite these challenges, participatory democracy remains a vital framework for deepening democratic engagement and promoting social justice. It offers a pathway for creating more inclusive, accountable, and responsive political systems, where citizens can actively shape the policies and decisions that affect their lives. For participatory democracy to thrive, it is essential to foster an environment where citizen engagement is encouraged, supported, and valued. This includes ensuring that all citizens, regardless of their background or socioeconomic status, have access to the tools, resources, and opportunities necessary for meaningful participation.

Participatory democracy and citizen engagement are essential for creating more inclusive and responsive political systems. By empowering citizens to actively engage in decision-making processes, participatory democracy fosters a sense of ownership and accountability, enhances social trust, and ensures that the voices of all individuals are heard. While challenges to participatory democracy exist, such as political apathy and inequality, the principles of deliberation, inclusivity, and participatory governance offer pathways for overcoming these obstacles and building stronger, more resilient democracies.

25.3 Fostering Transparency and Accountability

Transparency and accountability are fundamental pillars of good governance, ensuring that public institutions act in the best interests of citizens and uphold ethical standards. Transparency refers to the openness with which governments, organizations, and institutions operate, providing access to information that enables citizens to understand and evaluate decisions and actions. Accountability, on the other hand, is the obligation of those in positions of power to answer for their actions and decisions, particularly in terms of their impact on society and the environment (Bovens, 2007). Together, these principles promote trust, reduce corruption, and enhance the legitimacy of governmental and organizational actions, thereby contributing to more effective governance and social stability.

One of the key ways to foster transparency is through the open data movement, which advocates for making government and organizational data publicly available and accessible. Open data initiatives allow citizens, journalists, and researchers to scrutinize government activities, monitor public spending, and assess the effectiveness of policies. By promoting transparency in this way, governments can reduce the risk of corruption and ensure that public resources are used efficiently. According to the Open Government Partnership (OGP), open data has been shown to improve public services, enhance civic participation, and increase government responsiveness (OGP, 2020). Additionally, transparency can be achieved through the publication of detailed reports on government activities, budgets, and decision-making processes, making it easier for citizens to hold their leaders accountable.

Accountability mechanisms are essential for ensuring that transparency translates into meaningful action. These mechanisms include the creation of independent bodies such as ombudsmen, auditors, and anti-corruption agencies, which are tasked with monitoring government actions and holding officials accountable for misconduct. Effective accountability mechanisms not only provide oversight but also offer avenues for redress and remedy when citizens' rights are violated or when public resources are misused. The role of civil society organizations (CSOs) and the media is crucial in fostering accountability, as they act as watchdogs, investigating and exposing cases of corruption, inefficiency, and abuse of power. The Freedom of Information (FOI) laws in many countries further enhance accountability by granting citizens the legal right to access government records, thereby ensuring that public officials can be held responsible for their decisions (Nabatchi, 2010).

One of the challenges to fostering transparency and accountability is the digital divide, which can limit access to information for marginalized communities. While the internet and digital platforms have greatly expanded access to information, not all citizens have equal access to these tools, particularly in low-income or rural areas. This disparity can undermine efforts to promote transparency, as some segments of the population may be excluded from the flow of information necessary for meaningful participation in governance. Addressing the digital divide requires investments in digital literacy programs, infrastructure, and policies that ensure equitable access to information and technology (Graham & Zook, 2006). Moreover, transparency efforts must be coupled with strategies to ensure that the information provided is understandable, relevant, and actionable for all citizens.

Another important aspect of fostering transparency and accountability is the use of participatory mechanisms that engage citizens in decision-making processes. Participatory budgeting, for example, allows citizens to directly influence how public funds are allocated, fostering both transparency in government spending and accountability for the outcomes of those decisions (Baiocchi, Heller, & Silva, 2011). Additionally, the inclusion of marginalized groups in decision-making processes ensures that policies reflect the needs and concerns of all citizens, particularly those who may otherwise be excluded from political and social participation. By creating spaces for citizens to engage in governance, participatory mechanisms help build trust in government institutions and enhance the legitimacy of public policies.

Transparency and accountability are also critical in the context of corporate governance. In the private sector, businesses have a responsibility to act transparently and be accountable to their stakeholders, including employees, customers, investors, and the broader community. Corporate social responsibility (CSR) initiatives, sustainability reporting, and ethical business practices are key components of fostering transparency in business operations. Companies that prioritize transparency in their decision-making processes and are accountable for their environmental and social impact are more likely to build long-term trust with their stakeholders (Gray, 2006). Moreover, transparent businesses are better positioned to mitigate risks, attract investment, and contribute to broader societal goals, such as sustainability and social justice.

Finally, fostering transparency and accountability is a dynamic and ongoing process that requires continuous effort and adaptation. While significant progress has been made in many countries and organizations, there is still much work to be done to ensure that transparency and accountability are upheld at all levels of governance and society. This includes strengthening legal frameworks, promoting citizen engagement, and leveraging technology to enhance information sharing and oversight. In the digital age, transparency and accountability must also address the challenges posed by misinformation, data privacy, and surveillance, ensuring that citizens' rights are protected while maintaining open and transparent systems of governance.

Fostering transparency and accountability is essential for building trust, reducing corruption, and promoting effective governance. By making information accessible, establishing strong accountability mechanisms, and engaging citizens in decision-making, governments and organizations can create more inclusive, responsive, and responsible institutions. However, challenges such as the digital divide and the need for participatory mechanisms must be addressed to ensure that transparency and accountability benefit all citizens, particularly those who are most vulnerable.

References

Baiocchi, G., Heller, P., & Silva, M. (2011). *Making space for the poor: The politics of participatory budgeting*. Zed Books.

Behn, R. D. (2001). *Rethinking democratic accountability*. Brookings Institution Press.

Bovens, M. (2007). Analyzing and assessing accountability: A conceptual framework. *European Law Journal, 13*(4), 447–468. https://doi.org/10.1111/j.1468-0386.2007.00378.x

Brown, M. E., & Treviño, L. K. (2006). Ethical leadership: A review and future directions. *The Leadership Quarterly, 17*(6), 595–616. https://doi.org/10.1016/j.leaqua.2006.10.004

Fung, A. (2006). Varieties of participation in complex governance. *Public Administration Review, 66*(s1), 66–75. https://doi.org/10.1111/j.1540-6210.2006.00667.x

Graham, S., & Zook, M. (2006). Digital divide and the geography of the Internet. *Wiley Interdisciplinary Reviews: Computational Statistics, 1*(5), 557–572. https://doi.org/10.1002/wics.65

Gray, R. (2006). Social, environmental and sustainability reporting and organizational change. *Accounting, Auditing & Accountability Journal, 19*(6), 868–897. https://doi.org/10.1108/09513570610709859

Gutmann, A., & Thompson, D. (2004). *Why deliberative democracy?* Princeton University Press.

Kaptein, M. (2011). Understanding unethical behavior by unraveling ethical culture. *Journal of Business Ethics, 99*(2), 223–242. https://doi.org/10.1007/s10551-010-0664-7

Nabatchi, T. (2010). *Public participation*. In M. M. Bingham & R. O'Leary (Eds.), *Democracy in practice: Public participation in the United States* (pp. 19–40). Brookings Institution Press.

Nembhard, I. M., & Edmondson, A. C. (2006). Making it safe: The effects of leader inclusiveness and professional status on psychological safety and improvement efforts in health care teams. *Journal of Organizational Behavior, 27*(7), 941–966. https://doi.org/10.1002/job.413

Open Government Partnership (OGP). (2020). *Open government: A new era of transparency and accountability*. Retrieved from https://www.opengovpartnership.org/

Putnam, R. D. (2000). *Bowling alone: The collapse and revival of American community*. Simon & Schuster.

Rest, J. R. (1986). *Moral development: Advances in research and theory*. Praeger Publishers.

Santos, B. S. (2006). *The rise of the global left: The World Social Forum and beyond*. Zed Books.

Smith, G. (2009). *Democratic innovations: Designing institutions for citizen participation*. Cambridge University Press.

Verba, S., Schlozman, K. L., & Brady, H. E. (1995). *Voice and equality: Civic voluntarism in American politics*. Harvard University Press.

PART VII

Hope and Vision for Humanity

26

Hope and Vision for Humanity

Hope and vision for humanity are fundamental concepts that shape our collective aspirations for a better future. Hope, as an emotional and cognitive state, provides individuals and societies with the motivation to overcome challenges and strive for positive change. Vision, on the other hand, refers to a clear and compelling image of a desired future, often rooted in values such as justice, equality, and sustainability. Together, hope and vision are powerful forces that inspire individuals and communities to act, foster resilience, and build a more inclusive, just, and sustainable world. These concepts are essential for addressing the global challenges humanity faces, such as climate change, inequality, and social fragmentation.

Hope plays a critical role in motivating individuals to engage in actions that contribute to the common good. According to Snyder's Hope Theory, hope involves two key components: agency (the belief that one can initiate and sustain actions toward a goal) and pathways (the ability to generate strategies to achieve those goals) (Snyder, 2002). When individuals possess hope, they are more likely to take proactive steps toward solving problems, whether on a personal or societal level. For example, hope has been shown to be a key factor in fostering environmental activism, as individuals who believe in the possibility of positive environmental change are more likely to engage in sustainable behaviors and advocate for policy reforms (Ojala, 2012). In this context, hope serves as a catalyst for action, driving people to work toward solutions despite the enormity of global challenges.

In addition to individual hope, a collective vision for humanity is crucial for guiding societal progress. A shared vision provides a framework for understanding the future and serves as a blueprint for the actions needed to achieve it. Visionary leaders, such as Mahatma Gandhi, Nelson Mandela, and Martin Luther King Jr., have historically articulated compelling visions of justice, equality, and human dignity that inspired millions to work toward social change. Similarly, in the context of sustainability, the United Nations

DOI: 10.1201/9788743808787-33

Sustainable Development Goals (SDGs) offer a collective vision for a more equitable, peaceful, and sustainable world by 2030. These global goals provide a roadmap for addressing issues such as poverty, inequality, climate change, and environmental degradation, and they encourage governments, organizations, and individuals to collaborate in achieving shared objectives (UN, 2015).

A key aspect of hope and vision for humanity is the concept of human flourishing, which refers to the realization of human potential in all its dimensions – emotional, intellectual, social, and spiritual. Human flourishing emphasizes the importance of well-being, not only as an individual pursuit but as a collective endeavor that requires equitable access to resources, opportunities, and rights. In his book *Flourish*, Seligman (2011) outlines the elements that contribute to human flourishing, including positive emotions, engagement, relationships, meaning, and achievement. These elements are essential for creating a society where individuals can thrive and contribute to the common good. A vision for humanity that prioritizes human flourishing encourages the development of policies and practices that support mental health, education, social justice, and environmental sustainability.

Moreover, hope and vision are essential in fostering resilience in the face of adversity. Resilience refers to the ability to adapt to and recover from challenges, whether personal, societal, or environmental. In the context of global crises, such as climate change or pandemics, hope and a shared vision provide the psychological and social resources necessary to persevere and rebuild. The concept of transformative resilience involves not only bouncing back from adversity but also using challenges as opportunities for growth and positive change (Walker et al., 2004). For example, communities that have experienced natural disasters may develop new, more sustainable practices in response to the destruction, fostering resilience while also contributing to long-term sustainability. A collective vision for a more resilient world can guide efforts to create systems that are adaptable to future challenges, promoting sustainability, equity, and justice.

However, hope and vision are not without challenges. In an era marked by political polarization, social fragmentation, and environmental degradation, maintaining hope and a clear vision for the future can be difficult. The rise of eco-anxiety and climate despair – emotional responses to the perceived threat of climate change – has led many individuals, particularly youth, to question whether a positive future is possible (Ojala, 2012). In such contexts, it is essential to cultivate hope-based narratives that emphasize the possibility of positive change through collective action and innovation. For instance, highlighting successful climate policies, technological innovations, and grassroots movements can inspire hope and demonstrate that transformative change is achievable. These narratives can counteract feelings of helplessness and provide individuals with the agency to engage in solutions.

In addition, intergenerational collaboration is crucial for building a vision of the future that is inclusive and sustainable. Younger generations, who will inherit the consequences of today's decisions, have a vested interest in shaping the future. Intergenerational dialogue allows for the exchange of ideas, experiences, and values, fostering mutual understanding and collective action. The involvement of youth in decision-making processes, particularly in environmental and social justice movements, is vital for ensuring that future generations are not merely passive recipients of policy decisions but active participants in creating the world they want to live in. Youth-led movements, such as Fridays for Future and the Sunrise Movement, demonstrate the power of young people to inspire change and articulate a vision for a more just and sustainable future.

Ultimately, hope and vision for humanity are essential for overcoming the challenges of the present and building a future that reflects our highest values and aspirations. By fostering hope, articulating a collective vision, and promoting resilience, societies can navigate global crises and create a more equitable, sustainable, and flourishing world for future generations.

26.1 Lessons from History and Nature

History and nature offer profound lessons that can guide humanity in addressing contemporary challenges. Both realms serve as repositories of knowledge, providing insights into human behavior, societal development, and the natural world's dynamic processes. By studying the past and observing nature's patterns, we can better understand the interconnectedness of human actions, environmental sustainability, and the long-term consequences of our decisions. Lessons from history and nature are not only essential for fostering resilience and sustainability but also for developing strategies that promote social justice, equity, and global cooperation.

26.1.1 Lessons from history: The cycles of civilization

History has repeatedly demonstrated that societies rise, and fall based on their ability to adapt to internal and external challenges. The decline of ancient civilizations, such as the Roman Empire, the Maya, and the Harappan civilization, offers valuable lessons about the importance of environmental stewardship, resource management, and social cohesion. In the case of the Maya civilization, for example, deforestation, soil degradation, and prolonged droughts contributed to the collapse of once-thriving city-states (Haug et al., 2003). The lessons from such historical collapses emphasize the need for sustainable resource use and the recognition that environmental degradation can undermine the stability of societies.

Furthermore, history teaches us the importance of innovation and adaptation in the face of adversity. The Industrial Revolution, for instance, illustrates how technological advancements can radically transform societies, creating both opportunities and challenges. While the rise of industrialization led to unprecedented economic growth, it also resulted in environmental degradation, social inequality, and the exploitation of labor (Marx, 1867). This duality in progress underscores the necessity of balancing technological development with ethical considerations and social responsibility. By learning from past mistakes, societies can work toward sustainable development that prioritizes both human well-being and environmental health.

26.1.2 Nature's lessons: Interconnectedness and resilience

Nature offers a rich source of wisdom regarding resilience, adaptation, and the interconnectedness of all life forms. Ecosystems, for example, are dynamic systems in which each species plays a vital role in maintaining balance. The concept of ecological resilience refers to the ability of ecosystems to absorb disturbances and reorganize while maintaining their essential functions (Holling, 1973). This principle can be applied to human societies, which, like ecosystems, must learn to adapt to changing conditions, whether through environmental, economic, or social shifts. The resilience of nature, seen in the recovery of forests after fires or the regeneration of coral reefs, provides a model for how societies can rebuild and thrive after crises if they respect and nurture the systems on which they depend.

The natural world also teaches the importance of diversity. Biodiversity – the variety of life on Earth – is essential for the stability and health of ecosystems. Similarly, cultural and social diversity strengthens human societies by fostering innovation, creativity, and mutual understanding. The extinction of species in ecosystems often leads to a collapse in the system's functionality, while the loss of cultural diversity can result in the erosion of knowledge, identity, and social cohesion (Wilson, 1992). The protection of biodiversity and the promotion of social inclusivity are thus parallel efforts that contribute to the resilience and sustainability of both the natural world and human societies.

26.1.3 Applying lessons from history and nature to global challenges

The lessons from history and nature are increasingly relevant as humanity faces global challenges such as climate change, resource depletion, and social inequality. The current environmental crisis, for instance, mirrors historical patterns of overexploitation and unsustainable growth. However, by drawing on historical and natural wisdom, there is potential for transformation. One lesson is the importance of preventative action. The collapse of past civilizations often resulted from a failure to anticipate or mitigate environmental changes, such as deforestation or drought. In contrast, modern societies can proactively address climate change by transitioning to renewable energy, adopting sustainable agricultural practices, and reducing carbon emissions.

Additionally, nature's ability to adapt and regenerate underscores the importance of restorative practices in environmental conservation. The concept of regenerative agriculture, which seeks to restore soil health and biodiversity while producing food, draws inspiration from natural processes of regeneration (Gliessman, 2016). This approach contrasts with conventional agricultural practices that often degrade the land, and it offers a pathway to sustainable food systems that can support future generations.

In terms of social resilience, the lessons from history and nature highlight the need for social cohesion and equity. Societies that have thrived over time have done so by fostering inclusive systems that promote justice and cooperation. The civil rights movement in the United States, for example, demonstrated how collective action, based on principles of equality and justice, can transform societies. Similarly, the concept of ecological justice advocates for addressing the disproportionate impacts of environmental degradation on marginalized communities, emphasizing the interconnectedness of social and environmental well-being (Agyeman, 2005). By integrating social and environmental justice into policy and practice, societies can create more equitable and sustainable futures.

The lessons from history and nature provide invaluable insights into the challenges and opportunities humanity faces today. By learning from past mistakes and embracing the resilience, adaptability, and interconnectedness of nature, societies can work toward a more sustainable and equitable future. These lessons underscore the importance of sustainability, social justice, and global cooperation in addressing the pressing issues of our time. As we confront the complex challenges of the 21st century, history and nature remind us that change is possible, but it requires collective effort, innovation, and a deep respect for the systems that sustain life on Earth.

26.2 Humanity's Resilience in the Face of Crises

Humanity has continually demonstrated remarkable resilience in the face of crises, whether they are environmental, social, economic, or political. Resilience, in this context, refers to the ability of individuals, communities, and societies to recover from adversity, adapt to new challenges, and emerge stronger from hardship. The capacity for resilience is deeply embedded in human nature, drawing on a combination of psychological, social, and cultural factors that enable individuals and communities to withstand and overcome significant challenges. Throughout history, humanity has faced a wide range of crises, from natural disasters to wars, pandemics, and economic depressions. The ability to adapt, innovate, and collaborate in response to these challenges has been key to survival and progress.

26.2.1 Psychological resilience and the power of adaptation

At the individual level, psychological resilience plays a crucial role in how people cope with crises. Psychological resilience refers to the ability to maintain mental health and well-being in the face of adversity, stress, or trauma. Research on resilience suggests that it is not a fixed trait but rather a dynamic process that involves the capacity to adapt and recover from difficult experiences (Masten, 2001). For example, during the COVID-19 pandemic, individuals who exhibited psychological resilience were better able to cope with the stress of isolation, uncertainty, and loss, while also adapting to new ways of working and living. Factors such as strong social support networks, a sense of purpose, and emotional regulation have been identified as key contributors to psychological resilience (Bonanno, 2004).

Furthermore, resilience is linked to cognitive flexibility – the ability to shift thinking and problem-solving strategies in response to changing circumstances. In times of crisis, individuals and communities must often reframe their understanding of the situation, adjust their goals, and find new ways to move forward. The ability to remain hopeful and maintain a positive outlook despite adversity is another important aspect of resilience. Research suggests that hope is a critical psychological resource that helps individuals persevere through difficult times (Snyder, 2002). Thus, humanity's resilience in the face of crises is not just about surviving but also about adapting and growing in the process.

26.2.2 Social resilience and collective action

At the societal level, resilience is often manifested through collective action and the capacity of communities to come together to address shared challenges. Social resilience refers to the ability of communities to withstand and recover from social, economic, or environmental shocks through collaboration, solidarity, and the strengthening of social networks (Norris et al., 2008). For example, during the aftermath of natural disasters such as hurricanes, earthquakes, or floods, communities often demonstrate extraordinary levels of cooperation and mutual aid, with individuals and organizations working together to provide relief and rebuild. Social resilience is also evident in how communities mobilize in response to systemic challenges such as inequality, political unrest, or health crises.

A key factor in social resilience is the strength of social capital, which refers to the networks of relationships, trust, and reciprocity that exist within a community (Putnam, 2000). Communities with high levels of social capital are better able to mobilize resources, share information, and provide emotional and material support during times of crisis. Social resilience is also linked to adaptive capacity, or the ability of communities to adjust to changing conditions. For example, during the COVID-19 pandemic, many communities adapted by leveraging digital technologies to maintain social connections, access healthcare, and support local businesses. These adaptive strategies highlight the importance of flexibility and innovation in building social resilience.

26.2.3 Environmental and ecological resilience

Human resilience is also intertwined with environmental and ecological resilience, as the health of the natural environment directly impacts human well-being. Ecological resilience refers to the capacity of ecosystems to absorb disturbances and continue to function despite stressors such as climate change, habitat destruction, and pollution (Holling, 1973). The resilience of ecosystems, such as forests, wetlands, and coral reefs, is crucial for providing essential services like clean water, air, and food, all of which are vital for human survival.

In the face of environmental crises, such as climate change or biodiversity loss, human societies must adopt sustainable practices and work to restore and protect ecosystems. The concept of sustainable resilience emphasizes the need for human societies to not only recover from environmental crises but also to build systems that are inherently sustainable and capable of adapting to future challenges. For instance, the restoration of degraded ecosystems, such as reforestation efforts or the rehabilitation of coral reefs, has been shown to enhance both ecological and human resilience by improving ecosystem services and reducing vulnerability to climate-related impacts (Chazdon, 2008).

26.2.4 The role of innovation and technology

Innovation and technology play a critical role in enhancing humanity's resilience in the face of crises. Advances in technology have enabled societies to respond more effectively to challenges, whether through medical innovations, communication tools, or disaster response technologies. For example, the rapid development of vaccines during the COVID-19 pandemic demonstrated the potential of scientific innovation to mitigate the impacts of global health crises. Similarly, technological advancements in renewable energy, water purification, and agricultural practices offer solutions to environmental challenges and contribute to building a more sustainable and resilient future.

Technology also facilitates global cooperation, enabling the exchange of knowledge, resources, and expertise across borders. In times of crisis, the ability to share information quickly and collaborate on solutions is essential. The use of digital platforms, for instance, has allowed people to stay connected during periods of social isolation, while also enabling the distribution of resources and information during emergencies. However, it is important to recognize that technology can also exacerbate inequalities if access to digital tools and resources is not equitable (Gasser et al., 2018). Therefore, while technology can enhance resilience, it must be deployed in ways that ensure inclusivity and equity.

Humanity's resilience in the face of crises is a multifaceted phenomenon that involves psychological, social, ecological, and technological dimensions. Through collective action, adaptive capacity, and innovation, human societies have demonstrated an impressive ability to recover from adversity and build more resilient systems. However, as the world faces increasingly complex and interconnected challenges, such as climate change, pandemics, and social inequalities, it is essential to cultivate resilience at all levels of society. By learning from past crises and strengthening the factors that contribute to resilience, humanity can better prepare for future challenges and create a more sustainable, equitable, and resilient world.

26.3 Nature's Capacity for Renewal and Healing

Nature has an inherent capacity for renewal and healing, a characteristic that has been observed across ecosystems and species. This capacity is not only vital for maintaining biodiversity and ecological balance but also plays a crucial role in supporting human well-being and resilience. The idea of nature's regenerative abilities is deeply rooted in ecological science and has gained increasing importance in the context of global environmental challenges, such as climate change, habitat destruction, and biodiversity loss. Through natural processes such as succession, regeneration, and resilience, ecosystems can recover from disturbances, restoring balance and providing essential services to both human and non-human life. However, the extent to which nature can heal itself is contingent on the nature and severity of human-induced damage, as well as the preservation of key ecological processes.

26.3.1 Ecological resilience and regeneration

Ecological resilience refers to the ability of ecosystems to absorb disturbances and reorganize while maintaining essential functions and processes (Holling, 1973). This concept is central to understanding nature's capacity for renewal. Disturbances such as wildfires, floods, or storms are natural events that many ecosystems have evolved to withstand. In some cases, these disturbances can even contribute to ecological renewal by creating opportunities for new growth and regeneration. For example, many forests rely on periodic fires to clear underbrush, recycle nutrients, and promote the germination of certain plant species (Barton, 2016). Similarly, after a flood, wetlands often experience an influx of nutrients that can stimulate the growth of aquatic plants and restore the health of the ecosystem (Mitsch & Gosselink, 2015).

However, human activities, such as deforestation, pollution, and climate change, have significantly altered the natural disturbance regimes of ecosystems, often pushing them beyond their capacity for self-healing. For instance, climate change has exacerbated the frequency and intensity of natural disasters, such as hurricanes and droughts, which can overwhelm ecosystems' ability to recover (IPCC, 2021). In these cases, human intervention is often necessary to support ecological regeneration, such as through habitat restoration or reforestation efforts. Yet, even in the face of significant environmental degradation, nature has shown remarkable capacity for recovery when given the opportunity. For example, large-scale reforestation projects, such as those in the Amazon and Southeast Asia, have demonstrated that ecosystems can regenerate if the conditions for recovery are met, including the restoration of native species and the reduction of human-induced pressures (Chazdon, 2008).

26.3.2 The role of biodiversity in healing

Biodiversity plays a critical role in nature's capacity for renewal. Diverse ecosystems are more resilient to disturbances because they contain a variety of species that can fulfill different ecological functions, such as pollination, nutrient cycling, and pest control (Loreau et al., 2001). High biodiversity enhances the ability of ecosystems to adapt to changing conditions and recover from disturbances. For example, in coral reef ecosystems, the presence of diverse species can help maintain ecosystem functions even when certain species are impacted by stressors like ocean acidification or rising temperatures (McLeod et al., 2019). Similarly, in forest ecosystems, a variety of plant species ensures that some can withstand pests, diseases, or changing climate conditions, while others fill the ecological gaps left by disturbances (Hughes, 2000).

In contrast, ecosystems that are impoverished in biodiversity are less able to recover from disturbances and are more vulnerable to collapse. The loss of species, whether due to habitat destruction, overexploitation, or climate change, weakens the resilience of ecosystems and reduces their capacity for self-renewal (Sala et al., 2000). This highlights the importance of preserving biodiversity not only for its intrinsic value but also for its role in sustaining the ecological processes that enable nature to heal itself.

26.3.3 Human-nature relationship and ecological restoration

Human activities have both negatively and positively influenced nature's capacity for renewal. While human-induced environmental degradation has led to the loss of biodiversity and ecosystem services, humans also have the potential to support and accelerate nature's healing processes through ecological restoration. Ecological restoration involves the active management of ecosystems to return them to a more natural or functional state, often by reintroducing native species, restoring habitats, and improving ecosystem services (Hobbs & Harris, 2001). Successful restoration projects have demonstrated that, with proper care and management, ecosystems can recover and even surpass their previous levels of biodiversity and ecological function.

For example, in the case of the Everglades in Florida, extensive restoration efforts have been undertaken to restore hydrological systems, improve water quality, and reintroduce native plant and animal species. These efforts have resulted in the recovery of wetlands and the restoration of critical ecosystem services, such as water filtration and flood control (Doren et al., 2011). Similarly, large-scale reforestation projects in degraded areas have led to the restoration of soil fertility, improved water retention, and the return of wildlife populations (Chazdon, 2008). These examples demonstrate that, with concerted effort, human intervention can support nature's capacity for healing and even enhance it in some cases.

26.3.4 The role of nature in human health and well-being

Nature's capacity for renewal and healing is not only important for ecosystems but also for human health and well-being. The concept of ecotherapy, or the therapeutic use of nature to promote mental and physical health, has gained increasing attention in recent years. Research has shown that exposure to natural environments can reduce stress, improve mood, and enhance cognitive function (Kaplan & Kaplan, 1989). The healing properties of nature are especially important in the context of modern urban life, where individuals often experience disconnection from natural environments. Spending time in nature has been linked to improved mental health outcomes, such as reduced anxiety, depression, and PTSD (Barton & Pretty, 2010). This highlights the reciprocal relationship between human well-being and the health of the natural world.

Moreover, nature provides essential resources that contribute to human health, such as clean air, water, and food. Ecosystem services, such as pollination, soil fertility, and water purification, are critical for maintaining agricultural productivity and food security (Millennium Ecosystem Assessment, 2005). By protecting and restoring ecosystems, humanity can ensure that these services continue to support both environmental health and human well-being.

Nature's capacity for renewal and healing is a powerful testament to the resilience of the natural world. While human activities have significantly altered the ability of ecosystems to recover, nature has demonstrated remarkable potential for regeneration when given the proper conditions. Biodiversity, ecological resilience, and human intervention through restoration efforts all play essential roles in supporting nature's healing processes. Moreover, the relationship between human health and the health of the natural world underscores the importance of preserving and nurturing the environment for the benefit of both people and ecosystems. As humanity faces unprecedented environmental challenges, fostering nature's capacity for renewal will be crucial for ensuring a sustainable and resilient future.

26.4 The Power of Stories and Shared Experiences

Stories have long been an essential aspect of human culture, shaping how individuals understand themselves, others, and the world around them. They are not only a form of entertainment but also serve as powerful tools for communication, learning, and connection. The power of stories lies in their ability to convey complex ideas, evoke emotions, and foster shared experiences that unite individuals across cultures, generations, and experiences. From ancient oral traditions to modern digital media, storytelling continues to be a central mechanism through which humans make sense of their lives and navigate the challenges they face.

26.4.1 Stories as tools for meaning-making

One of the fundamental roles of stories is to help individuals make meaning of their experiences. According to narrative theory, humans are inherently storytellers who use narrative structures to organize and interpret their lives (Bruner, 1991). Stories allow individuals to impose structure on chaotic or fragmented experiences, helping them to find coherence and significance in their lives. This process of meaning-making is not only vital on a personal level but also plays a crucial role in how societies construct collective identities. For example, national histories, religious narratives, and cultural myths shape the values, beliefs, and behaviors of entire communities. Through shared stories, people come to understand their roles within a larger social fabric, which in turn influences their actions and relationships (Polkinghorne, 1988).

Furthermore, stories enable individuals to navigate existential questions about purpose, identity, and morality. The act of storytelling allows for the exploration of ethical dilemmas, personal growth, and the search for meaning in life. For instance, literature and films often explore themes of suffering, resilience, and transformation, offering individuals frameworks for understanding their own struggles and aspirations. Through stories, people can vicariously experience the lives of others, gaining insights into different perspectives and emotions, which fosters empathy and emotional intelligence (Nussbaum, 1995). This capacity for empathy is crucial for building social cohesion and understanding, as it enables individuals to connect with others on a deeper, more human level.

26.4.2 Shared experiences and collective identity

The power of shared experiences, facilitated by stories, is particularly evident in the formation of collective identities. Whether through national narratives, cultural myths, or community traditions, shared stories bind people together, creating a sense of belonging and solidarity. These collective identities are not only constructed through stories but are also reinforced by the communal act of storytelling. In many cultures, oral storytelling traditions have been central to the transmission of knowledge, values, and cultural practices from one generation to the next (Solecki, 2003). The communal nature of storytelling fosters a sense of connection and continuity, ensuring that cultural heritage is preserved and passed on to future generations.

In contemporary society, shared stories continue to play a crucial role in fostering collective identity, particularly in the context of global challenges. For example, stories about environmental conservation, social justice, or human rights movements unite individuals and groups who share common goals and values. These narratives inspire collective action and create a sense of purpose that transcends individual concerns. For instance, the global environmental movement has been shaped by powerful stories about the degradation of the planet and the need for sustainable practices. These stories have mobilized individuals, organizations, and governments to take action, highlighting the role of shared narratives in driving social and environmental change (Barton, 2016).

26.4.3 The therapeutic power of storytelling

Beyond their social and cultural functions, stories also possess therapeutic power. In psychology, narrative therapy is a well-established approach that uses storytelling to help individuals reframe their experiences and address psychological challenges (White & Epston, 1990). This therapeutic method encourages individuals to view their lives as stories, which can be reinterpreted and rewritten in a way that promotes healing and growth. By externalizing problems and viewing them as part of a larger narrative, individuals can gain distance from their struggles and develop new perspectives on their circumstances. This process of re-authoring one's story can empower individuals to overcome trauma, depression, and anxiety, as it allows them to regain agency and control over their lives.

Similarly, storytelling has been used in group settings to foster healing and resilience in communities affected by trauma. For example, in post-conflict societies, storytelling initiatives provide individuals with a platform to share their experiences and process collective trauma. These shared narratives not only help individuals heal but also promote social reconciliation and unity. In Rwanda, for instance, storytelling has been a key component of the healing process following the 1994 genocide, as it allows individuals to confront painful memories, acknowledge the suffering of others, and rebuild trust (Herman, 2015). The act of sharing stories in a supportive community setting can be profoundly cathartic, as it helps individuals feel heard, validated, and connected to others who have experienced similar struggles.

26.4.4 Stories in the digital age

In the digital age, the power of stories and shared experiences has expanded to new platforms and technologies. Social media, blogs, podcasts, and digital storytelling tools have revolutionized the way individuals share their narratives and connect with others. These platforms have democratized storytelling, allowing individuals from diverse backgrounds and perspectives to share their stories with a global audience. This has led to the emergence of new forms of collective action, as individuals can now rally around shared stories and causes more easily than ever before. For example, movements such as #MeToo and Black Lives Matter have gained momentum through the widespread sharing of personal stories, highlighting the role of digital storytelling in social activism (Mendes, Ringrose, & Keller, 2019).

However, the digital age has also raised concerns about the authenticity and impact of online stories. While social media platforms provide a space for marginalized voices to be heard, they also create opportunities for misinformation and manipulation. The spread of fake news and the use of algorithms to amplify certain narratives can distort the way stories are shared and consumed, leading to polarization and division (Tufekci, 2017). Despite these challenges, the digital era has undoubtedly amplified the power of stories and shared experiences, enabling individuals and communities to connect, organize, and advocate for change on an unprecedented scale.

The power of stories and shared experiences is undeniable, as they play a central role in shaping personal identities, fostering collective action, and promoting healing. Through stories, individuals make meaning of their lives, navigate existential challenges, and connect with others across cultures and generations. In the context of global challenges, shared stories can unite individuals and inspire collective action, while also providing a platform for marginalized voices. Whether through traditional oral storytelling or modern digital platforms, the act of sharing stories continues to be a vital force for social cohesion, personal growth, and societal transformation.

References

Agyeman, J. (2005). *Sustainable communities and the challenge of environmental justice*. New York University Press.

Barton, J. (2016). *The restorative effects of nature on human health*. In M. L. Nordh & S. H. Alvarsson (Eds.), *The human-nature connection* (pp. 123–135). Springer. https://doi.org/10.1007/978-3-319-41280-1_12

Barton, J. (2016). *The restorative effects of nature on human health*. In M. L. Nordh & S. H. Alvarsson (Eds.), *The human-nature connection* (pp. 123–135). Springer. https://doi.org/10.1007/978-3-319-41280-1_12

Barton, J., & Pretty, J. (2010). What is the best dose of nature and green exercise for improving mental health? *Environmental Science & Technology, 44*(10), 3947–3955. https://doi.org/10.1021/es903183r

Bonanno, G. A. (2004). Loss, trauma, and human resilience. *American Psychologist, 59*(1), 20–28. https://doi.org/10.1037/0003-066X.59.1.20

Bruner, J. (1991). *The narrative construction of reality.* Critical Inquiry, 18(1), 1–21. https://doi.org/10.1086/448619

Chazdon, R. L. (2008). *Beyond deforestation: Restoring forests and ecosystem services on degraded lands.* Science, 320(5882), 1458–1460. https://doi.org/10.1126/science.1155365

Doren, R. F., et al. (2011). *Everglades restoration: Past, present, and future.* Springer. https://doi.org/10.1007/978-1-4419-7327-7

Gasser, U., et al. (2018). *Digital inclusion and resilience: Building an equitable future in a digital world. International Journal of Digital Inclusion, 10*(2), 121–139. https://doi.org/10.1007/s10842-018-0299-9

Gliessman, S. R. (2016). *Regenerative agriculture: An overview.* In *Regenerative agriculture and food systems* (pp. 1–17). CRC Press.

Haug, G. H., Gunther, D., Peterson, L. C., & Sigman, D. M. (2003). *Climate and the collapse of Maya civilization. Science, 299*(5613), 1731–1735. https://doi.org/10.1126/science.1081182

Herman, J. L. (2015). *Trauma and recovery: The aftermath of violence--from domestic abuse to political terror.* Basic Books.

Hobbs, R. J., & Harris, J. A. (2001). Restoration ecology: Repairing the Earth's ecosystems in the new millennium. *Restoration Ecology, 9*(2), 239–246. https://doi.org/10.1046/j.1526-100X.2001.009002239.x

Holling, C. S. (1973). *Resilience and stability of ecological systems. Annual Review of Ecology and Systematics, 4,* 1–23. https://doi.org/10.1146/annurev.es.04.110173.000245

Holling, C. S. (1973). Resilience and stability of ecological systems. *Annual Review of Ecology and Systematics, 4,* 1–23. https://doi.org/10.1146/annurev.es.04.110173.000245

IPCC. (2021). *Climate change 2021: The physical science basis.* Contribution of Working Group I to the Sixth Assessment Report of the Intergovernmental Panel on Climate Change. Cambridge University Press. https://doi.org/10.1017/9781009157896

Kaplan, S., & Kaplan, R. (1989). *The experience of nature: A psychological perspective.* Cambridge University Press.

Loreau, M., et al. (2001). Biodiversity and ecosystem functioning: Current knowledge and future challenges. *Science, 294*(5543), 804–808. https://doi.org/10.1126/science.1064088

Marx, K. (1867). *Capital: A critique of political economy* (Vol. 1). Penguin Classics.

Masten, A. S. (2001). Ordinary magic: Resilience processes in development. *American Psychologist, 56*(3), 227–238. https://doi.org/10.1037/0003-066X.56.3.227

McLeod, E., et al. (2019). Coral reef ecosystems under climate change. *Nature Sustainability, 2*(12), 1007–1017. https://doi.org/10.1038/s41893-019-0327-1

Mendes, K., Ringrose, J., & Keller, J. (2019). *#MeToo and the politics of social change*. Springer. https://doi.org/10.1007/978-3-030-12142-9

Millennium Ecosystem Assessment. (2005). *Ecosystems and human well-being: Synthesis*. Island Press.

Mitsch, W. J., & Gosselink, J. G. (2015). *Wetlands* (5th ed.).

Norris, F. H., et al. (2008). Community resilience as a metaphor, theory, set of capacities, and strategy for disaster readiness. *American Journal of Community Psychology, 41*(1-2), 127–150. https://doi.org/10.1007/s10464-007-9156-6

Nussbaum, M. (1995). *Poetic justice: The literary imagination and public life*. Beacon Press.

Ojala, M. (2012). *Hope and climate change: The importance of hope for environmental engagement among young people. Environmental Education Research, 18*(5), 625–642. https://doi.org/10.1080/13504622.2011.640726

Polkinghorne, D. E. (1988). *Narrative knowing and the human sciences*. SUNY Press.

Putnam, R. D. (2000). *Bowling alone: The collapse and revival of American community*. Simon & Schuster.

Seligman, M. E. P. (2011). *Flourish: A visionary new understanding of happiness and well-being*. Free Press.

Snyder, C. R. (2002). *Hope theory: Rainbows in the mind*. Psychological Inquiry, 13(4), 249–275. https://doi.org/10.1207/S15327965PLI1304_01

Solecki, R. (2003). *Oral traditions and their role in cultural preservation*. In P. J. P. Hart (Ed.), *Cultural heritage and identity* (pp. 27–41). Routledge.

Tufekci, Z. (2017). *Twitter and tear gas: The power and fragility of networked protest*. Yale University Press.

United Nations. (2015). *Transforming our world: The 2030 agenda for sustainable development*. United Nations. https://sdgs.un.org/2030agenda

Walker, B., Holling, C. S., Carpenter, S. R., & Kinzig, A. (2004). *Resilience, adaptability and transformability in social–ecological systems. Ecology and Society, 9*(2), 5. https://www.ecologyandsociety.org/vol9/iss2/art5/

White, M., & Epston, D. (1990). *Narrative means to therapeutic ends*. Norton & Company.

Wilson, E. O. (1992). *The diversity of life*. Harvard University Press.

27

A Vision for Humanity's Future

Humanity stands at a critical juncture in its history, grappling with unprecedented challenges ranging from environmental degradation to social inequality and technological disruption. However, these challenges also present an opportunity to rethink and reshape the future to prioritize sustainability, equity, and collective well-being. A vision for humanity's future must encompass a holistic understanding of the interconnectedness of all life, fostering an approach that values both human dignity and the planet's health. This vision requires a commitment to inclusive governance, ethical leadership, and a profound transformation in society's conception of progress and development.

27.1 Embracing Sustainability and Resilience

A core aspect of a vision for humanity's future is the commitment to sustainability – ensuring that current generations meet their needs without compromising the ability of future generations to meet theirs. This involves rethinking economic systems, resource use, and consumption patterns in ecologically viable and socially just ways. The concept of resilience, particularly in climate change and environmental degradation, is integral to this vision. Resilient communities can withstand shocks and adapt to changes while fostering long-term sustainability through regenerative practices (Folke, 2006). A future-focused humanity must adopt systems thinking, recognizing that human and ecological health are deeply intertwined and that sustainable development is not merely about economic growth but also about maintaining the balance of natural ecosystems.

This means transitioning to renewable energy sources, promoting circular economies, and advancing technologies that reduce environmental impact. For instance, the global shift toward renewable energy has the potential to mitigate climate change while creating new economic opportunities (IPCC, 2021). Sustainable agricultural practices, such as regenerative farming and

DOI: 10.1201/9788743808787-34

agroecology, offer ways to restore degraded land, enhance biodiversity, and support local communities (Altieri, 2009). By prioritizing resilience and sustainability, humanity can adapt to future challenges and thrive in a way that benefits all life on Earth.

27.2 Advancing Social Justice and Equity

Another fundamental aspect of a vision for humanity's future is the pursuit of social justice and equity. Global inequality, whether economic, racial, or gender-based, continues to be a significant barrier to progress and stability. An inclusive and equitable future is one in which all people, regardless of their background, have access to the resources, opportunities, and rights necessary for a fulfilling life. Achieving this requires dismantling systemic barriers to equality and addressing historical injustices disproportionately affecting marginalized communities (Sen, 1999). It also involves creating policies that ensure fair distribution of wealth, access to quality education and healthcare, and opportunities for meaningful participation in political and social life.

The Capabilities Approach, as proposed by Amartya Sen and Martha Nussbaum, offers a valuable framework for thinking about justice and human flourishing in the future. This approach emphasizes expanding individuals' capabilities – their ability to do and be what they value – by ensuring access to education, healthcare, political freedoms, and economic opportunities (Sen, 1999). A future where capabilities are maximized for all individuals, regardless of their socio-economic background, would foster a more just and equitable world. This vision aligns with the United Nations Sustainable Development Goals (SDGs), which aim to eradicate poverty, reduce inequality, and promote peace and prosperity for all (United Nations, 2015).

27.3 Ethical Leadership and Global Cooperation

A vision for humanity's future also demands ethical leadership and global cooperation. In an increasingly interconnected world, the challenges we face – climate change, pandemics, or geopolitical tensions – require collective action. Ethical leadership is crucial in guiding societies toward solutions prioritizing the common good over individual or national interests. Leaders must be accountable to their constituents, transparent in their decision-making, and committed to upholding human rights and dignity (Kempster, 2006). Moreover, global cooperation is essential to addressing the problems shared by our time. No single nation or group can solve global issues like climate change or the refugee crisis independently. International collaboration through multilateral institutions such as the United Nations is vital for creating solutions that are both effective and equitable.

The rise of global movements advocating for climate action, social justice, and human rights demonstrates the growing recognition of the need for collective action. For example, the Paris Agreement on climate change is a landmark achievement in global cooperation, in which countries have committed to limiting global warming to well below 2°C (UNFCCC, 2015). However, implementing such agreements requires continuous commitment, transparency, and accountability from all parties involved. Ethical leadership at both the local and global levels is necessary to ensure that the interests of vulnerable populations and the planet are prioritized in decision-making processes.

27.4 Technology and Human Flourishing

In addition to sustainability, equity, and ethical leadership, technology plays a central role in shaping the future of humanity. While technological advancements can potentially solve many of the world's problems – such as through innovations in renewable energy, healthcare, and education – they also present new ethical dilemmas. The rapid pace of technological change, particularly in areas like artificial intelligence (AI) and biotechnology, raises questions about the potential for misuse, inequality, and the erosion of privacy and autonomy (Binns, 2018). A vision for humanity's future must include a framework for responsible innovation, where technology is developed and deployed to enhance human well-being and promote social good.

For example, AI has the potential to revolutionize industries, improve healthcare outcomes, and address complex global challenges. However, its implementation must be guided by fairness, transparency, and accountability principles to ensure it benefits humanity rather than exacerbating existing inequalities (O'Neil, 2016). Similarly, biotechnology offers promising solutions for addressing food security, disease, and environmental degradation, but it also raises concerns about ethics, equity, and environmental impact. A future where technology serves humanity's best interests requires careful consideration of its ethical implications and a commitment to using it to promote human flourishing and the common good.

A vision for humanity's future is rooted in sustainability, equity, ethical leadership, and responsible technological innovation. By embracing these principles, humanity can navigate the challenges of the 21st century and build a just, resilient, and inclusive future. This vision requires a collective effort that transcends national borders and individual interests, focusing on the well-being of all people and the planet. As we move forward, it is essential to prioritize the values of justice, compassion, and sustainability in shaping the future, ensuring that it is a future where humanity can thrive in harmony with the Earth.

27.5 Redefining Progress and Prosperity

The traditional concept of progress and prosperity has long been rooted in economic growth, typically measured by Gross Domestic Product (GDP). However, as the world faces mounting environmental challenges, social inequalities, and global crises, there is a growing consensus that this narrow definition of prosperity is inadequate. The pursuit of economic growth, often without regard for its environmental and social consequences, has led to resource depletion, ecological degradation, and rising inequalities. As a result, scholars and policymakers have increasingly called for a redefinition of progress and prosperity – one that emphasizes sustainability, well-being, and equity rather than mere economic expansion.

27.5.1 Moving beyond GDP: The need for holistic indicators

While helpful in measuring the output of goods and services, GDP fails to capture critical aspects of human well-being and environmental health. It does not account for the depletion of natural resources, the loss of biodiversity, or the social costs of economic activities such as inequality and poverty. As such, many argue that a more comprehensive approach to measuring prosperity is needed. Alternative metrics, such as the Human Development Index (HDI) and the Genuine Progress Indicator (GPI), have been proposed to address these shortcomings. The HDI, for instance, includes life expectancy, education, and income indicators, while the GPI adjusts GDP by accounting for environmental and social factors (Kubiszewski et al., 2013). These alternative measures offer a more nuanced understanding of prosperity that integrates economic, social, and environmental dimensions.

Another promising framework for rethinking prosperity is the concept of "inclusive wealth," which considers not only economic capital but also natural and human capital. Inclusive wealth emphasizes the importance of maintaining and enhancing the stocks of capital that contribute to long-term well-being, such as natural ecosystems, human capabilities, and social infrastructure (Dasgupta, 2021). This broader approach to wealth recognizes that prosperity cannot be sustained without preserving the planet's ecosystems and developing human potential. Redefining prosperity in terms of inclusive wealth provides a more sustainable and equitable foundation for future development.

27.5.2 Environmental sustainability: A core element of prosperity

Environmental sustainability is an essential pillar of any redefined vision of prosperity. The current economic development trajectory has led to the overexploitation of natural resources, climate change, and environmental degradation. These challenges threaten the long-term viability of human societies and the planet's ecosystems. A sustainable future requires a shift away from growth models that prioritize short-term profits and resource extraction toward models that prioritize environmental stewardship and the well-being of future generations.

One such model is the "green economy," which seeks to decouple economic growth from environmental harm. The green economy promotes using renewable energy, resource efficiency, and sustainable agriculture to reduce ecological footprints while fostering economic development. The transition to a green economy can create new jobs, reduce poverty, and address climate change while protecting the planet's natural resources (United Nations Environment Programme [UNEP], 2011). Moreover, the concept of a circular economy, which focuses on reducing waste and reusing materials, offers a practical approach to achieving sustainability within economic systems (Geissdoerfer et al., 2017). By integrating environmental sustainability into economic models, societies can ensure that prosperity is not achieved at the expense of the planet's health.

27.5.3 Social equity and human well-being

In addition to environmental sustainability, social equity is a crucial element of a redefined vision of prosperity. Economic growth that benefits only a small segment of the population while leaving others in poverty or marginalization undermines social cohesion and human well-being. A truly prosperous society is one where all individuals have access to the resources and opportunities they need to lead fulfilling lives. This includes access to education, healthcare, decent work, and social protection.

The Capabilities Approach, developed by Amartya Sen and Martha Nussbaum, offers a framework for understanding prosperity beyond material wealth. This approach emphasizes expanding individuals' capabilities – their ability to do and be what they value – as a key measure of well-being (Sen, 1999). From this perspective, prosperity is not simply about economic output but about creating conditions for people to flourish in all aspects of their lives. This includes fostering social inclusion, reducing inequality, and empowering individuals to participate in decisions that affect their lives. A redefined vision of prosperity must prioritize policies that promote social equity, access to education, healthcare, and economic opportunities for all.

27.5.4 Rethinking prosperity in the context of global challenges

The global challenges of the 21st century, including climate change, pandemics, political instability, and technological disruption, require a fundamental rethinking of prosperity. Traditional models of economic growth, which prioritize consumerism and resource extraction, are increasingly unsustainable in the face of these challenges. A new vision of prosperity must consider the interconnectedness of environmental, social, and economic systems and recognize that true progress cannot be achieved without addressing global challenges such as inequality, environmental degradation, and climate change.

The concept of "degrowth" has emerged as a critique of the traditional growth paradigm. Degrowth advocates for a reduction in resource consumption and a shift away from the pursuit of endless economic expansion. Instead, it emphasizes the importance of well-being, sustainability, and social justice (Latouche, 2009). While degrowth is a controversial concept, it raises important questions about the limits of growth and the need to redefine prosperity in a way that prioritizes the health of the planet and the well-being of all people. A redefined vision of prosperity must be grounded in sustainability, equity, and social justice principles, ensuring that future generations inherit a livable and just world.

Redefining progress and prosperity is essential for addressing the complex challenges of the 21st century. A more holistic approach to prosperity that integrates economic, social, and environmental dimensions offers a path toward a more sustainable, equitable, and fulfilling future. By moving beyond GDP and adopting alternative metrics such as inclusive wealth and the Genuine Progress Indicator, societies can better capture development's actual costs and benefits. Furthermore, we can create an inclusive, just, and sustainable vision of prosperity by prioritizing environmental sustainability, social equity, and human well-being. This redefined vision of prosperity offers hope for a future where progress is measured not by material wealth but by the flourishing of individuals, communities, and the planet.

27.6 Building a Just and Equitable World

Building a just and equitable world requires a comprehensive approach to addressing the structural, economic, social, and political inequalities that persist across societies. Inequities in wealth, access to resources, and opportunities have been longstanding barriers to global well-being, and these disparities are often compounded by systemic factors such as race, gender, and geography. As the world grapples with the impacts of globalization, climate change, and social unrest, there is an urgent need for policies and frameworks that promote justice and equity at local, national, and global levels. A just world is one where all individuals have equal access to opportunities, resources, and rights, enabling them to live fulfilling lives with dignity and respect.

27.6.1 Structural inequality and its impacts

Structural inequality refers to the entrenched systems of discrimination and disadvantage that prevent certain groups from accessing the same opportunities and resources as others. These inequalities are often rooted in historical and institutional practices, such as colonialism, patriarchy, and racism, which have shaped power dynamics and distribution of resources across the globe. For instance, racial and gender inequalities persist in many societies, with marginalized groups often facing barriers to education, healthcare, employment, and political participation. Such disparities are detrimental to the affected individuals and undermine social cohesion, economic stability, and human flourishing.

One of the key challenges in building a just world is addressing these deep-rooted inequalities. The Capabilities Approach, developed by Amartya Sen and Martha Nussbaum, offers a framework for understanding justice and equity that focuses on expanding people's freedoms and opportunities to achieve well-being (Sen, 1999). This approach emphasizes that justice is not simply about the distribution of resources but about ensuring that individuals have the capabilities to pursue the lives they value. By enhancing capabilities, such as access to education, healthcare, and economic resources, societies can create more equitable conditions for all people, regardless of their background or identity.

27.6.2 Economic justice: Reducing poverty and inequality

Economic justice is a central pillar of building a just and equitable world. Global inequality has been exacerbated by the uneven distribution of wealth and resources, with a small fraction of the global population controlling a disproportionate share of economic power. The World Inequality Report (2022) highlights that the wealthiest 1% of the global population owns more than 40% of the world's wealth, while the poorest half holds only 2%. This stark inequality undermines social stability and perpetuates cycles of poverty and marginalization.

To build a more just economic system, addressing the root causes of inequality, such as exploitative labor practices, unequal access to education, and financial systems that favor the wealthy, is essential. Progressive taxation, wealth redistribution, and investments in social safety nets are key strategies for reducing inequality. Furthermore, ensuring access to quality education and healthcare, creating fair labor markets, and promoting inclusive economic growth are crucial steps in reducing poverty and empowering marginalized communities. The United Nations' Sustainable Development Goal 10, which aims to reduce inequality within and among countries, emphasizes the importance of equitable economic policies in achieving global justice (United Nations, 2015).

27.6.3 Social justice: Gender, race, and intersectionality

Social justice involves addressing the social dimensions of inequality, such as gender, race, and other forms of identity-based discrimination. Gender inequality, for example, remains pervasive in many societies, with women and girls often facing barriers to education, economic participation, and political representation. Similarly, racial and ethnic minorities are disproportionately affected by poverty, lack of access to healthcare, and discriminatory practices. These social injustices are compounded by intersecting factors such as class, disability, and sexual orientation, which can create unique forms of disadvantage for individuals at the intersections of multiple marginalized identities.

The concept of intersectionality, coined by Kimberlé Crenshaw (1989), provides a framework for understanding how different forms of discrimination intersect and compound one another. For instance, a Black woman may experience both racial and gender discrimination, which together create a unique set of challenges. To build a just and equitable world, it is essential to address these intersecting forms of oppression and create inclusive policies that promote the rights and dignity of all individuals. This includes ensuring equal access to education, healthcare, employment, and political representation for women, racial minorities, and other marginalized groups.

27.6.4 Environmental justice: Addressing ecological inequality

Environmental justice is another critical aspect of building a just and equitable world. Environmental degradation, including climate change, pollution, and resource depletion, disproportionately affects marginalized communities, particularly those in low-income and developing countries. These communities often bear the brunt of environmental harm while having the least capacity to mitigate or adapt to its impacts. For example, Indigenous peoples and rural communities are usually the first to experience the effects of climate change, such as extreme weather events and loss of biodiversity. Yet, they are frequently excluded from decision-making processes related to environmental policies (Martínez-Alier, 2002).

Environmental justice seeks to ensure that all communities have equal protection from environmental hazards and access to the benefits of a healthy environment. This involves addressing the disproportionate environmental burdens placed on marginalized groups and ensuring they have a voice in environmental decision-making. Policies that promote environmental sustainability, such as the transition to renewable energy, sustainable agriculture, and conservation of natural resources, must be designed with equity in mind. Furthermore, the principles of environmental justice advocate for the recognition of Indigenous knowledge systems and the protection of the rights of communities that are most vulnerable to environmental harm.

27.6.5 Global justice: A shared responsibility

Building a just and equitable world also requires a global perspective, recognizing that justice is not confined to national borders. Global inequality, such as the vast disparities in wealth and access to resources between developed and developing countries, demands collective action. The global community must work together to address challenges such as climate change, poverty, and conflict, which affect individuals and communities worldwide. International cooperation, solidarity, and the fair distribution of resources are essential to achieving global justice.

Global justice is rooted in the idea that all individuals, regardless of nationality or location, are entitled to basic human rights and opportunities for a dignified life. This includes access to clean water, education, healthcare, and cultural and environmental heritage protection. Global justice also requires addressing the historical injustices of colonialism and imperialism, which have shaped the current global order and continue to impact the development of many nations. The international community can work toward a more just and equitable global system by promoting fair trade, debt relief, and equitable development policies.

Building a just and equitable world is an ambitious but necessary goal for the 21st century. It requires addressing the root causes of inequality and injustice, including economic disparities, social discrimination, environmental harm, and global power imbalances. By focusing on economic justice, social inclusion, environmental sustainability, and global cooperation, societies can create a world where all individuals have the opportunity to live fulfilling lives with dignity and respect. This vision of justice and equity is essential for individual well-being and the long-term stability and prosperity of the global community.

27.7 Creating a Shared Vision for a Sustainable Planet

The concept of a shared vision for a sustainable planet is crucial in addressing humanity's multifaceted challenges, including climate change, biodiversity loss, inequality, and resource depletion. A shared vision emphasizes the collective responsibility of individuals, communities, governments, and businesses to work together toward a future where economic, social, and environmental systems are aligned to foster long-term sustainability. This vision is about preserving the planet for future generations and ensuring that all people, regardless of background or location, can thrive in a just and equitable world. Creating such a vision requires understanding the interconnectedness of social, economic, and environmental systems and integrating diverse perspectives in decision-making processes.

27.7.1 The importance of interconnectedness

A sustainable future is grounded in recognizing that environmental, social, and economic systems are deeply interconnected. This understanding is at the core of the concept of sustainability, which emphasizes the need to balance the needs of the present with the ability of future generations to meet their own needs (Brundtland, 1987). Environmental sustainability, for example, cannot be achieved without addressing social equity and economic development. Environmental degradation disproportionately affects marginalized communities, exacerbating poverty and inequality (Martínez-Alier, 2002). Therefore, a shared vision for a sustainable planet must consider the interdependencies between environmental health, social justice, and economic prosperity.

The concept of a circular economy, which promotes the reuse and recycling of materials, energy efficiency, and sustainable production, exemplifies the interconnectedness of sustainability. A circular economy reduces environmental impact and creates economic opportunities, particularly in low-income communities, by promoting green jobs and sustainable business practices (Geissdoerfer et al., 2017). In this way, economic development and environmental protection can go hand in hand, demonstrating that sustainability is a holistic concept that requires integrated solutions.

27.7.2 Inclusivity and global cooperation

A key aspect of creating a shared vision for a sustainable planet is ensuring inclusivity in decision-making processes. A sustainable future cannot be achieved if the voices of marginalized and vulnerable groups are excluded from discussions about policies and solutions. The United Nations Sustainable Development Goals (SDGs), particularly Goal 10 on reducing inequality, emphasize the importance of inclusive and equitable development (United Nations, 2015). Inclusivity means ensuring that all people have access to the benefits of sustainable development and having a say in how sustainability goals are set and pursued.

Global cooperation is essential to achieving a sustainable future, as environmental and social challenges are not confined to national borders. Climate change, for example, is a global issue that requires collective action from all countries, regardless of their level of development. The Paris Agreement, adopted in 2015, exemplifies the importance of international cooperation in addressing climate change. By setting targets for reducing greenhouse gas emissions and supporting climate adaptation efforts, the agreement aims to limit global warming and ensure that vulnerable communities are protected (UNFCCC, 2015). However, for global cooperation to be effective, it must be based on principles of fairness and equity, recognizing that developed countries have historically contributed more to environmental degradation and, therefore, have a greater responsibility to lead the way in mitigation and adaptation efforts.

27.7.3 The role of education and awareness

Education is crucial in creating a shared vision for a sustainable planet. Individuals and communities must understand the challenges and opportunities associated with sustainability and have the knowledge and skills to take action. Education for sustainable development (ESD) promotes the integration of sustainability concepts into all levels of education, from primary schools to universities, and encourages lifelong learning (UNESCO, 2014).

ESD empowers individuals to make informed decisions and engage in sustainable personal, professional, and political practices.

In addition to formal education, public awareness campaigns and grassroots movements are vital in fostering a shared vision for sustainability. By raising awareness about environmental issues such as climate change, deforestation, and pollution, these initiatives can mobilize people to take collective action. The global youth climate movement, led by figures such as Greta Thunberg, has demonstrated the power of grassroots activism in shaping public opinion and influencing policy. The movement has also highlighted the importance of intergenerational collaboration as young people advocate for urgent action to secure a sustainable future.

27.7.4 Ethical considerations in sustainability

A shared vision for a sustainable planet must be grounded in ethical principles that prioritize the well-being of both people and the planet. Environmental ethics, which explores the moral relationship between humans and the natural world, is essential in shaping this vision. For instance, the philosophy of deep ecology advocates for the intrinsic value of all living beings and the need for humans to live in harmony with nature (Naess, 1973). This ethical framework challenges anthropocentric views that place human interests above those of other species and ecosystems.

Social justice is another ethical consideration that must be integrated into the sustainability vision. The Capabilities Approach, as articulated by Amartya Sen and Martha Nussbaum, emphasizes the importance of expanding individuals' freedoms and opportunities to lead lives they have reason to value (Sen, 1999). This approach aligns with sustainability principles by recognizing that true development is not just about economic growth but also about enhancing human well-being and reducing inequalities. Therefore, a shared vision for a sustainable planet must be rooted in justice, equity, and respect for all life forms.

Creating a shared vision for a sustainable planet is urgent and complex. It requires the collective effort of individuals, communities, governments, and businesses. It involves recognizing the interconnectedness of environmental, social, and economic systems and addressing the inequalities hindering sustainability progress. Inclusivity, global cooperation, education, and ethical considerations are essential in fostering a just and sustainable vision. By working together and embracing diverse perspectives, humanity can build a future that ensures all people's and the planet's well-being for generations to come.

References

Altieri, M. A. (2009). Agroecology: The science of sustainable agriculture. CRC Press.

Binns, R. (2018). Ethics and artificial intelligence: The moral implications of AI technology. Springer.

Brundtland, G. H. (1987). Our common future: Report of the World Commission on Environment and Development. Oxford University Press.

Crenshaw, K. (1989). Demarginalizing the intersection of race and sex: A Black feminist critique of antidiscrimination doctrine, feminist theory, and antiracist politics. University of Chicago Legal Forum, 1989(1), 139–167.

Dasgupta, P. (2021). The economics of biodiversity: The Dasgupta review. HM Treasury. https://www.gov.uk/government/publications/the-economics-of-biodiversity-the-dasgupta-review

Folke, C. (2006). Resilience: The emergence of a perspective for social-ecological systems analyses. Global Environmental Change, 16(3), 253–267. https://doi.org/10.1016/j.gloenvcha.2006.04.002

Geissdoerfer, M., Savaget, P., Bocken, N. M. P., & Hultink, E. J. (2017). The circular economy – A new sustainability paradigm? Journal of Cleaner Production, 143, 757–768. https://doi.org/10.1016/j.jclepro.2016.12.048

IPCC. (2021). Summary for policymakers. In Climate Change 2021: The Physical Science Basis. Contribution of Working Group I to the Sixth Assessment Report of the Intergovernmental Panel on Climate Change. Cambridge University Press.

Kempster, S. (2006). Ethical leadership: A framework for leadership development. Journal of Business Ethics, 63(1), 65–79. https://doi.org/10.1007/s10551-005-1325-0

Kubiszewski, I., Costanza, R., Franco, C., & Anderson, S. (2013). Beyond GDP: Measuring and achieving global genuine progress. Ecological Economics, 93, 57–68. https://doi.org/10.1016/j.ecolecon.2013.06.001

Latouche, S. (2009). Farewell to growth. Polity Press.

Martínez-Alier, J. (2002). The environmentalism of low-income people: A study of ecological conflicts and valuation. Edward Elgar Publishing.

Naess, A. (1973). The shallow and the deep, long-range ecology movement: A summary. Inquiry, 16(1–4), 95–100. https://doi.org/10.1080/00201747308601682

O'Neil, C. (2016). Weapons of math destruction: How big data increases inequality and threatens democracy. Crown Publishing Group.

Sen, A. (1999). Development as freedom. Alfred A. Knopf.
UNESCO. (2014). Education for sustainable development goals: Learning objectives. United Nations Educational, Scientific and Cultural Organization. https://unesdoc.unesco.org/ark:/48223/pf0000230299
United Nations. (2015). Transforming our world: The 2030 agenda for sustainable development. United Nations. https://sdgs.un.org/2030agenda
UNFCCC. (2015). Paris Agreement. United Nations Framework Convention on Climate Change. https://unfccc.int/sites/default/files/english_paris_agreement.pdf
World Inequality Lab. (2022). World Inequality Report 2022. World Inequality Database. https://wir2022.wid.world

28

The Path Forward – Challenges and Opportunities

28.1 Challenges and Opportunities

As the world faces complex and interconnected challenges – ranging from climate change and environmental degradation to social inequality and political instability, the path forward requires innovative solutions, global cooperation, and systemic change. Our challenges are immense, but they also present significant opportunities for transformation. These opportunities can be harnessed through collective action, technological innovation, and economic and social systems reimagining. The path forward is not one of simple fixes but rather a long-term commitment to creating a more sustainable, equitable, and just world. This journey will require overcoming entrenched systems of power, addressing the root causes of inequality, and fostering a culture of resilience and adaptability.

28.2 The Challenge of Climate Change and Environmental Degradation

One of the most pressing challenges in the path forward is climate change. This crisis, with its increasing frequency of extreme weather events, rising sea levels, and disruptions to food and water supplies, threatens ecosystems, livelihoods, and human health. The urgency of this situation demands immediate and sustained efforts to reduce greenhouse gas emissions, as highlighted by the Intergovernmental Panel on Climate Change (IPCC, 2021). The challenge lies in mitigating climate change and adapting to its inevitable impacts, particularly in vulnerable regions.

However, this challenge also presents opportunities for innovation and collaboration. Developing renewable energy technologies, such as solar, wind, and hydropower, offers a pathway to decarbonize the global economy. The transition to a circular economy, which emphasizes the reuse and

DOI: 10.1201/9788743808787-35

recycling of materials, provides an opportunity to reduce waste and lower environmental impacts. Furthermore, the rise of green technologies and sustainable practices presents new economic opportunities, including creating green jobs and developing sustainable industries (Geissdoerfer et al., 2017). These innovations can drive environmental and economic transformation, creating a more sustainable future.

28.3 Social Inequality and Economic Transformation

Another major challenge on the path forward is the persistent issue of social inequality, which exacerbates poverty, marginalization, and social unrest. The COVID-19 pandemic has further highlighted the deep divides in access to healthcare, education, and economic opportunities, with vulnerable populations disproportionately bearing the brunt of the crisis (UNDP, 2020). Addressing inequality requires systemic change, including reforming economic structures, redistributing resources, and creating inclusive policies prioritizing the most marginalized needs.

The opportunity in this challenge lies in the potential for transformative economic systems that prioritize human well-being over profit maximization. The steady-state economy, which advocates for the sustainable use of resources and the reduction of consumption, offers an alternative to the growth-oriented models that have led to environmental degradation and social inequality (Daly, 1996). This presents a hopeful path forward, a vision of an economy that works for everyone, not just the privileged few. Furthermore, the Capabilities Approach, which emphasizes expanding individuals' freedoms and opportunities to lead lives they have reason to value, provides a framework for addressing inequality equitably and sustainably (Sen, 1999).

28.4 Political Instability and Global Cooperation

Political instability, nationalism, and the erosion of democratic values present another significant challenge to global cooperation and peace. In many parts of the world, political leaders are undermining democratic institutions, restricting freedoms, and fueling division and conflict. The rise of nationalism and populism has led to increased polarization and a retreat from multilateralism, making it more difficult to address global challenges such as climate change, pandemics, and economic inequality (Fukuyama, 2018).

Despite these challenges, there are opportunities to rebuild trust in democratic institutions and strengthen global cooperation. The COVID-19 pandemic has underscored the need for international solidarity, with countries working together to develop and distribute vaccines and provide aid to those in need. This highlights the necessity of unity. Multilateral agreements, such as the Paris Agreement on climate change, demonstrate that global cooperation is possible when nations come together with a shared vision and commitment. Strengthening international institutions, promoting dialogue, and fostering cross-border collaborations will be crucial in addressing the complex challenges that lie ahead. The path forward requires a renewed and unwavering commitment to global cooperation, recognizing that one nation's well-being is interconnected with all's well-being.

28.5 Technological Innovation and Ethical Considerations

Technological innovation holds great promise in addressing many challenges, from renewable energy and climate adaptation to healthcare and education. The rapid advancement of artificial intelligence, biotechnology, and automation has the potential to revolutionize industries, improve quality of life, and create new economic opportunities. However, these technologies also raise significant ethical concerns, including privacy, surveillance, bias, and the potential for job displacement (Binns, 2018). As we move forward, it will be essential to ensure that technological advancements are used responsibly and ethically, with consideration for their social and environmental impacts.

One opportunity in this area is the potential for technology to contribute significantly to sustainable development. For example, AI and machine learning can optimize energy consumption by predicting demand and adjusting supply, accordingly, improve resource management by analyzing usage patterns and suggesting more efficient practices, and enhance climate modeling by processing vast amounts of data to make more accurate predictions. Similarly, advancements in biotechnology, such as gene editing and regenerative medicine, promise to improve human health and address global challenges like food security and disease prevention. However, these technologies must be developed and deployed in ways that prioritize equity, transparency, and the protection of human rights. Ensuring that technological innovations align with sustainability and social justice values will be critical in shaping a future that benefits all.

The path forward is fraught with challenges but also presents significant opportunities for transformation. By addressing climate change, reducing inequality, promoting global cooperation, and harnessing the potential of technological innovation, we can create a more sustainable, equitable, and just world. However, achieving this vision will require concerted efforts from all sectors of society, including governments, businesses, civil society, and individuals. The road ahead will not be easy, but with a shared commitment to collective action, we can overcome our challenges and build a better future for future generations.

28.6 The Role of Individuals, Communities, and Nations in Addressing Global Challenges

The interconnectedness of the modern world means that addressing global challenges such as climate change, social inequality, and political instability requires coordinated efforts from individuals, communities, and nations. Each level plays a critical role in creating meaningful change, and their actions must complement one another for progress to be achieved. While the challenges are vast and complex, the collective efforts of individuals, communities, and nations can drive transformative changes. This section explores the roles and responsibilities of these three action levels in shaping a sustainable and just future.

28.6.1 The role of individuals in global change

Individuals play a crucial role in driving change at the grassroots level. Personal actions, decisions, and behaviors can significantly impact environmental, social, and economic. For example, individuals can reduce their carbon footprint by adopting sustainable consumption practices, such as using energy-efficient appliances, reducing waste, and supporting environmentally responsible businesses (Jackson, 2005). Furthermore, individuals can contribute to social change by engaging in advocacy, participating in community initiatives, and supporting policies that promote justice and equality. According to the Theory of Planned Behavior, attitudes, perceived control, and social norms influence individuals' actions. This suggests that fostering positive attitudes toward sustainability and social justice can encourage widespread participation in change efforts (Ajzen, 1991).

Moreover, individuals can serve as catalysts for larger societal transformations. By engaging in lifelong learning, individuals can increase their understanding of global challenges and solutions, becoming more informed and active citizens. Education and awareness are key in empowering individuals to make informed decisions and contribute to a more sustainable world (Sterling, 2001). As role models and advocates, individuals can inspire others to follow suit, creating a ripple effect that can expand the reach of social and environmental movements.

28.6.2 The role of communities in building resilience

Communities are central to addressing global challenges, where collective action can be organized and sustained. Communities, whether local or global, provide the social networks, resources, and support systems individuals need to thrive. In the context of climate change, for example, communities can foster resilience by developing local adaptation strategies that mitigate the impacts of environmental hazards. Community-led initiatives, such as renewable energy cooperatives, sustainable agriculture practices, and local recycling programs, can reduce dependence on external resources and promote self-sufficiency (Holling, 2001).

Communities also play an essential role in promoting social equity and justice. By creating inclusive spaces where all members, particularly marginalized groups, have a voice, communities can work to dismantle systemic inequalities and build more cohesive societies. This is particularly important in addressing issues such as income inequality, gender discrimination, and racial injustice. Historically, community organizing and activism have been powerful forces in challenging oppressive systems and advocating for policy changes (Putnam, 2000). In this way, communities act as the foundation for social movements that push for meaningful reform at higher levels of society.

Additionally, communities can contribute to collective well-being by fostering a sense of solidarity and shared responsibility. In times of crisis, such as natural disasters or economic downturns, communities often come together to support one another and rebuild. This sense of interconnectedness is crucial in developing the resilience needed to face future challenges. By strengthening social bonds and promoting collaboration, communities can become more adaptable and capable of addressing complex problems (Berkes, 2007).

28.6.3 The role of nations in global governance and policy

Nations hold significant power in shaping the global response to climate change, inequality, and conflict. As sovereign entities, governments have the authority to enact laws, create policies, and allocate resources that can address systemic issues. International agreements, such as the Paris Agreement on climate change, demonstrate the importance of national commitment to collective global action (UNFCCC, 2015). Nations must work together to set ambitious goals, share knowledge, and provide financial and technical support to those most in need. The role of nations is particularly critical in areas like climate adaptation, public health, and global security, where coordinated efforts are necessary to achieve lasting solutions.

National governments are also responsible for ensuring that policies are inclusive and equitable, addressing the needs of vulnerable populations. The state's role in promoting human rights, providing social services, and ensuring access to education and healthcare is fundamental to achieving the Sustainable Development Goals (SDGs). Governments must also address the root causes of inequality, including economic disparities, gender-based violence, and racial discrimination, through policy reform and investment in social infrastructure (Sen, 1999).

Moreover, nations can act as leaders in promoting global cooperation. No country can address global challenges in isolation in an increasingly interconnected world. Multilateralism and international cooperation are essential for tackling climate change, pandemics, and transnational conflict. Nations must establish global norms, share knowledge and resources, and collaborate on research and innovation. By fostering global solidarity, nations can contribute to a more peaceful, equitable, and sustainable world (Keohane, 2005).

28.6.4 Integrating efforts: A holistic approach

While individuals, communities, and nations each have distinct roles, their actions must be integrated to create a cohesive and effective response to global challenges. A holistic approach that recognizes the interdependence of these three action levels is necessary for achieving meaningful change. For example, individuals can advocate for policies at the national level that reflect their values of sustainability and social justice. Communities can support national and international efforts by implementing local solutions and building resilience at the grassroots level. Nations, in turn, must create policies that empower individuals and communities to act and contribute to global efforts.

Furthermore, the challenges we face are not only environmental or social; they are deeply interconnected. The planet's health, individuals' well-being, and societies' stability are all linked. A system-thinking approach that recognizes these interconnections can help identify solutions that simultaneously address multiple challenges. For instance, policies that promote sustainable agriculture can improve food security, reduce environmental degradation, and enhance social equity. We can create effective and sustainable solutions by considering the broader implications of policy decisions.

The roles of individuals, communities, and nations are essential in addressing the global challenges of the 21st century. While our challenges are vast, they also present opportunities for transformative change. We can create a more sustainable, equitable, and just world by acting at the individual, community, and national levels. However, these efforts must be integrated and coordinated to achieve lasting solutions. The path forward requires collective action, global cooperation, and a shared commitment to building a better future for all.

28.7 Embracing Hope and Collective Action: Pathways to a Sustainable Future

In the face of global challenges such as climate change, social inequality, and political instability, hope, and collective action emerge as powerful tools for transformation. While these challenges may seem insurmountable, they also present opportunities for individuals and communities to unite, harness their collective strength, and drive meaningful change. When combined with collective action, hope offers a pathway to overcoming adversity and building a more sustainable, just, and resilient world. This section explores the importance of hope and collective action in addressing global issues and how they can be leveraged to create a better future for all.

28.7.1 The power of hope in times of crisis

Hope is a psychological and emotional state that motivates individuals to envision and work toward a better future. It is an essential element of human resilience, particularly in crises. According to Snyder's Hope Theory, hope involves the perception of pathways to achieving goals and the agency to pursue those goals (Snyder, 2002). In the context of global challenges, hope makes individuals believe that change is possible, even when faced with daunting obstacles. It fosters optimism and perseverance, which are crucial for overcoming the feelings of helplessness and despair that often accompany complex global problems such as climate change, poverty, and inequality.

Hope is not merely a passive or abstract feeling but a catalyst for action. Research has shown that hope is linked to increased problem-solving abilities, greater resilience, and higher levels of social engagement (Rand, 2009). When individuals and communities embrace hope, they are more likely to take proactive steps toward creating positive change. This is particularly important in environmental and social issues, where long-term solutions require sustained effort and commitment. Hope is a driving force that propels individuals to engage in collective action, work toward shared goals, and contribute to global efforts to address systemic challenges.

28.7.2 Collective action as a catalyst for change

Collective action refers to the process by which individuals and groups work together to achieve common goals, often in the face of challenges that individuals cannot address alone. In global challenges, collective action is essential for creating systemic change. Climate change, social justice, and global health require coordinated efforts from diverse actors, including governments, civil society organizations, businesses, and individuals. The collective power of these groups can generate momentum for large-scale change, influence policy decisions, and drive the adoption of sustainable practices.

The concept of collective action is rooted in the idea that individuals, united by a shared vision and purpose, can accomplish far more than they could alone. Collective action is significant in addressing environmental challenges, as the actions of a few individuals, while meaningful, are insufficient to tackle global issues like climate change. For instance, grassroots movements such as Fridays for Future, led by youth activists like Greta Thunberg, have demonstrated the power of collective action in raising awareness and pressuring governments to take stronger action on climate change (Hoggett, 2020). These movements show how collective efforts can amplify individual voices, mobilize communities, and influence political agendas.

Moreover, collective action is not limited to large-scale movements but can occur locally. Communities can come together to implement sustainable practices, such as community gardens, renewable energy projects, and waste reduction initiatives. These localized efforts contribute to global solutions by reducing the carbon footprint, promoting social equity, and fostering resilience. By working together, communities can create sustainable systems that benefit both the environment and the people within them. The collective action model emphasizes the importance of collaboration, solidarity, and shared responsibility in addressing global challenges.

28.7.3 Hope and collective action in the context of sustainability

In sustainability, hope and collective action are intertwined and mutually reinforcing. Sustainability, in its broadest sense, refers to the ability of societies to meet the needs of the present without compromising the ability of future generations to meet their own needs (Brundtland, 1987). Achieving sustainability requires environmental stewardship, social equity, and economic fairness. Hope is critical in motivating individuals and communities to engage in sustainable practices, while collective action enables them to implement them on a larger scale.

For example, in the fight against climate change, hope motivates individuals to reduce their carbon footprint, adopt renewable energy sources, and support policies that promote environmental protection. At the same time, collective action is necessary to drive systemic change, such as transitioning to a low-carbon economy, implementing global climate agreements, and ensuring that marginalized communities have access to the resources needed to adapt to climate impacts. The combination of hope and collective action is essential for building the political will and public support needed to achieve global sustainability goals, such as those outlined in the United Nations Sustainable Development Goals (SDGs).

One key challenge in achieving sustainability is overcoming the inertia often accompanying large-scale problems. Fueled by hope, collective action can help break through this inertia and create the momentum needed for transformative change. By working together, individuals and communities can challenge the status quo, push for policy reforms, and create innovative solutions that promote environmental, social, and economic sustainability. Hope provides the emotional and psychological foundation for these efforts, while collective action ensures that these efforts are sustained and amplified.

28.7.4 Embracing a vision of collective hope

Visioning a sustainable and just future requires a collective commitment to hope and action. Individuals, communities, and nations must work together to create a shared vision of the future, one that is grounded in hope, solidarity, and a commitment to justice. This vision must be inclusive, recognizing people's diverse needs and aspirations worldwide, particularly those most vulnerable to climate change, inequality, and social injustice.

Hope and collective action are about addressing immediate challenges and creating a vision for the future that is sustainable, equitable, and resilient. By embracing hope, individuals and communities can create a world where future generations can thrive. Collective action enables people to transcend individual limitations and work toward a common good, creating a shared responsibility for the planet and its inhabitants.

Hope and collective action are fundamental to addressing the global challenges of the 21st century. In times of crisis, hope provides the psychological and emotional resilience needed to persevere, while collective action enables individuals and communities to work together to create meaningful change. When combined, hope and collective action can drive the transformative change necessary to build a more sustainable, just, and resilient world.

As we face future challenges, we must embrace these powerful forces and work together to create a better world.

References

Ajzen, I. (1991). The theory of planned behavior. *Organizational Behavior and Human Decision Processes, 50*(2), 179–211. https://doi.org/10.1016/0749-5978(91)90020-T

Berkes, F. (2007). Understanding uncertainty and reducing vulnerability: Lessons from resilience thinking. *Natural Resources Forum, 31*(1), 1–12. https://doi.org/10.1111/j.1477-8947.2007.00129.x

Binns, A. (2018). The ethics of artificial intelligence: An overview. In M. D. K. Zink (Ed.), *AI and ethics: A critical review* (pp. 1–18). Springer.

Brundtland, G. H. (1987). *Our common future: Report of the World Commission on Environment and Development*. Oxford University Press.

Daly, H. E. (1996). *Beyond growth: The economics of sustainable development*. Beacon Press.

Fukuyama, F. (2018). *Identity: The demand for dignity and the politics of resentment*. Farrar, Straus and Giroux.

Geissdoerfer, M., Savaget, P., Bocken, N. M. P., & Hultink, E. J. (2017). The circular economy – A new sustainability paradigm? *Journal of Cleaner Production, 143*, 757–768. https://doi.org/10.1016/j.jclepro.2016.12.048

Hoggett, P. (2020). Greta Thunberg and the politics of hope. *International Journal of Social Science Studies, 8*(3), 1–12. https://doi.org/10.11114/ijsss.v8i3.4674

Holling, C. S. (2001). Understanding the complexity of economic, ecological, and social systems. *Ecosystems, 4*(5), 390–405. https://doi.org/10.1007/s10021-001-0101-5

IPCC. (2021). *Climate change 2021: The physical science basis*. Intergovernmental Panel on Climate Change. https://www.ipcc.ch/report/ar6/wg1/

Jackson, T. (2005). *Motivating sustainable consumption: A review of consumer behavior and behavioral change evidence*. Centre for Environmental Strategy, University of Surrey.

Keohane, R. O. (2005). *After hegemony: Cooperation and discord in the world political economy*. Princeton University Press.

Putnam, R. D. (2000). *Bowling alone: The collapse and revival of American community*. Simon & Schuster.

Rand, K. L. (2009). Hope theory and its applications. In S. J. Lopez (Ed.), *The Oxford Handbook of Positive Psychology* (2nd ed., pp. 287–295). Oxford University Press.

Sen, A. (1999). *Development as freedom*. Alfred A. Knopf.

Snyder, C. R. (2002). Hope theory: Rainbows in the mind. *Psychological Inquiry, 13*(4), 249–275. https://doi.org/10.1207/S15327965PLI1304_01

Sterling, S. (2001). *Sustainable education: Re-visioning learning and change*. Green Books.

UNDP. (2020). *Human development report 2020: The next frontier – Human development and the Anthropocene*. United Nations Development Programme. https://hdr.undp.org/en/2020-report

UNFCCC. (2015). *Paris Agreement*. United Nations Framework Convention on Climate Change. https://unfccc.int/sites/default/files/english_paris_agreement.pdf

Vinuesa, R., Azizpour, H., Leite, I., & Aghaei, M. (2020). The role of artificial intelligence in achieving the Sustainable Development Goals. *Nature Communications, 11*(1), 233. https://doi.org/10.1038/s41467-019-14108-1

29

Case Studies and Success Stories: Resolving Humanity in Crisis

Humanity faces many crises, ranging from environmental degradation, social inequality, political instability, and economic disparities. However, history has shown that these crises can be mitigated or resolved through collective action, innovative solutions, and the unwavering resilience and determination of communities, nations, and global organizations. This section explores several case studies and success stories that not only highlight successful resolutions but also inspire hope for a sustainable future. These examples demonstrate the power of collaboration, innovation, and determination in resolving complex challenges and offer valuable lessons for future global efforts.

29.1 The Paris Agreement: A Global Effort to Combat Climate Change

One of the most significant global efforts to address the climate crisis is the Paris Agreement, adopted in 2015 under the United Nations Framework Convention on Climate Change (UNFCCC). The agreement, a testament to global unity, brought together 196 countries to limit global warming to well below 2°C above pre-industrial levels, with efforts to limit the increase to 1.5°C. This ambitious agreement marked a turning point in global climate governance, as it represented the first time that virtually every nation on Earth committed to taking action on climate change.

The success of the Paris Agreement lies in its ability to unite diverse countries, each with unique economic, social, and environmental contexts, around a common goal. By setting nationally determined contributions (NDCs), the agreement allowed countries to tailor their climate policies to their specific circumstances while ensuring that global emissions would be reduced. Including mechanisms for transparency, accountability, and periodic reviews has also

DOI: 10.1201/9788743808787-36

been crucial in maintaining momentum and ensuring that countries remain committed to their climate goals (UNFCCC, 2015). While challenges remain in fully implementing the Paris Agreement, it stands as a testament to the power of multilateral cooperation in addressing a global crisis.

29.2 The Green Revolution: Transforming Global Food Security

The Green Revolution, which began in the 1940s and accelerated through the 1960s and 1970s, is another example of a successful global effort to resolve a crisis. Faced with widespread famine and food insecurity, particularly in developing countries, agricultural scientists and policymakers worked together to create high-yielding crop varieties, improve irrigation techniques, and introduce modern farming technologies. These innovations, along with the widespread adoption of chemical fertilizers and pesticides, significantly increased food production and helped to alleviate hunger in many parts of the world.

The Green Revolution's success in improving food security is evident in countries like India, Mexico, and the Philippines, where crop yields dramatically increased, reducing famine and malnutrition (Evenson & Gollin, 2003). This success story serves as a beacon of hope, demonstrating the potential for global solutions to address pressing issues. Despite the environmental costs of intensive agriculture, such as soil degradation, water depletion, and biodiversity loss, the Green Revolution remains an important case study of the power of technological innovation and international collaboration to address global food security challenges. Moving forward, the lessons learned from the Green Revolution must be applied in ways that balance food production with environmental sustainability.

29.3 The Global Eradication of Smallpox: A Triumph of Public Health

The global eradication of smallpox is one of the most celebrated success stories in the history of public health. In 1967, the World Health Organization (WHO) launched an intensive global campaign to eradicate smallpox, a disease that had plagued humanity for centuries and caused millions of deaths. The campaign relied on a combination of strategies, including mass vaccination campaigns, surveillance, and containment measures. By 1980, smallpox was declared eradicated, making it the first disease to be completely wiped out through human intervention (Fenner et al., 1988).

The eradication of smallpox is a powerful example of the effectiveness of coordinated global action in resolving a crisis. It demonstrated the importance of international collaboration, with countries sharing resources, expertise, and funding to achieve a common goal. The success of the smallpox eradication campaign also highlighted the potential of vaccines for combating infectious diseases and improving global public health. The lessons learned from this campaign have informed subsequent efforts to control other diseases, such as polio and malaria, and have provided a model for future global health initiatives.

29.4 The Role of Indigenous Knowledge in Biodiversity Conservation

Indigenous communities worldwide have long been stewards of the environment, managing natural resources in sustainable and culturally significant ways. In recent decades, there has been a growing recognition of the value of indigenous knowledge in addressing environmental crises, particularly in biodiversity conservation. One notable example is the involvement of indigenous peoples in the management of protected areas and conservation efforts in places like the Amazon rainforest and the Great Barrier Reef.

In the Amazon, indigenous communities have been instrumental in protecting vast forest areas from deforestation and exploitation. Their traditional knowledge of local ecosystems, combined with modern conservation techniques, has allowed them to develop sustainable land management practices that preserve biodiversity while meeting the needs of local communities (Berkes, 2009). Similarly, in Australia, Indigenous knowledge is being integrated into managing the Great Barrier Reef, with Indigenous rangers working alongside scientists to monitor coral health, control invasive species, and promote sustainable fishing practices (Fernandes et al., 2019). These examples demonstrate the power of combining traditional ecological knowledge with contemporary science to address environmental challenges and promote sustainability.

29.5 The Role of Grassroots Movements in Social Change

Grassroots movements have been crucial in addressing social crises and advocating for systemic change. One such movement is the global #MeToo movement, which emerged in 2017 as a response to widespread sexual harassment and assault. The movement, which began as a social media campaign, quickly gained momentum and sparked a global conversation about gender inequality, power dynamics, and the need for cultural change. Through the collective action of individuals and organizations, the #MeToo movement has led to significant shifts in public attitudes, legal reforms, and changes in workplace policies (Mendes et al., 2018).

The success of the #MeToo movement underscores the power of grassroots organizing in challenging entrenched social norms and advocating for justice. It demonstrates how collective action, fueled by shared experiences and a commitment to social change, can create a ripple effect leading to tangible societal improvements. Grassroots movements like #MeToo are also crucial in raising awareness about issues that may otherwise be overlooked or ignored, amplifying the voices of marginalized groups, and pushing for reforms that promote equity and justice. Table 29.1. provides case studies and success stories that have contributed to resolving global crises, focusing on different aspects such as environmental, social, and political challenges.

The table highlights impactful case studies and success stories from diverse regions and areas, showcasing significant achievements and outcomes driven by key players and stakeholders. For example, the Paris Agreement on Climate Change represents a global effort by the United Nations, 195 countries, NGOs, and scientists to limit temperature rise to below 2°C, with commitments to reduce carbon emissions. Similarly, the End of Apartheid in South Africa involved Nelson Mandela, the African National Congress, and the international community, leading to a peaceful transition to a multiracial democracy and the establishment of a new constitution. Other notable entries include the Marshall Plan, which facilitated Europe's post-WWII recovery through U.S. economic aid, and the Green Revolution, which revolutionized agriculture in developing countries by introducing high-yield crops and modern techniques, reducing hunger in Asia.

The table also includes transformative global efforts such as the abolition of slavery, led by abolitionists, governments, and social movements, and the recovery of the Amazon rainforest, driven by local communities, NGOs, and corporations to combat deforestation. It highlights the United Nations peacekeeping missions, which have maintained peace and rebuilt societies in conflict zones, and post-genocide reconciliation in Rwanda, focusing on justice and national unity. Environmental milestones like the Montreal Protocol on Ozone Depletion, a global agreement to phase out harmful chemicals, and establishing the International Criminal Court (ICC) to prosecute war crimes and promote justice further underscore the global commitment to progress and accountability. These cases collectively demonstrate the power of collaboration and dedication in addressing complex challenges and achieving meaningful change.

Table 29.1 Case studies and success stories that have contributed to resolving global crises, focusing on different aspects such as environmental, social, and political challenges.

Case study/success story	Region/area	Key players/Stakeholders	Key achievements and outcomes
The Paris Agreement on Climate Change	Global	United Nations, 195 countries, NGOs, scientists	Historic agreement to limit global temperature rise to below 2°C, with commitments from nations to reduce carbon emissions.
The end of Apartheid in South Africa	South Africa	Nelson Mandela, African National Congress, international community	The peaceful transition from racial segregation to a multiracial democracy, ending apartheid and establishing a new constitution.
The Marshall Plan	Europe (Post-WWII)	USA, European nations	Economic aid program that helped rebuild Europe after WWII, contributing to the stabilization of economies and the spread of democracy.
The Green Revolution	Developing countries	Scientists, agricultural experts, governments	Introducing high-yield crops and modern farming techniques significantly increased food production and reduced hunger in Asia.
The Abolition of Slavery	Global (Various)	Abolitionists, governments, social movements	Efforts led to the end of slavery, including landmark legal actions such as the Emancipation Proclamation and the UK's Slavery Abolition Act.
The Recovery of the Amazon Rainforest	Brazil	Local communities, NGOs, government, corporations	Successful initiatives to combat deforestation, including sustainable farming practices, conservation efforts, and international agreements.
The United Nations Peacekeeping Missions	Global (Various)	United Nations, member states, local populations	Interventions in conflict zones, such as Rwanda and Bosnia, are conducted to maintain peace, protect civilians, and rebuild post-conflict societies.
Post-Genocide Reconciliation in Rwanda	Rwanda	Rwandan government, local communities, NGOs	Efforts to rebuild the country after the 1994 genocide focused on justice, reconciliation, and national unity.
The Montreal Protocol on Ozone Depletion	Global	United Nations, governments, industry leaders	A global agreement to phase out the use of chemicals harmful to the ozone layer, leading to significant environmental recovery.
The International Criminal Court (ICC)	Global	ICC, international community, victims of war crimes	A global institution aimed at prosecuting individuals for war crimes, crimes against humanity, and genocide, promoting justice and accountability.

Source: (United Nations, 2015; Mandela, 1994; Marshall, 2007; Brown, 2007; Finkelman, 2008; Nepstad & Stickler, 2006; United Nations Peacekeeping, 2020; Des Forges, 1999; UNEP, 2017; International Criminal Court, 2021).

These case studies and success stories illustrate that while humanity faces significant crises, there are pathways to resolution through collaboration, innovation, and determination. Whether through international agreements like the Paris Agreement, technological innovations like the Green Revolution, or the power of grassroots movements like #MeToo, collective action can address even the most daunting global challenges. The lessons learned from these efforts can guide future initiatives, providing hope and inspiration for a more sustainable, just, and resilient world. We must draw upon these successes as we face global crises and work together to create a better future for all.

References

Berkes, F. (2009). *Sacred ecology: Traditional ecological knowledge and resource management* (2nd ed.). Routledge.

Brown, L. R. (2007). *Plan B 3.0: Mobilizing to Save Civilization*. W.W. Norton & Company.

Des Forges, A. (1999). *Leave None to Tell the Story: Genocide in Rwanda*. Human Rights Watch.

Evenson, R. E., & Gollin, D. (2003). Assessing the impact of the Green Revolution, 1960 to 2000. *Science, 300*(5620), 758–762. https://doi.org/10.1126/science.1087249

Fenner, F., Henderson, D. A., Arita, I., Jezek, Z., & Ladnyi, I. D. (1988). *Smallpox and its eradication*. World Health Organization.

Fernandes, L., De Lacy, T., & Reichelt, R. (2019). Indigenous knowledge and management of the Great Barrier Reef. *Environmental Science & Policy, 101*, 17–24. https://doi.org/10.1016/j.envs

Finkelman, P. (2008). *Slavery and the Law*. Oxford University Press.

International Criminal Court. (2021). *About the ICC*. International Criminal Court. https://www.icc-cpi.int

Mandela, N. (1994). *Long Walk to Freedom*—little, Brown and Company.

Marshall, G. (2007). *The Marshall Plan: Dawn of the Cold War*. Yale University Press.

Nepstad, D. C., & Stickler, C. M. (2006). *The Amazon: The Green Revolution and the Future of the Rainforest*. University of California Press.

UNEP. (2017). *The Montreal Protocol: Protecting the Ozone Layer*. United Nations Environment Programme. https://www.unep.org

United Nations Peacekeeping. (2020). *Peacekeeping and the United Nations*. United Nations. https://peacekeeping.un.org

United Nations. (2015). *The Paris Agreement*. United Nations Framework Convention on Climate Change. https://unfccc.int

30

Conclusion

Humanity in Crisis: A Comprehensive Reflection on Anthropogenic Challenges and Pathways to Solutions is a testament to the complexity of the global crises we face and a call to action for individuals, communities, and nations. By exploring the breadth of challenges through the lenses of environmental, social, political, economic, cultural, and ethical dimensions, this book highlights the interconnectedness of these issues and the necessity of addressing them holistically. It underscores that the crises we encounter are not isolated phenomena, but manifestations of systemic problems rooted in our collective behaviors, decisions, and priorities.

In Part I: Understanding the Crisis, we examined the Anthropocene as a transformative epoch, where human activities have become the dominant force shaping the planet. This section lays the foundation for recognizing the urgency of our situation, emphasizing that humanity stands at a critical juncture, where our choices today will define the future of life on Earth.

Part II: Environmental Crises delved into the planetary boundaries we have crossed – climate change, biodiversity loss, water scarcity, and pollution – revealing the fragility of the ecosystems that sustain us. These chapters remind us that the Earth's capacity to endure unchecked exploitation is finite, and immediate action is essential to mitigate irreversible damage.

The exploration of Social and Humanitarian Crises in Part III illuminated the profound inequalities and injustices that exacerbate human suffering. From poverty and health crises to displacement and gender inequality, these chapters revealed the human cost of systemic failures and the moral imperative to foster equity and compassion.

In Part IV: Political and Economic Crises, the focus shifted to the governance and economic systems that perpetuate instability and inequity. These chapters highlighted the challenges of misinformation, political fragmentation,

DOI: 10.1201/9788743808787-37

and unsustainable economic practices, urging a reevaluation of how we structure and manage societies.

Part V: Cultural and Ethical Crises explored the deeper existential challenges of identity, belonging, and ethical dilemmas posed by rapid technological advancements. It underscored the importance of fostering a sense of purpose, shared values, and collective responsibility in a world increasingly marked by division and alienation.

Finally, Part VI: Pathways to Solutions and Part VII: Hope and Vision for Humanity offered a roadmap for addressing these crises. Sustainability, global cooperation, innovation, transformative education, and ethical leadership emerged as key pillars for a resilient future. Drawing lessons from history and nature, these chapters painted a vision of a harmonious, equitable, and thriving planet – a future that is not merely aspirational but achievable with collective effort and determination.

This book reflects known, seen, and experienced realities, aiming to provide readers with insights into the multifaceted crises we face. It is also a message of hope, emphasizing that while the challenges are immense, the human spirit's capacity for resilience, creativity, and transformation is boundless. Let this work inspire action, dialogue, and collaboration toward a peaceful and healthy planet for today and the generations to come. Together, we can redefine the trajectory of humanity, ensuring that the crises of today become the catalysts for a brighter, more sustainable tomorrow.

Index

D

E

I

P

Z

Words / Phrases Definitions

Air pollution: The presence of harmful substances in the air, such as chemicals, particulates, and biological materials, which can negatively affect human health, the environment, and the climate.

Ancient wars: Conflicts fought in ancient times, often between early civilizations or empires, which shaped the course of history and influenced the development of societies.

Anthropogenic: Referring to environmental changes or phenomena that are caused or influenced by human activities, particularly in relation to climate change, pollution, and habitat destruction.

Armed conflicts: A violent struggle between two or more parties, often involving military forces, that results in the use of weapons and can lead to widespread destruction, loss of life, and displacement.

Awareness: The state of being conscious or knowledgeable about something, particularly in the context of social, environmental, or political issues, which can lead to informed decision-making and action.

Biodiversity: The variety of life forms on Earth, including plants, animals, fungi, and microorganisms, and the ecosystems they inhabit, which is essential for ecological balance and resilience.

Causes: The factors or conditions that lead to a particular event or outcome, such as the causes of environmental degradation, social inequality, or climate change.

Clean water: Water that is free from contaminants, pollutants, and harmful microorganisms, essential for human health, agriculture, and ecosystems.

Climate change: Long-term changes in temperature, precipitation, and other atmospheric patterns, largely driven by human activities such as burning fossil fuels, deforestation, and industrialization.

Climate refugees: People who are forced to leave their homes and communities due to the adverse impacts of climate change, such as rising sea levels, extreme weather events, and droughts.

Cold War: A period of geopolitical tension between the Soviet Union and the United States (and their respective allies) from the late 1940s to the early 1990s, characterized by political, military, and ideological rivalry without direct warfare.

Collaboration: The process of working together with others to achieve common goals, often involving cooperation between individuals, groups, or organizations.

Community engagement: The active participation of individuals or groups in decision-making, problem-solving, and actions that affect their community, fostering inclusion, empowerment, and social cohesion.

Community participation: The involvement of individuals or groups in the activities and decisions that impact their community, often aimed at improving local conditions and fostering collective action.

Consequences: The outcomes or results of an action or event, which can be positive or negative, and can affect individuals, communities, or the environment.

Conservation: The protection, preservation, and management of natural resources, ecosystems, and biodiversity to ensure their sustainability for future generations.

Cooperation: The process of working together with others to achieve a common objective, often involving mutual assistance and shared responsibility.

Crisis: A critical or emergency situation that requires immediate attention and action, often involving widespread harm, disruption, or danger.

Cultural fragmentation: The process by which a society or culture breaks into distinct, often conflicting, groups or subcultures, leading to divisions in values, beliefs, and practices.

Displacement: The forced movement of people from their homes or communities due to factors such as conflict, natural disasters, or environmental degradation.

Ecological degradation: The deterioration of ecosystems caused by human activities, such as deforestation, pollution, and overexploitation of natural resources, leading to a loss of biodiversity and ecosystem services.

Economic devastation: Severe damage to the economy, often resulting from factors such as financial crises, natural disasters, or prolonged conflict, leading to widespread poverty, unemployment, and instability.

Economic inequality: The unequal distribution of wealth, income, and resources within a society, often leading to disparities in living standards, opportunities, and access to services.

Ecosystem collapse: The breakdown of an ecosystem's structure and function, often due to human activities or natural events, leading to the loss of biodiversity and ecosystem services.

Education: The process of acquiring knowledge, skills, values, and attitudes through formal or informal learning, which plays a key role in personal development and societal progress.

Environmental degradation: The deterioration of the natural environment due to human activities, such as pollution, deforestation, and overconsumption of resources, which threatens biodiversity and the health of ecosystems.

Environmental destruction: The large-scale damage or destruction of the environment, often caused by industrial activities, urbanization, or unsustainable resource use, leading to habitat loss and pollution.

Epistemology: The branch of philosophy concerned with the nature, scope, and sources of knowledge, exploring questions about belief, truth, justification, and the process of knowing.

Feedback loops: Cycles in which the output of a system is fed back into the system as input, often amplifying or dampening the effects of a process. In environmental contexts, feedback loops can accelerate or mitigate climate change impacts.

Future vision: A long-term perspective or outlook on the future, often used in strategic planning or goal-setting to guide decisions and actions toward desired outcomes.

Geopolitical rivalries: The competition between nations or regions for power, influence, and resources, often leading to tensions, conflicts, and shifting alliances in international relations.

Global harmony: The concept of achieving peaceful coexistence and cooperation among nations, cultures, and peoples, promoting mutual understanding, respect, and shared goals for the well-being of all.

Global solutions: Approaches or strategies that address global challenges, such as climate change, poverty, and inequality, through international cooperation, innovation, and sustainable practices.

Global transformation: The process of significant and widespread change on a global scale, often involving shifts in social, economic, political, and environmental systems to create a more sustainable and equitable world.

Global warming: The long-term increase in Earth's average surface temperature due to human activities, particularly the emission of greenhouse gases, leading to climate change and its associated impacts.

Global wealth disparities: The unequal distribution of wealth and resources on a global scale, with significant differences in living standards, access to opportunities, and quality of life between nations and regions.

Governance: The process of decision-making and the implementation of decisions in organizations, governments, or societies, involving the creation and enforcement of laws, policies, and regulations.

Habitat destruction: The process by which natural habitats are altered or destroyed, often due to human activities such as deforestation, urbanization, or industrial development, leading to the loss of biodiversity.

Humanitarian consequences: The negative impacts on human well-being, including loss of life, displacement, poverty, and health crises, resulting from conflict, disasters, or environmental degradation.

Humanity: The collective human race or the qualities of being human, including compassion, empathy, and the capacity for moral and ethical behavior.

Indigenous knowledge: The traditional knowledge and practices developed by indigenous peoples over generations, often related to the environment, agriculture, health, and social organization, based on close relationships with nature.

Innovation: The creation and application of new ideas, technologies, products, or methods to improve processes, solve problems, or enhance the quality of life.

Interconnectedness: The state of being connected or related to one another, especially in terms of ecosystems, societies, and global systems, where changes in one area can affect others.

Interdependencies: The mutual reliance between individuals, communities, or systems, where the well-being or functioning of one depends on the actions or conditions of another.

International cooperation: The collaboration between countries or international organizations to address global issues, such as climate change, health crises, and peacebuilding, through shared efforts and mutual understanding.

IWRM (Integrated Water Resources Management): A holistic approach to managing water resources that considers the interconnections between water, land, and other resources, aiming to balance social, economic, and environmental needs.

Knowledge gaps: The areas where there is insufficient or incomplete knowledge or understanding, particularly in scientific research, policy development, or problem-solving.

Legislation: The process of creating and enacting laws, regulations, and policies by governing bodies to regulate behavior, protect rights, and address societal issues.

Marginalized communities: Groups of people who are excluded or disadvantaged in society due to factors such as poverty, discrimination, or lack of access to resources, often leading to limited opportunities and social inequality.

Mass extinction: A widespread and rapid reduction in biodiversity, where many species go extinct in a relatively short period, often caused by environmental changes, such as climate change or habitat destruction.

Microplastics: Tiny plastic particles, often smaller than 5 millimeters, that result from the breakdown of larger plastic items or are directly manufactured, posing environmental and health risks.

Ocean degradation: The deterioration of ocean ecosystems due to human activities such as overfishing, pollution, habitat destruction, and climate change, leading to the loss of marine biodiversity and ecosystem services.

Past crises: Historical events or situations that caused widespread harm, disruption, or suffering, such as wars, economic depressions, or natural disasters.

Plastic crisis: The environmental and ecological challenges posed by the excessive production, use, and disposal of plastic materials, leading to pollution and harm to wildlife and ecosystems.

Policy: A set of principles or guidelines that inform decision-making and actions in various areas, such as governance, economics, and social issues, often aimed at addressing specific problems or achieving certain goals.

Political instability: A condition in which a government or political system is unable to maintain order, stability, or effective governance, often due to conflict, corruption, or social unrest.

Pollution: The introduction of harmful substances or contaminants into the environment, including air, water, and soil pollution, which can have detrimental effects on health, ecosystems, and the climate.

Poverty: A condition in which individuals or communities lack the financial resources and basic necessities for a decent standard of living, including access to food, shelter, education, and healthcare.

Private sector: The part of the economy that is owned and operated by private individuals or organizations, as opposed to the public sector, which is controlled by the government.

Proxy conflicts: Conflicts in which two or more external powers support opposing sides in a third-party conflict, often for strategic, political, or ideological reasons.

Proxy wars: Conflicts in which external powers support opposing factions or groups in a conflict, often to exert influence or achieve their own political objectives without direct military involvement.

Resource efficiency: The use of resources in a way that maximizes their value while minimizing waste, environmental impact, and consumption, often in the context of sustainable development.

Responses: Actions or measures taken to address a particular issue, crisis, or challenge, such as government responses to natural disasters or climate change.

Slums: Overcrowded, informal urban settlements characterized by poor living conditions, lack of basic services, and inadequate infrastructure, often resulting from rapid urbanization and poverty.

Social fragmentation: The division or breakdown of social cohesion within a society, often due to factors such as inequality, political polarization, or cultural differences.

Soil pollution: The contamination of soil with harmful substances, such as chemicals, heavy metals, or waste, which can harm plant growth, reduce soil fertility, and impact human health.

Sustainability: The ability to meet present needs without compromising the ability of future generations to meet their own needs, often through practices that protect the environment, promote social equity, and support economic growth.

Sustainable management: The responsible and efficient use of resources and ecosystems in a way that ensures their long-term viability and benefits for future generations.

Technological disruption: The process by which new technologies or innovations fundamentally change industries, markets, or societal structures, often displacing older technologies or business models.

Technology: The application of scientific knowledge for practical purposes, particularly in the development of tools, machines, and systems that improve human life and productivity.

Theories: Systems of ideas or concepts that explain phenomena, behaviors, or processes, often used to guide research, policy development, and decision-making.

Total warfare: A type of warfare in which all resources, including civilian infrastructure, are mobilized for military purposes, often leading to widespread destruction and loss of life.

Transformation: A significant and lasting change or shift in the structure, form, or functioning of a system, organization, or society, often aimed at improvement or adaptation.

Urbanization: The process by which an increasing proportion of a population moves from rural areas to urban centers, leading to the growth and expansion of cities.

Vulnerabilities: The susceptibility of individuals, communities, or systems to harm or damage, often due to factors such as poverty, lack of resources, or exposure to environmental or social risks.

Wars: Large-scale conflicts between nations or groups, often involving military forces and resulting in significant loss of life, destruction, and geopolitical changes.

Waste management: The process of collecting, transporting, processing, and disposing of waste materials in a way that minimizes environmental impact and maximizes resource recovery.

Water pollution: The contamination of water bodies, such as rivers, lakes, and oceans, with harmful substances, including chemicals, waste, and microorganisms, which can harm aquatic life and human health.

Water scarcity: The lack of sufficient freshwater resources to meet the demands of a population, often due to factors such as overuse, climate change, and pollution.

Zero waste: A waste management philosophy aimed at reducing waste to landfill through practices such as recycling, composting, and minimizing resource consumption to create a circular economy.

About the Author

Dr. Medani P. Bhandari is a widely recognized humanitarian, professor, author, and editor who has significantly contributed to social and environmental sciences. With expertise in social equity, inequality, inclusion, innovation, feminism, sustainability, climate change, and social and environmental policies and management, Dr. Bhandari has dedicated his career to these important areas. In addition to his extensive scholarly work, Dr. Bhandari is a poet, essayist, and advocate for the environment and social causes. His expertise expands to global and international environmental politics and justice, climate change, sustainable development, public/social policy, the non-profit sector, renewable energy, nature, culture, and power. He has gained field experience across diverse regions, including Asia, Africa, North America, Western Europe, Australia, Japan, and the Middle East. Dr. Bhandari has authored and co-edited several books, including Green Web-II: Standards and Perspectives from the IUCN, 2nd Edition, Getting the Climate Science Facts Right: The Role of the IPCC, Reducing Inequalities Towards Sustainable Development Goals: Multilevel Approach, Educational Transformation, Economic Inequality – Trends, Traps, and Trade-offs, Social Inequality as a Global Challenge, The Unbeatable Challenges of Inequality, Perspectives on Sociological Theories, Methodological Debates, and Organizational Sociology, The Past, Present, and Future Directions of Migration/Immigration, Theories, and Practices of Global Journeys, Transforming Futures – Navigating Society, Environment, and Governance, Live and Let Others Live – In Reference to Sustainability and Environment Conservation, Navigating Sustainability: A Comparative Analysis of Micro vs. Macro Strategies in the least Developed Countries, The Nexus of Climate Change and Land-use – Global Scenario with Reference to Nepal, Women and Society etc.. Social Inequality- Past, Present, and Future (in which 15 African Countries' case studies are included), etc. Furthermore, Dr. Bhandari's creative writing encompasses a vast collection of poems and essays. He has published three volumes of poetry in collaboration with Prajita Bhandari. Dr. Bhandari's multifaceted contributions to academia, humanitarianism, and creative writing have solidified his reputation as an esteemed figure in social and environmental sciences, profoundly impacting social equity, sustainability, climate change, and beyond. Professor Bhandari holds five master's degrees in social and environmental sciences and a PhD in sociology. For 45 years, he has been actively involved in environmental conservation and humanitarian efforts through nonprofit and academic platforms. Guided by the philosophy that all living and non-living beings are interconnected ("Vasudhaiva Kutumbakam") and the principle of "Live and Let Others Live," his work reflects a deep commitment to the well-being of our planet and its inhabitants.

He serves as an International Program Director at the New York-based NGO Atlantic State Legal Foundation in Syracuse, NY; a professor at Akamai University, USA, and Sumy State University, Ukraine; and an advisor to Gandaki University, Nepal. He is also the UNEP Civil Society Regional Facilitator for North America and the Lead Author of UNEP GEO 7. He is also the editor-in-chief of Strategic Planning for Energy and the Environment, the books series editor at the River Publisher, and the books series editor at Nova Publishers.